The Invertebrates:
Function and Form

The Invertebrates:
Function and Form

A Laboratory Guide

SECOND EDITION

Irwin W. Sherman
Vilia G. Sherman

*University of California, Riverside
and
The Marine Biological Laboratory,
Woods Hole*

MACMILLAN PUBLISHING CO., INC.
New York
COLLIER MACMILLAN PUBLISHERS
London

Macmillan Publishing Co., Inc.
866 Third Avenue, New York, New York 10022

Collier Macmillan Canada, Ltd.

Library of Congress Cataloging in Publication Data

Sherman, Irwin W
    The invertebrates.

1. Invertebrates—Laboratory manuals. I. Sherman,
Vilia G., joint author. II. Title.
QL362.S54 1976   592'.0028   75-14481
ISBN 0-02-409841-8

Printing: 7 8   Year: 2

To Our Parents

Preface
To the Second Edition

Professor Frank A. Brown was once approached by a student who intended to take his course in invertebrate zoology. The student wished to know whether Dr. Brown's course would be different from the course he had taught the previous year. Dr. Brown replied that since the invertebrates had not evolved very much during the year, the course would be substantially the same.

Invertebrate structure and function have not changed since the first edition of this laboratory guide was published, either. Invertebrate function and form are so varied and diverse, however, that no guide can cover everything there is to know about them. The selection of material for the second edition is not substantially different from the first, but changes have been made: errors have been corrected; new line drawings have been added and others modified; the references* have been updated; dissection directions have been expanded; and the section on the Mollusca has been almost completely rewritten. Many other sections were reorganized for pedagogical or practical reasons. The sources of living material, listed in Appendix B, are keyed by number to the specific organisms that must be ordered for each exercise.

The revision has been guided by many helpful suggestions from instructors and students. During the last few years, some instructors indicated that they have employed an organism-by-organism approach in their laboratories while using certain exercises in the guide. Large classes housed in limited laboratory facilities cannot always perform work on several organisms simultaneously, especially when these organisms must be retained for successive laboratory periods. Although we feel it is a pity to sacrifice the system-by-system, functional approach, practical realities must often overcome the desires of the instructor. In large phyla, where many specimens are involved, a classical specimen-by-specimen approach must sometimes be taken, where all the work on one specimen is completed before starting on another one. To facilitate the use of the manual in such situations, this edition includes page numbers with the table of contents at the beginning of each chapter. It should therefore

---

*We recommend adding to the laboratory reference shelf *Practical Invertebrate Zoology,* edited by R. P. Dales, Sidgwick and Jackson, London, 1969.

be easy for students to find and use the relevant portions of each exercise applicable to an individual organism under study, repeating the process for each organism.

We hope that this revised edition will continue to stimulate a functional, system-by-system approach to the study of invertebrates. As before, contributions and suggestions for improvement will be appreciatively received.

I. W. S.
V. G. S.

Preface
To the First Edition

This manual has been designed for the laboratory portion of a course in invertebrate zoology at the level of the second or third year of college. It is assumed that the student has had an introductory course in biology including material on cells, organisms, and physiological processes; an understanding of evolutionary theory will also be helpful.

Much of the material included here was developed in connection with a course taught by the senior author at the University of California, Riverside, for several years. No claim for originality is made; many of the experiments and directions have been freely modified from those of other authors and from colleagues at the Marine Biological Laboratory and other institutions.

Classical approaches to laboratory courses in invertebrate zoology have usually required the student to perform lengthy and laborious dissections of a variety of invertebrate animals. The traditional complaint heard from students taking such a course is that there is an overwhelming amount of seemingly irrelevant and confusing classification, terminology, and anatomical detail; the student tends to become discouraged, to lose sight of the fact that the animals he is examining were once living beings, and to find it difficult to relate structure to function. This manual is based on the premise that form and function are inseparable considerations in understanding the biology of invertebrates. The exercises have been designed in an attempt to provide a guide that allows students to see not only how invertebrates are constructed, but how they work.

The field of invertebrate zoology is huge; at least 90% of all described species are invertebrates and they show extreme diversity. The material must usually be covered in a few short courses. The means employed in understanding this diversity vary with the interests of the teacher, the background of the student, the location and facilities of the school, and the role the course plays in the curriculum. We do not pretend that this manual provides all the answers or is the only approach that may be taken. In the hope of increasing the flexibility and widening the use of the manual, more material is given than can possibly be encompassed in a single course. We have tried to design the exercises from the point of view of both content and layout so that each section is relatively autonomous. The instructor may thus choose the exercises that best fit his approach to the subject, the background of the student, and the time and material available.

Because it is often easier for students to comprehend functional organization and adaptations that occur within rather than between phyla, the exercises have been organized on a phylogenetic basis. However, the manual can be adapted for treatment of invertebrate zoology according to functional systems by extracting the relevant sections from each exercise. This approach runs the risk of luring students into doing things with animals before their organization is understood. The phylogenetic approach is practical and economical because it does not require that the same array of organisms be brought into the laboratory each week for the study of a different functional system. Its major drawback is that students must return to previously studied animals as each system is treated within each phylum. The students will grumble that their lives are complicated by the "busy work" of repetitive handling of organisms. Nevertheless, we have found that this serves as a pedagogical aid, for frequent exposure to the same animals promotes familiarity and reinforces knowledge. It is our hope that the organization provides a cohesive framework for the laboratory period and, in conjunction with the questions in each section and references at the end of each exercise, may provide a springboard for additional investigations.

In designing such a manual, a course must be steered between the Scylla and Charybdis of anatomy and physiology. The exercises may therefore not be as comprehensive as some instructors will like, nor is there as much anatomy and/or physiology as others will require. We chose to omit anatomical details from many of the sections and therefore advise that the following volumes be available in the laboratory: *The Invertebrates,* Volumes I–VI, by L. H. Hyman, McGraw-Hill, 1940–1968; *Practical Invertebrate Anatomy* by W. S. Bullough, St. Martin's Press, 1962; *Selected Invertebrate Types,* edited by F. A. Brown, Jr., Wiley, 1950; *Dissection Guides,* Volume V, *Invertebrates* by H. G. Q. Rowett, Holt, Rinehart and Winston, 1957. Phylogenetic relationships are stressed where pertinent, but taxonomy is given less emphasis than is usually the case in invertebrate zoology courses. A brief classification and list of characteristics of the organisms studied and their relatives are provided in the appendix to each exercise.

Ten out of the twenty-five to thirty recognized phyla have been selected for study. These we consider the major phyla, based on the number of individuals and number of species comprising them. An unusually large portion of the flow of solar energy passes through the bodies of the members of these ten phyla, so that they may also be designated as major groups from that standpoint. They make up the bulk of the animal biomass in natural communities, and therefore representatives are reasonably common and easily obtainable for laboratory study.

Selection of materials has been primarily based on specimen availability. The majority, if not all, of the animals can be ordered from commercial biological supply houses and local collectors. A list of sources of living materials as well as equipment needed for each laboratory is provided at the end of each exercise. A set of instructions for making up solutions and media may be found at the end of the manual. Because a large proportion of invertebrates are marine, emphasis on these forms is inevitable. Living marine invertebrates can be sent thousands of miles by air freight and, when placed in marine aquaria utilizing refrigerated natural or artificial seawater, can usually be maintained for weeks or even months. There is no need for inland schools to be restricted to preserved materials, freshwater invertebrates, and seasonal studies. If facilities are not available for holding living invertebrates, this manual should not be used.

Because living materials are the very heart of these exercises, the following suggestions may be helpful:

1. Place orders at least 3 weeks in advance of your needs and specify shipment by air freight.

2. Have live specimens delivered midweek; when specimens are ordered for the beginning of the week, they may arrive early and spend the weekend on a loading dock.

3. Have as standard equipment in the laboratory at least one 25-gallon refrigerated seawater aquarium. A 25-gallon tank usually serves the needs of forty students.

In all cases, investigations of functional morphology can be made with the simplest of laboratory apparatus, the most complex items needed being a colorimeter and a kymograph. Neurophysiological experiments requiring complicated equipment are not treated; these are better handled in courses in invertebrate or comparative physiology.

Note on the Curriculum and Selection of Exercises

Considerably more material than can be covered in a single course has been included in this laboratory manual. Instructors will find that they must choose sections of the exercises that meet the background of their own students and the time and material available. The choice has been left to the discretion of the instructor; however, the following guidelines may be helpful.

Each of the exercises is of a different length, dictated by the nature of the phylum, availability of organisms, and techniques employed in their study. A flexible schedule is necessary when considerable numbers of living specimens are utilized, and the alert teacher will always have supplemental materials available (e.g., slides and preserved specimens) so that the laboratory period may be put to effective use in the event of mishaps.

Experience in the use of the manual has indicated that it can be handled effectively in twice-weekly laboratory sessions of 3 hours duration each. The semester system usually allows 90 hours of laboratory and the two-quarter sequence allows 120 hours of laboratory; in shorter-semester courses as little as 60 hours may be available. The following outlines are suggested:

	Semester Two 3-hr labs per week (90 hr) Lab hours	Two quarters Two 3-hr labs per week (120 hr) Lab hours	Semester One 4-hr lab per week (60 hr) Lab hours
Protozoa	9	12	8
Porifera	3	6	4
Cnidaria	9	12	8
Platyhelminthes	6	9	4
Aschelminthes	6	9	4
Annelida	9	12	8
Arthropoda	15	18	8
Mollusca	9	12	8
Echinodermata	12	12	8
Protochordata	6	6	—
Open (field trips, examinations, etc.)	6	12	—

Schools vary greatly in their special offerings, and some may prefer to omit the Protozoa, Arthropoda, and Protochordata because they may be offered in other courses. In general, most schools treat Porifera, Cnidaria, Platyhelminthes, Aschelminthes, Annelida, Mollusca, and Echinodermata in courses on invertebrates.

Students can usually cover about five to six pages of the manual per 4-hour session. Some, depending on background and preparation, can cover more ground and should be encouraged to do so. Many of the exercises have been designed so that students can work in pairs. Each exercise begins with a brief introduction to the group citing the major features of the phylum. The next section gives an anatomical survey of some representatives and often can be handled by using preserved materials and/or slides rather than living specimens, although the latter are preferable. The remainder of the sections in each exercise can stand, more or less, as independent units. Each starts with a description of the system, followed by directions for experimental work. Directions for practical work are distinguished by a small symbol ▶. The student should be encouraged to read the entire exercise for the day *before* coming to the laboratory, even though all the material will not be covered. He then can proceed immediately to the practical work, with the background information clear in his mind.

Topics covered and sections selected for study depend on the emphasis desired by the instructor as well as factors already mentioned. In general, experiments on the neuromuscular system, behavior, and reproduction are extremely time consuming, and these sections can be omitted without seriously affecting the continuity of the laboratory. These aspects of invertebrate biology are quite effective as individual student projects.

Alternatively, the instructor may wish to take a different approach and study topics such as feeding, circulation, digestion, and the neuromuscular system on a comparative and physiological basis between phyla. The pitfall of this approach is that the student sees only the system and not its relation to the whole organism. This treatment can be successful only if students have sufficient background in anatomy and phylogeny.

Acknowledgments

This manual is a compilation of exercises that demonstrate the relation of form to function in the invertebrates. We are indebted to many of the workers and students who have played a role in developing these exercises and in introducing them to students and teachers throughout the world. We have been particularly fortunate in having the able assistance of numerous colleagues and friends who have freely provided technical and theoretical knowledge, suggested numerous improvements, and given useful criticism as well as encouragement. We are especially grateful to John Anderson, Allison Burnett, Fred and Norma Diehl, Robert Eaton, Eric B. Edney, Allahverdi Farmanfarmaian, Frank M. Fisher, Milton Fingerman, Jonathan P. Green, George G. Holz, Jr., Meredith Jones, Robert K. Josephson, James Lash, and W. D. Russell-Hunter. Any errors that remain and the organization of the manual are our own responsibility.

We hope that this manual will be found useful and would appreciate receiving contributions of experiments and suggestions for improvement.

I. W. S.
V. G. S.

RIVERSIDE, CALIFORNIA

Foreword
To the Student

We recognize very well that the best way to get to know the invertebrates is to set oneself up by the sea where the fauna and flora are abundant, to collect and examine the organisms, and above all to observe them in their natural habitats, noting reactions and changing patterns of distribution and behavior. In this way, real understanding of the enormous range of adaptive features shown by invertebrates can be revealed. Unfortunately, for most courses in invertebrate zoology extended field studies are not possible, so other approaches must be sought.

This manual is not designed to replace the lecture material or your text assignments. Neither is it a comprehensive survey of invertebrates, and it certainly will not give you all details of invertebrate anatomy and physiology. The manual is intended as a guide, and its main contention is that examination of form and its relation to function among the invertebrates is the most satisfactory approach toward understanding these animals. In brief surveys of invertebrate biology such as are encompassed in courses of a semester or a quarter, one cannot hope to examine even a part of the invertebrate species known, so we have arbitrarily adopted a few forms for detailed study on a comparative basis and by phylum. Through examination of living and preserved materials, it is hoped that you will see how structural organization relates to function, and that there will be raised in your mind many biological questions that can be answered by the study of invertebrates.

The field of invertebrate zoology is vast, and no one can hope to have all the information at his fingertips. To make the most of the limited laboratory time available, the following hints are offered:

1. Read the exercises *before* coming to the laboratory and try to understand the procedures thoroughly. Directions for practical work are distinguished from descriptive material by a small symbol ▶, and you should be able to proceed directly to these on coming to the laboratory, with the background material clear in your mind.

2. Many of the exercises contain more than you alone can accomplish during the normal laboratory period. Your instructor may tell you which sections should be omitted and may wish you to work in pairs. Exercise choice in the sorts of experiments you undertake, and exchange information with your fellow students. Organize your time in advance so that assigned work can be completed in the time available. Some portions of the exercises require prolonged or

periodic observation and must be set up at the beginning of the laboratory period. These are usually indicated by an asterisk and a footnote wherever they occur.

3. Use the abbreviated classification of the phylum and list of the characteristics of the animals to be studied and their related forms; these are found at the end of each exercise. A list of equipment needed for each laboratory period is also provided, and you should assemble everything you need for one period before beginning work.

4. Do not become unduly frustrated with details—much of the material will be repeated and in time become familiar. In the laboratory, follow the directions and see for yourself how the animal functions. If information is lacking in the manual, consult the references listed at the end of each exercise and those available in the laboratory. Check your observations with those reported in the literature. Not all scientists agree, and your well-thought-out opinions, although different from those of others, may in fact be correct. Do not be afraid to state them in the face of opposition, or because they seem ridiculous or oversimple. Remember that the best way to turn "I don't know" into "I know" is to admit ignorance and to investigate for yourself.

5. Questions are interspersed throughout the text and vary in difficulty. Some can conveniently be answered only by deductive reasoning, and some may require outside reading from text or references before they can be answered. Some are unanswerable and will stimulate research and discussion in the laboratory.

Attempt to answer all the questions. The purpose of the manual is to stimulate your curiosity. If this is accomplished, the authors will consider themselves well rewarded.

Contents

PROTOZOA*

Contents

The diagnostic feature of Protozoa (*proto,* first; *zoon,* animal; Gk.) is that they are **unicellular.** All their activities are performed within the framework of a single cell, and the body is not divided up into tissues and organs. The functional units of the body, e.g., nuclei, mitochondria, are called **organelles.** These are small aggregates of macromolecules arranged in specialized configurations and they perform specific tasks; they are analogous to **metazoan (multicellular) organs.** Some authors argue that the body of a protozoan is, in a functional sense at least, better compared with the entire metazoan organism than with any of its component cells and, therefore, that the Protozoa are acellular, not unicellular. This concept, although valuable in terms of function, is not tenable at the cytological level. Electron microscope studies of Protozoa clearly demonstrate that in their ultrastructure they are entirely comparable with individual metazoan cells.

*Students unfamiliar with or requiring a refresher course in the use of the microscope should refer to the review at the end of this exercise before proceeding.

1

Animals of microscopic dimensions such as the Protozoa are able to perform many of their functions without any specialized organelles. In respiratory gaseous exchange, elimination of wastes, and intracellular distribution of materials, they rely almost exclusively on simple diffusion. Protozoa are the simplest forms of animal life, and some members of this phylum have plantlike characteristics; that is, they possess cellulose cell walls and chlorophyll and can perform photosynthesis. Such organisms are the modern representatives of the common stock that gave rise to both plants and animals.

Although Protozoa are often referred to as "primitive" or "lowly" forms of life, they often exhibit an amazing degree of differentiation at the cellular level. The Protozoa have evolved to the cells we see today by a long series of complex changes that have fitted each one for life in its particular environment. They are a widespread group ecologically, being found in fresh, marine, and brackish water, and are successful as symbionts of every grade. Thinking of these organisms as being primitive or ancestral is as naive and erroneous as thinking that men are the same as their apelike ancestors.

Note on the Preparation of Wet Mounts

During this exercise it will be necessary to prepare several temporary wet mounts. These should be saved for use in various parts of the exercise and may be exchanged with other students to reduce the amount of time spent in preparing material. Temporary mounts can be sealed by touching a *very* small drop of mineral oil to the edge of a cover slip, permitting it to come in contact with the water. The oil spreads out as an invisible film over the surface of the water and forms an effective seal that lasts for many hours.

1–1. LOCOMOTION

All Protozoa live in a watery environment, and their methods of locomotion are related to swimming or crawling activity. Some are sessile or sedentary and may have their movements restricted to body contractions. In some instances, movement may be passive, or not associated with any organelles (subphylum Sporozoa).

The actual surface of Protozoa, as in other cells, is limited by a **three-ply plasma** or **cell membrane** (Figure 1.1). Although we cannot study the cell surface by means of the electron microscope in this course, we can look at some of the properties of the membrane in locomotion and at the associated elaborate specializations formed at the cell surface. Methods of locomotion and the organelles involved (e.g., cilia, flagella, pseudopodia) are important in the classification of Protozoa (see p. 34).

a. Flagella and Cilia

Flagella and cilia are filamentous projections from the surface of protozoan or metazoan cells that accomplish locomotion or produce a current. One of the striking features of the flagellum and cilium is that, at the level of the electron microscope, both organelles have a similar fundamental structure of nine outer and two central fibrils (Figure 1.2). The entire filament (**axoneme**) is covered by the cell membrane (**sheath**). The central fibers terminate at a basal plate at the body surface, and the nine peripheral fibrils continue internally to contribute to the structure of the **kinetosome (basal body)**. However, cilia are usually shorter than flagella, occur together in greater numbers, and are most often arranged in rows, although these differences

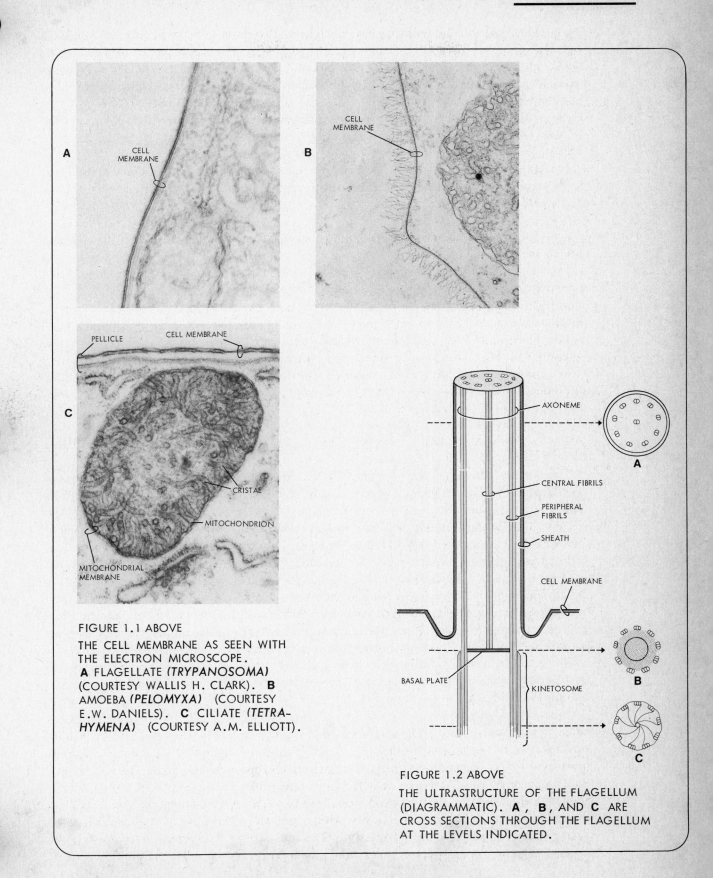

FIGURE 1.1 ABOVE

THE CELL MEMBRANE AS SEEN WITH THE ELECTRON MICROSCOPE. **A** FLAGELLATE *(TRYPANOSOMA)* (COURTESY WALLIS H. CLARK). **B** AMOEBA *(PELOMYXA)* (COURTESY E.W. DANIELS). **C** CILIATE *(TETRA-HYMENA)* (COURTESY A.M. ELLIOTT).

FIGURE 1.2 ABOVE

THE ULTRASTRUCTURE OF THE FLAGELLUM (DIAGRAMMATIC). **A**, **B**, AND **C** ARE CROSS SECTIONS THROUGH THE FLAGELLUM AT THE LEVELS INDICATED.

are not always absolute. Some functional distinction can be made between cilia and flagella. A typical flagellum produces a movement of fluid along the length of the flagellar axis and moves fluid continuously throughout its beat. The cilium produces a movement of fluid at right angles to the long axis of the cilium when it is in the middle of its active stroke; the cilium has an active phase during which movement of environmental fluid occurs and a recovery phase in which there is little or no movement of fluid. Groups of cilia generally beat together in coordinated fashion, whereas flagella usually beat more independently of each other. Note that there are exceptions to this, e.g., *Volvox*.

Protozoa with cilia are classified as ciliates (subphylum Ciliophora), and those with flagella at some stage in their life history are classified as flagellates and amoeboflagellates (subphylum Sarcomastigophora).

▶   i. FLAGELLUM.   Cultures of the freshwater flagellates *Chlamydomonas, Euglena,* and *Volvox* will be provided (Figures 1.3, 1.4, and 1.5). Place a drop of a culture on a slide and add a drop of 10% methyl cellulose, which will restrict the rapid movements of the organisms and allow you to study movement in "slow motion." Place a cover slip on the drop and observe the movement of the flagellum, which can be more easily seen by reducing the light. Does the flagellar beat move from tip to base (point of insertion in the body) or in the reverse direction? To see the flagellum even more clearly, stain with Lugol's iodine by adding a drop of the iodine solution to the edge of the cover slip and allowing the stain to be drawn under by capillarity.

▶   In the colonial flagellate *Volvox,* notice that the flagellar beat is coordinated from cell to cell to produce an oriented motion of the entire colony. What is the distinction between a colony and an individual?

▶   Note the number of flagella present in each of the specimens. (These are important in taxonomy; see Classification of Protozoa, p. 34.) The flagellar apparatus may be seen more easily in prepared slides of *Euglena* and *Volvox.*

The number and arrangement of flagella may be quite complicated in some groups of zooflagellates (Hypermastigida). This can best be seen in the flagellates that inhabit the termite gut.

▶   Take a termite and grasp the middle of the body with a pair of forceps. Run the curved end of another pair of forceps from the middle of the insect's body toward the posterior so that the gut contents are squeezed out onto a microscope slide. Cover with a cover slip. Note that the gut contents are full of a bewildering array of flagellates. A number of genera are represented, of which *Trichonympha* is probably the best known (Figure 1.6). The flagellar apparatus is quite complicated, and greater cytological detail may be seen in thin films of the gut contents that have been allowed to dry in air and then stained with Giemsa stain. Is the flagellar beat coordinated? Why are these locomotor organelles considered to be flagella and not cilia?

The flagellum may be modified in such a way that it is attached along the length of the cell to form an **undulating membrane.** Some authors feel that the undulating membrane is an adaptation for swimming in the viscous blood.

▶   Obtain a drop of blood from the tail vein of a rat infected with the nonpathogenic trypanosome *Trypanosoma lewisi* (Figure 1.7) by snipping the tip of the tail with a sharp pair of scissors. Before the drop of blood clots, cover with a cover slip and observe the vigorous movements of the trypanosomes (best pinpointed from the erratic motion of the erythrocytes). Study stained slides of *Trypanosoma lewisi* or prepare thin films of the blood of the rat and stain with Giemsa stain. (For method, see Appendix A.) Note the relationship of the flagellum to the basal body (Figure 1.7). Closely related organisms *Trypanosoma rhodesiense* and *Trypanosoma gambiense* are the causative agents of African sleeping sickness in man. Examine demonstration slides of these organisms.

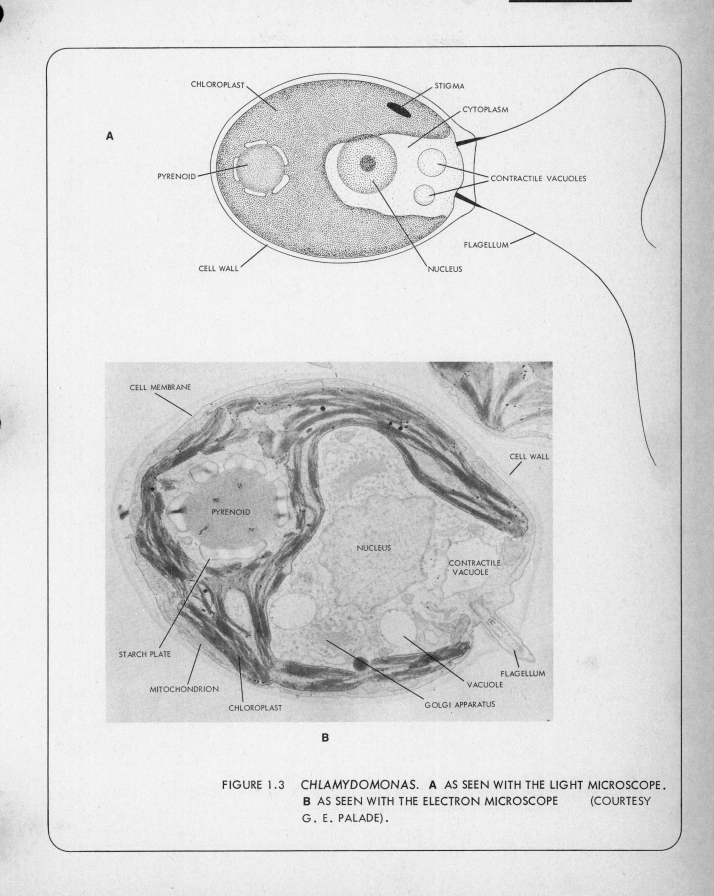

FIGURE 1.3   *CHLAMYDOMONAS.* **A** AS SEEN WITH THE LIGHT MICROSCOPE. **B** AS SEEN WITH THE ELECTRON MICROSCOPE   (COURTESY G. E. PALADE).

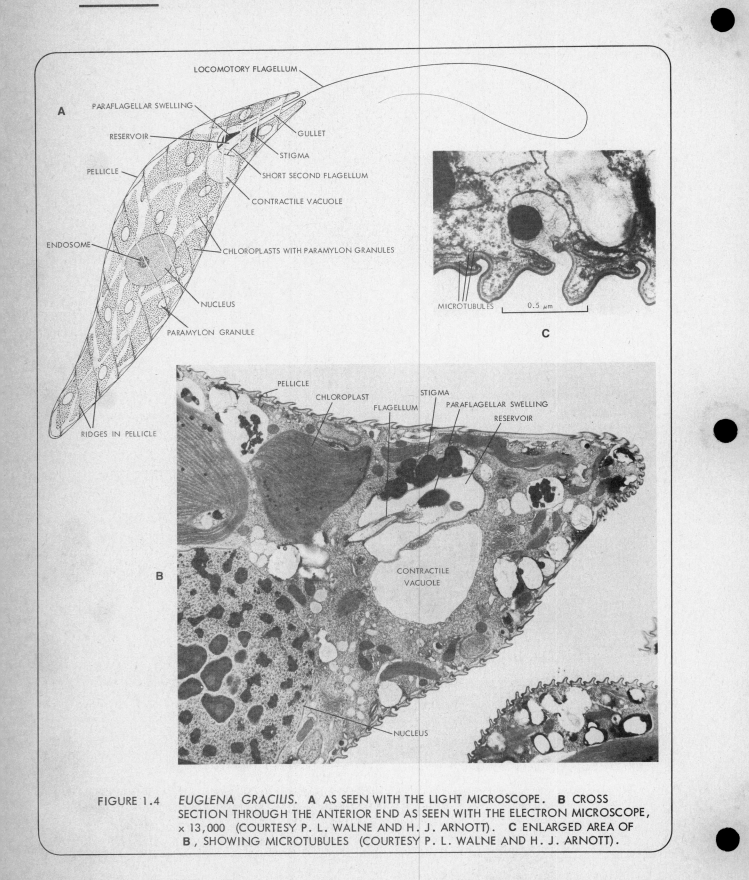

LOCOMOTORY FLAGELLUM

A

PARAFLAGELLAR SWELLING

RESERVOIR

GULLET

STIGMA

PELLICLE

SHORT SECOND FLAGELLUM

CONTRACTILE VACUOLE

ENDOSOME

CHLOROPLASTS WITH PARAMYLON GRANULES

NUCLEUS

PARAMYLON GRANULE

RIDGES IN PELLICLE

MICROTUBULES   0.5 μm

C

PELLICLE   CHLOROPLAST   STIGMA

FLAGELLUM   PARAFLAGELLAR SWELLING

RESERVOIR

CONTRACTILE VACUOLE

B

NUCLEUS

FIGURE 1.4   *EUGLENA GRACILIS.* **A** AS SEEN WITH THE LIGHT MICROSCOPE. **B** CROSS SECTION THROUGH THE ANTERIOR END AS SEEN WITH THE ELECTRON MICROSCOPE, × 13,000 (COURTESY P. L. WALNE AND H. J. ARNOTT).   **C** ENLARGED AREA OF **B**, SHOWING MICROTUBULES (COURTESY P. L. WALNE AND H. J. ARNOTT).

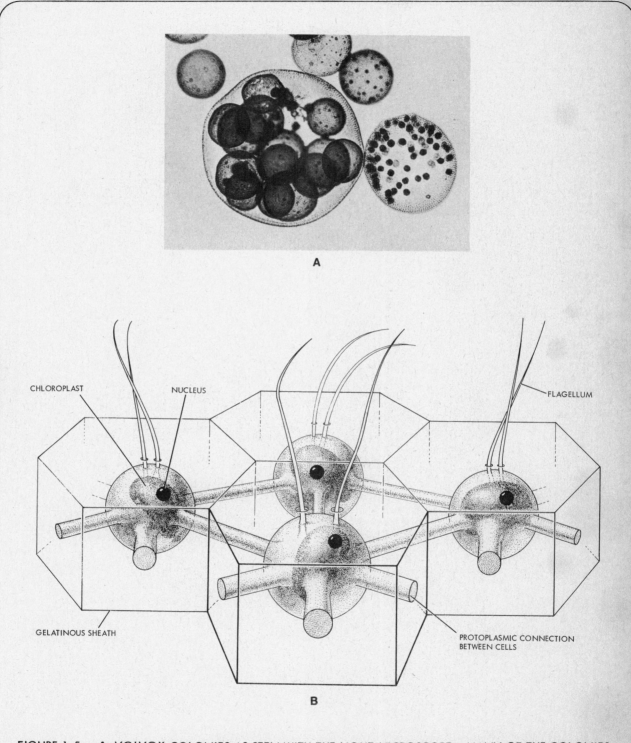

FIGURE 1.5   **A** *VOLVOX COLONIES AS SEEN WITH THE LIGHT MICROSCOPE. MANY OF THE COLONIES SHOW DAUGHTER COLONIES INSIDE THEM (COURTESY CAROLINA BIOLOGICAL SUPPLY COMPANY).* **B** *DIAGRAMMATIC REPRESENTATION OF THE CELLS THAT COMPRISE THE VOLVOX COLONY.*

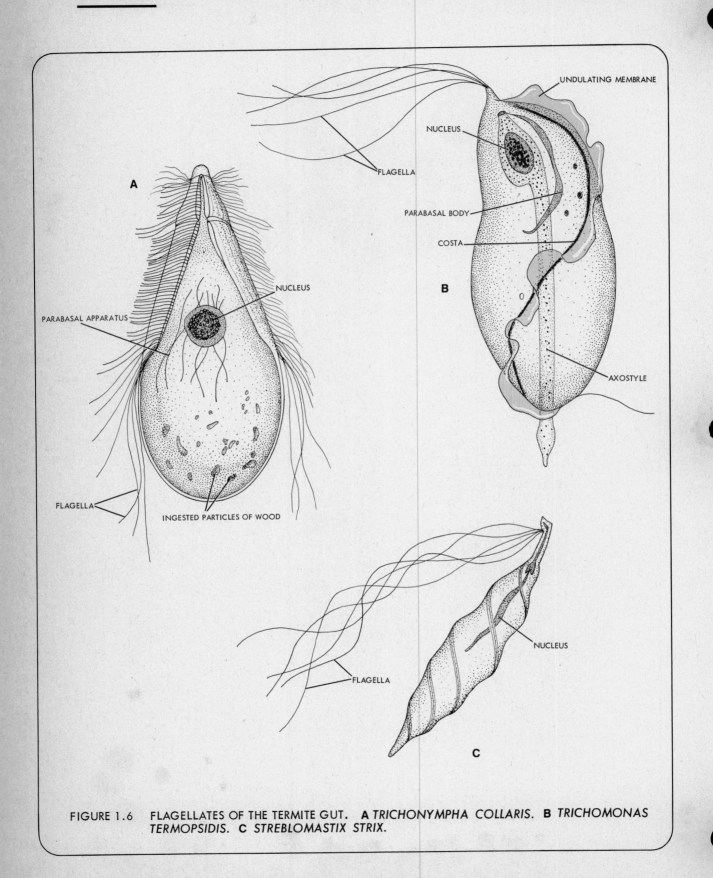

FIGURE 1.6   FLAGELLATES OF THE TERMITE GUT.   **A** *TRICHONYMPHA COLLARIS.* **B** *TRICHOMONAS TERMOPSIDIS.* **C** *STREBLOMASTIX STRIX.*

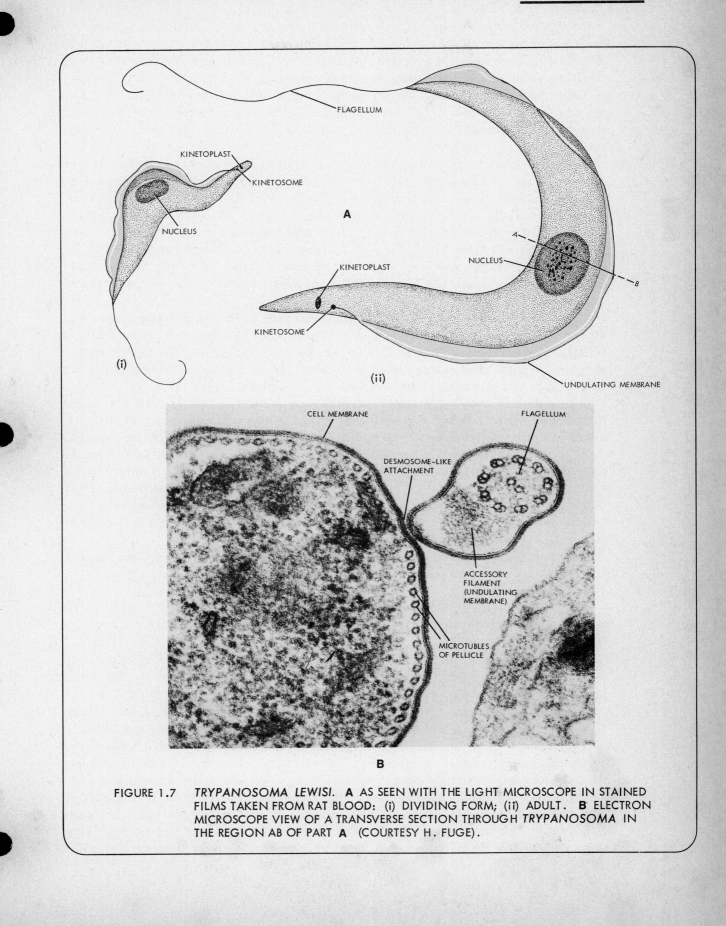

FIGURE 1.7 *TRYPANOSOMA LEWISI.* **A** AS SEEN WITH THE LIGHT MICROSCOPE IN STAINED FILMS TAKEN FROM RAT BLOOD: (i) DIVIDING FORM; (ii) ADULT. **B** ELECTRON MICROSCOPE VIEW OF A TRANSVERSE SECTION THROUGH *TRYPANOSOMA* IN THE REGION AB OF PART **A** (COURTESY H. FUGE).

▶ ii. CILIUM. Obtain drops of cultures of the freshwater ciliates *Paramecium, Spirostomum,* and *Blepharisma.* Make wet mounts and examine the distribution of cilia on the surface of the body of each species (Figures 1.8, 1.9, and 1.10). Reduce the light for better resolution. Do ciliates swim in a manner similar to the flagellates previously examined? How is swimming in ciliates and flagellates related to organellar beat, coordination, and position on the body? Is it strictly related to the number of cilia or flagella present? Is a certain portion of the ciliate body always directed forward? Does the animal rotate on its long axis? Diagram the swimming motion.

▶ Observation of the surface structure of *Paramecium* can be improved by mixing a drop of culture with nigrosin stain. Hold blotting paper against the slide edge to drain off any excess stain and place the slide near a lamp until the film has dried. Observe under high power. A detailed view of the surface architecture is shown in Figure 1.8**B.** Study demonstrations of ciliates stained by the silver technique of Chatton-Lwoff (Figure 1.11), which clearly shows the longitudinal rows of kinetosomes. The **pellicle (cortex)** of *Paramecium* is composed of all the membranous structures associated with the cell surface, usually consisting of membrane-enclosed sacs and microtubules (Figure 1.8**C**). Note the relationship of the pellicle to the cell membrane.

Cilia may form **membranelles** (Figure 1.12**B**), **cirri** (Figure 1.13**B**), or **undulating membranes.** Membranelles are composed of two to three rows of cilia that move together in such a fashion that the organelle beats like a paddle. Cirri are composed of bundles of cilia that taper at the tip; there is no common membrane surrounding the cirrus. The undulating membrane of ciliates is a simple, single row of cilia that function together as a membrane.

▶ Make wet mounts of drops from cultures of the pond dwellers *Stentor* (Figure 1.12**A**) and/or *Blepharisma* (Figure 1.10) and observe the membranelles in operation. The body of *Stentor* is trumpet-shaped and the flat, upper oral surface is a broadly expanded funnel, the **peristome**; on the edge of the peristome is a spiraled zone of membranelles that runs to the cell mouth **(cytostome).** Notice that here the cilia do not usually function in locomotion, but rather create a current of water for food gathering. How does *Stentor* swim? The peristomal region of *Blepharisma* is elongated, with a projecting undulating membrane. A wet mount of the freshwater hypotrich *Euplotes* will demonstrate how cirri function as tiny legs in locomotion (Figure 1.13**A**).

Can you make any generalizations about the relationship between symmetry and locomotion, based on your observations of the organisms you have so far examined?

b. Amoeboid Movement

The principal locomotor organelle of amoebae is the **pseudopodium** (''false foot''). The body form of most amoebae is constantly changing, and a variety of types of pseudopodia can be seen among the representatives of this group.

▶ Observe the unrestricted motion of specimens of the freshwater *Amoeba proteus* (Figure 1.14A), i.e., not under a cover slip. Then cover the specimens with a cover slip and illuminate the preparation with the iris diaphragm of the substage condenser closed down. Observe the **hyaline cap** at the tip of each pseudopodium, the movement of granules in the endoplasm, and the clear ectoplasm. Carefully observe the movement of granules in the advancing tip, and notice the way in which they are arrested at the sides of the pseudopodium. *Amoeba* may be observed in side view by fixing two cover slips on either side of a microscope slide with the aid of petroleum jelly, and placing a drop of culture in the space between the two cover slips (Figure 1.14**B**). Tilt the microscope tube 90° so that the amoebae crawl along the edge of the

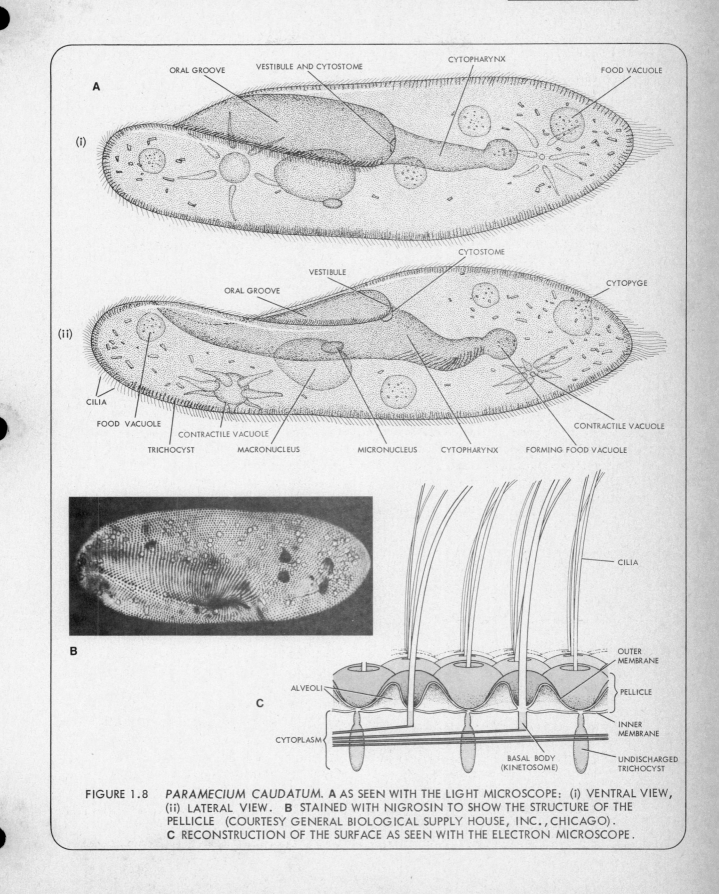

FIGURE 1.8   *PARAMECIUM CAUDATUM.* **A** AS SEEN WITH THE LIGHT MICROSCOPE: (i) VENTRAL VIEW, (ii) LATERAL VIEW.   **B** STAINED WITH NIGROSIN TO SHOW THE STRUCTURE OF THE PELLICLE (COURTESY GENERAL BIOLOGICAL SUPPLY HOUSE, INC., CHICAGO). **C** RECONSTRUCTION OF THE SURFACE AS SEEN WITH THE ELECTRON MICROSCOPE.

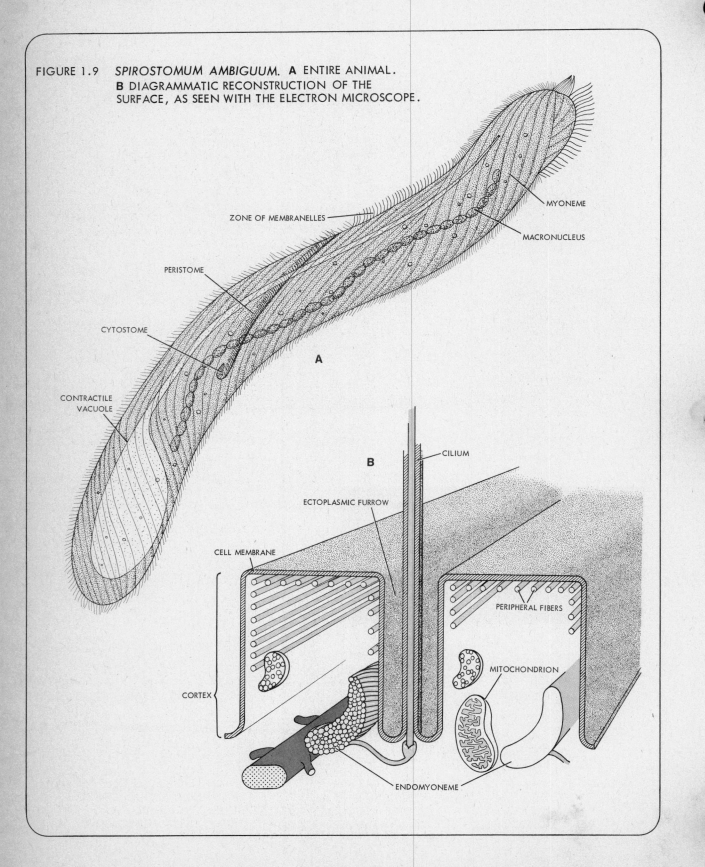

FIGURE 1.9   *SPIROSTOMUM AMBIGUUM.* **A** ENTIRE ANIMAL.
B DIAGRAMMATIC RECONSTRUCTION OF THE
SURFACE, AS SEEN WITH THE ELECTRON MICROSCOPE.

ZONE OF MEMBRANELLES

MYONEME

MACRONUCLEUS

PERISTOME

CYTOSTOME

CONTRACTILE
VACUOLE

A

B

CILIUM

ECTOPLASMIC FURROW

CELL MEMBRANE

PERIPHERAL FIBERS

CORTEX

MITOCHONDRION

ENDOMYONEME

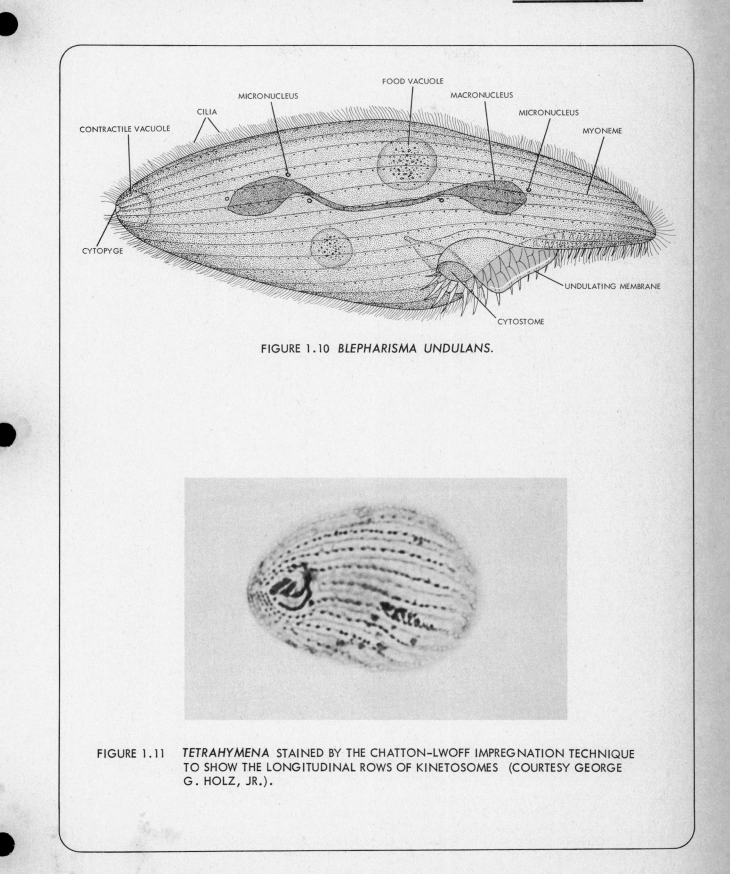

FIGURE 1.10 *BLEPHARISMA UNDULANS.*

FIGURE 1.11 *TETRAHYMENA* STAINED BY THE CHATTON–LWOFF IMPREGNATION TECHNIQUE TO SHOW THE LONGITUDINAL ROWS OF KINETOSOMES (COURTESY GEORGE G. HOLZ, JR.).

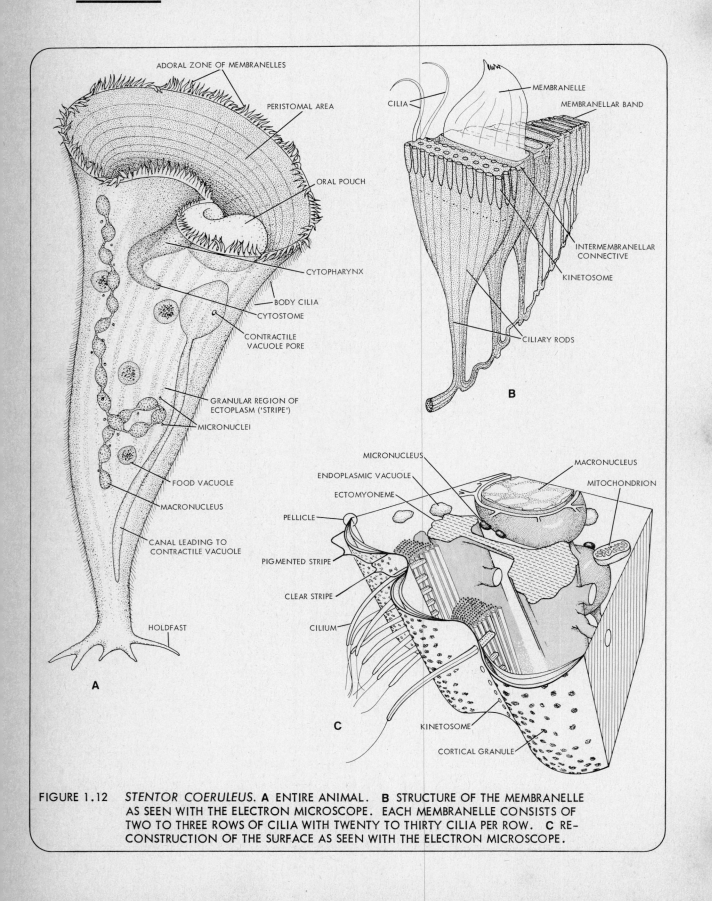

FIGURE 1.12　*STENTOR COERULEUS.* **A** ENTIRE ANIMAL. **B** STRUCTURE OF THE MEMBRANELLE AS SEEN WITH THE ELECTRON MICROSCOPE. EACH MEMBRANELLE CONSISTS OF TWO TO THREE ROWS OF CILIA WITH TWENTY TO THIRTY CILIA PER ROW. **C** RECONSTRUCTION OF THE SURFACE AS SEEN WITH THE ELECTRON MICROSCOPE.

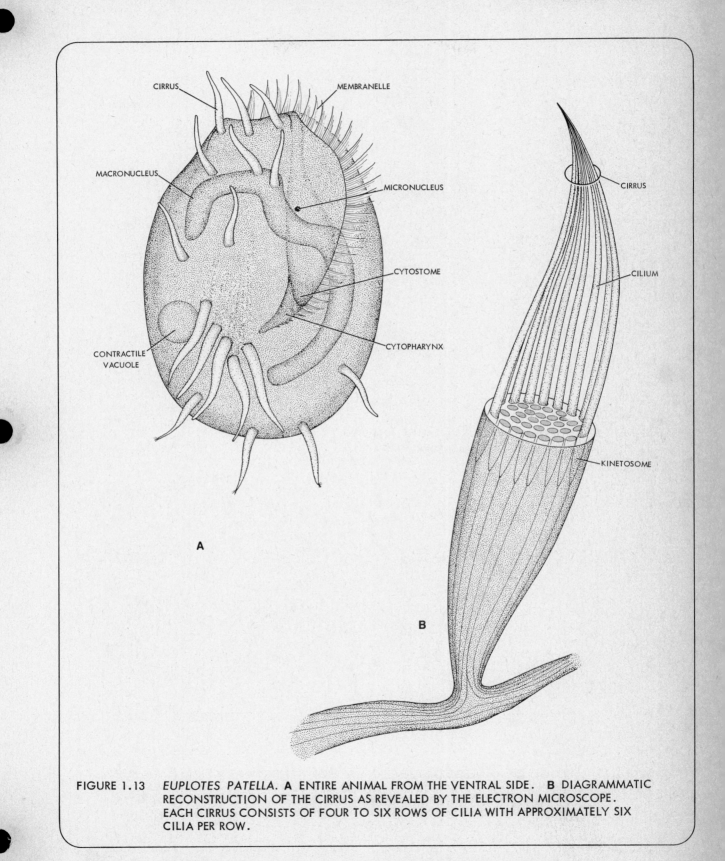

FIGURE 1.13   *EUPLOTES PATELLA.* **A** ENTIRE ANIMAL FROM THE VENTRAL SIDE.   **B** DIAGRAMMATIC
RECONSTRUCTION OF THE CIRRUS AS REVEALED BY THE ELECTRON MICROSCOPE.
EACH CIRRUS CONSISTS OF FOUR TO SIX ROWS OF CILIA WITH APPROXIMATELY SIX
CILIA PER ROW.

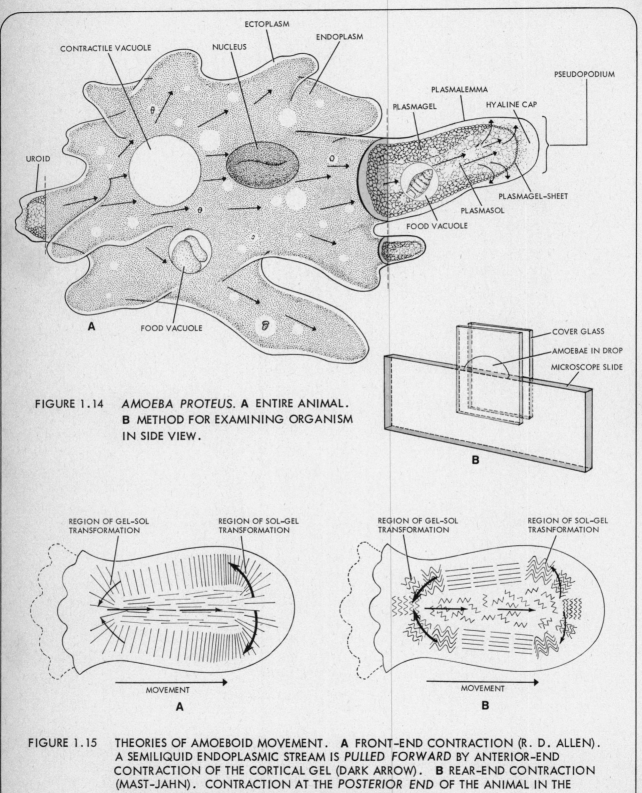

FIGURE 1.14 *AMOEBA PROTEUS.* **A** ENTIRE ANIMAL. **B** METHOD FOR EXAMINING ORGANISM IN SIDE VIEW.

FIGURE 1.15 THEORIES OF AMOEBOID MOVEMENT. **A** FRONT-END CONTRACTION (R. D. ALLEN). A SEMILIQUID ENDOPLASMIC STREAM IS *PULLED FORWARD* BY ANTERIOR-END CONTRACTION OF THE CORTICAL GEL (DARK ARROW). **B** REAR-END CONTRACTION (MAST-JAHN). CONTRACTION AT THE *POSTERIOR END* OF THE ANIMAL IN THE REGION WHERE ENDOPLASMIC GEL BECOMES MORE FLUID.

slide and not on the cover slips. Note the advancing pseudopodium. Does the entire animal contact the substrate surface? Is there a fixed anterior end? What is the position of the contractile vacuole relative to the advancing pseudopodium? Is there a characteristic shape to the posterior (**uroid**) end?

▶ The form of the pseudopodium may vary. Observe the difference between pseudopodia of *Pelomyxa* (called **lobopods**) and those of *Entamoeba terrapinae* and *Amoeba proteus*.

Amoeboid motion and the manner of production of pseudopodia has been a controversial area of investigation. Two major theories of amoeboid movement have emerged in recent years (Figure 1.15).

1–2. MOTILITY

Protozoa not only show movements in connection with locomotion (i.e., movement from place to place) but exhibit body contractions and intracellular cytoplasmic motion as well.

▶ Make wet mounts of *Euglena* and observe that considerable changes in body form are possible; these changes reflect a flexibility so characteristic of this organism as to have been named **euglenoid movement.** The ability to make contractions of this nature is related to the presence of **microtubules** (Figures 1.4 and 1.7), which are fibrils that underlie the cell membrane. Other Protozoa, particularly ciliates, employ organelles in body contraction known as **myonemes,** which are composed of bundles of microtubules. Make wet mounts of specimens from cultures of *Spirostomum, Stentor,* and *Vorticella* (Figures 1.9A, 1.12A, and 1.16) and observe the myonemes and the body movements produced by their contractility. In what direction do the myonemes run? What is the function of the contractions seen in these organisms? Note that ability to undergo deformational changes may not be associated with organelles in some species, e.g., *Amoeba*.

Protoplasmic streaming movements occur within cells. Endoplasmic streaming has already been noted in *Amoeba* and will be seen in the movement of food vacuoles in *Paramecium*. These streaming movements are usually not associated with organelles; in some cases they may be related to the presence of microtubules.

1–3. CELLULAR INCLUSIONS

a. Nuclear Structures

At the level of the electron microscope, the structure of the protozoan nucleus is similar to that of other cell nuclei. The nucleus is bounded by a double membrane that contains pores (Figure 1.19). The shape of the protozoan nucleus and the composition of its internal contents may, however, be quite variable. Flagellates and amoebae usually have a large ovoid or spherical nucleus, with a large Feulgen negative center or **endosome** that best corresponds to the **nucleolus.** Nuclei of this type are often described as **vesicular.** Ciliates typically have dimorphic nuclei: one or more large **macronuclei** and one or more small vesicular **micronuclei.** The macronucleus is compact, containing randomly distributed dense granules of Feulgen positive material, as well as scattered nucleoli. The macronucleus is concerned with vegetative functions (feeding, digestion, etc.); the micronucleus is involved in sexual recombination. The nucleus can best be seen after it has been stained.

►   i. FLAGELLATES AND SARCODINES. In order to see the nucleus in the living protozoan, examine wet mounts of samples from cultures of the colorless euglenoids *Astasia* or *Peranema* (Figure 1.20). The structural organization of the vesicular nucleus of these flagellates is similar to that of the nucleus of *Euglena* (Figure 1.4), but is more easily observed because there are no chloroplasts in the colorless forms to obscure your view. Close down the iris diaphragm for better resolution of the nucleus. Now stain the nucleus by placing a small drop of acidified methyl green near the edge of the cover slip so that it is slowly drawn beneath by capillary action. If the proper dilution of stain is obtained on the slide, the cells will be killed and the cytoplasm will take on a bluish tinge with the nucleus green. Compare your observations of stained material with those made on living cells.

►   Stain the nuclei of *Pelomyxa* and *Amoeba proteus* (Figure 1.14) in a similar manner. *Pelomyxa* is multinucleate. Try to visualize the nuclear form in three dimensions, i.e., is it spherical, biconcave, disc-shaped, or sausage-shaped?

►   ii. CILIATES. Obtain cultures of *Paramecium* (Figure 1.8) and make wet mounts; under high power, observe the form of the macronucleus. Stain with acidified methyl green as previously directed. Locate a micronucleus. Contrast the staining of the macronucleus with that of the micronucleus. Macronuclear form in ciliates may be quite variable. The macronucleus is elongate and sausage-shaped in *Vorticella* (Figure 1.16) but forms a string of "beads" in *Stentor* (Figure 1.12**A**) and *Spirostomum* (Figure 1.9) and is described as being **moniliform.** Observe the nuclei of these organisms in the living state, and stain as previously described. Try to see the morphology of the micronuclei in these ciliates.

b. Mitochondria and Kinetoplasts

►   Mitochondria are self-reproducing, DNA-containing, double-membrane-bound bodies that are the centers of oxidative metabolism (Figure 1.21). They may be vitally stained by placing a drop of *Paramecium* culture on a slide previously coated with Janus green B and neutral red. Try similar techniques with other living Protozoa. Are mitochondria located at specific sites within the cell?

►   Flagellates of the order Kinetoplastida, of which *Trypanosoma lewisi* is a representative, have a small structure, the **kinetoplast,** located close to the basal body (kinetosome) of the flagellum. Refer back to the prepared slides of *Trypanosoma lewisi* and *Trypanosoma brucei* to see the relationship between kinetosome and kinetoplast (Figure 1.7). The kinetoplast has been shown to be self-reproducing, DNA-containing, and double-membrane-bound and to have tubular extensions identical to mitochondria (Figure 1.21**B**). Thus, kinetoplast and mitochondrion are developmentally interrelated.

c. Plastids

The colors of Protozoa are as varied as those of the rainbow. Some are pink, red, green, yellow, brown, or blue, and the pigment is located in discrete bodies called **plastids.** If colorless, they are called **leucoplasts,** whereas those containing photosynthetic chlorophyll are called **chloroplasts.** The green color may be intense enough to obscure any other pigment present in the chloroplast or may be masked by other pigments (carotenoids, xanthophylls, biliproteins).

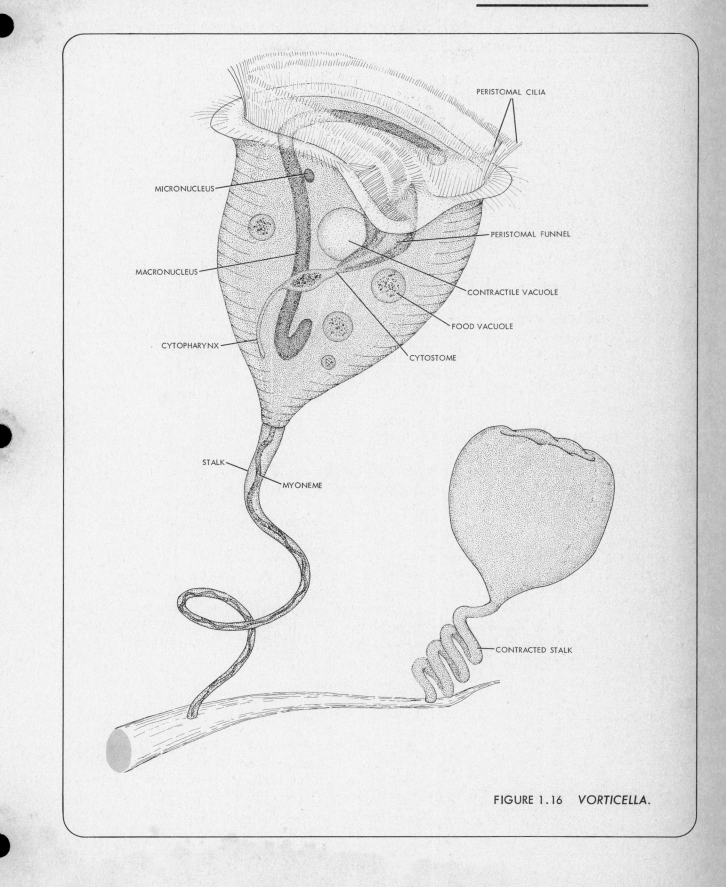

PERISTOMAL CILIA

MICRONUCLEUS

MACRONUCLEUS

PERISTOMAL FUNNEL

CONTRACTILE VACUOLE

FOOD VACUOLE

CYTOPHARYNX

CYTOSTOME

STALK

MYONEME

CONTRACTED STALK

FIGURE 1.16 *VORTICELLA*.

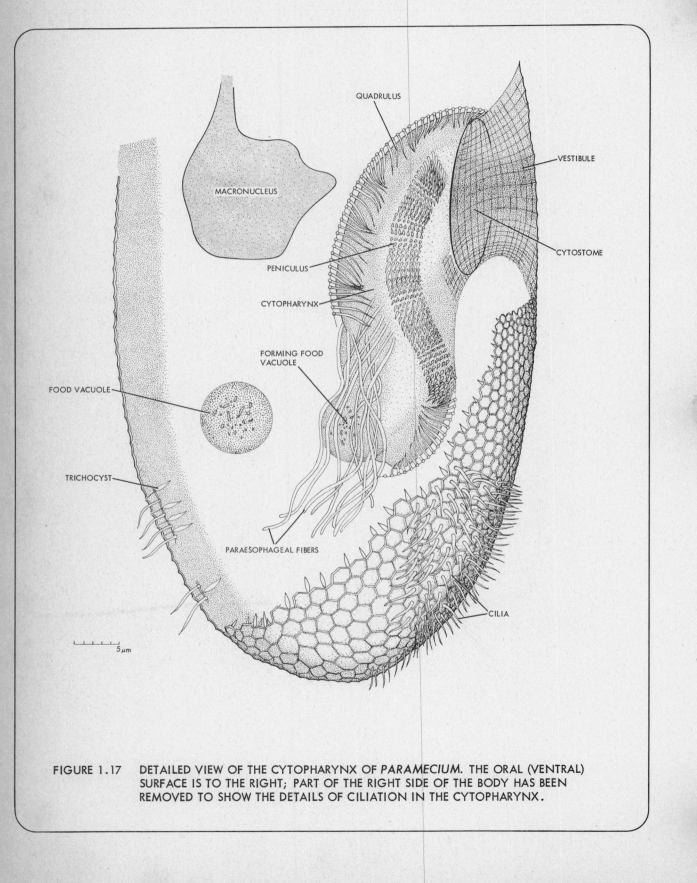

FIGURE 1.17   DETAILED VIEW OF THE CYTOPHARYNX OF *PARAMECIUM*. THE ORAL (VENTRAL) SURFACE IS TO THE RIGHT; PART OF THE RIGHT SIDE OF THE BODY HAS BEEN REMOVED TO SHOW THE DETAILS OF CILIATION IN THE CYTOPHARYNX.

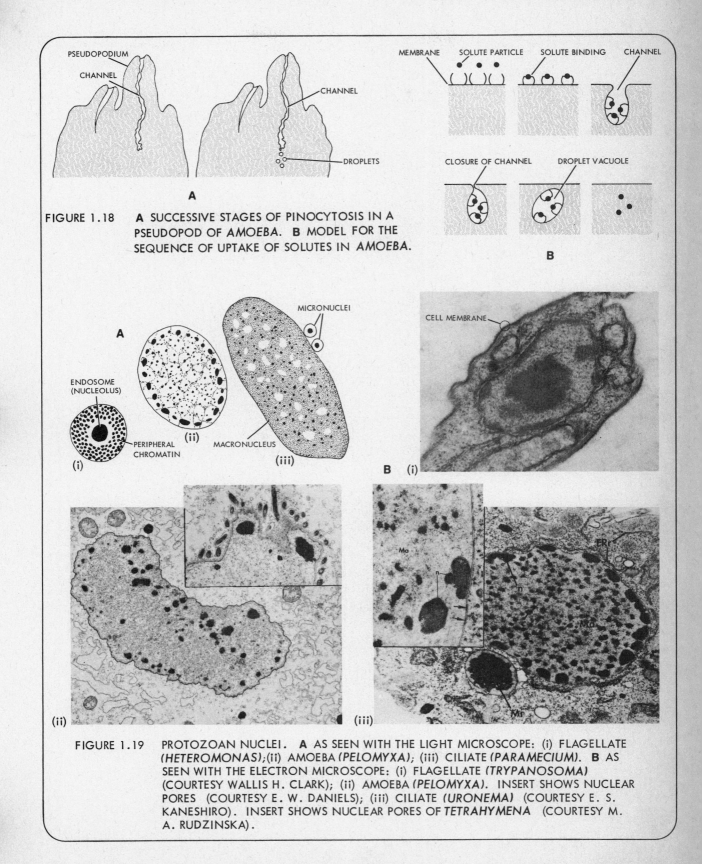

FIGURE 1.18 **A** SUCCESSIVE STAGES OF PINOCYTOSIS IN A PSEUDOPOD OF *AMOEBA.* **B** MODEL FOR THE SEQUENCE OF UPTAKE OF SOLUTES IN *AMOEBA.*

FIGURE 1.19 PROTOZOAN NUCLEI. **A** AS SEEN WITH THE LIGHT MICROSCOPE: (i) FLAGELLATE (*HETEROMONAS*); (ii) AMOEBA (*PELOMYXA*); (iii) CILIATE (*PARAMECIUM*). **B** AS SEEN WITH THE ELECTRON MICROSCOPE: (i) FLAGELLATE (*TRYPANOSOMA*) (COURTESY WALLIS H. CLARK); (ii) AMOEBA (*PELOMYXA*). INSERT SHOWS NUCLEAR PORES (COURTESY E. W. DANIELS); (iii) CILIATE (*URONEMA*) (COURTESY E. S. KANESHIRO). INSERT SHOWS NUCLEAR PORES OF *TETRAHYMENA* (COURTESY M. A. RUDZINSKA).

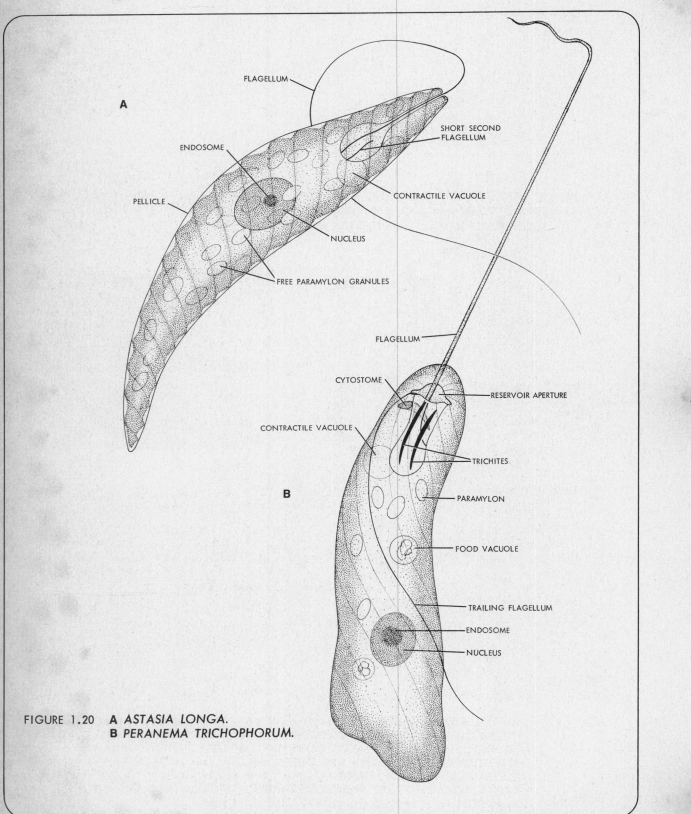

FLAGELLUM

A

SHORT SECOND
FLAGELLUM

ENDOSOME

CONTRACTILE VACUOLE

PELLICLE

NUCLEUS

FREE PARAMYLON GRANULES

FLAGELLUM

CYTOSTOME

RESERVOIR APERTURE

CONTRACTILE VACUOLE

TRICHITES

PARAMYLON

FOOD VACUOLE

B

TRAILING FLAGELLUM

ENDOSOME

NUCLEUS

FIGURE 1.20   **A** *ASTASIA LONGA.*
              B *PERANEMA TRICHOPHORUM.*

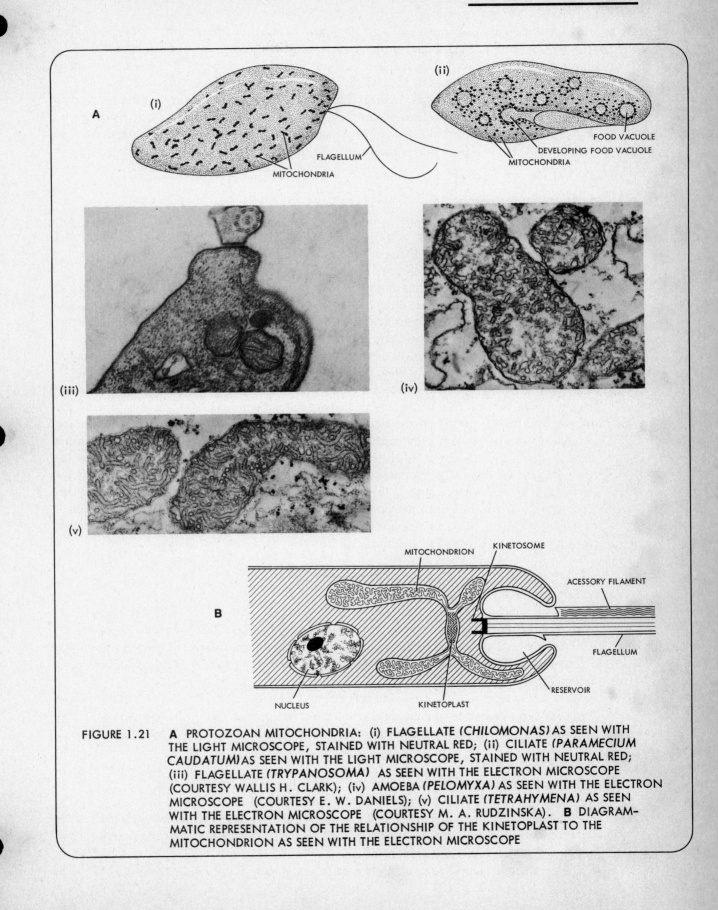

FIGURE 1.21   **A** PROTOZOAN MITOCHONDRIA: (i) FLAGELLATE *(CHILOMONAS)* AS SEEN WITH THE LIGHT MICROSCOPE, STAINED WITH NEUTRAL RED; (ii) CILIATE *(PARAMECIUM CAUDATUM)* AS SEEN WITH THE LIGHT MICROSCOPE, STAINED WITH NEUTRAL RED; (iii) FLAGELLATE *(TRYPANOSOMA)* AS SEEN WITH THE ELECTRON MICROSCOPE (COURTESY WALLIS H. CLARK); (iv) AMOEBA *(PELOMYXA)* AS SEEN WITH THE ELECTRON MICROSCOPE (COURTESY E. W. DANIELS); (v) CILIATE *(TETRAHYMENA)* AS SEEN WITH THE ELECTRON MICROSCOPE (COURTESY M. A. RUDZINSKA).   **B** DIAGRAMMATIC REPRESENTATION OF THE RELATIONSHIP OF THE KINETOPLAST TO THE MITOCHONDRION AS SEEN WITH THE ELECTRON MICROSCOPE

▶ Observe the shape and form of the chloroplasts in *Euglena* (Figure 1.4), *Chlamydomonas* (Figure 1.3), and *Volvox* (Figure 1.5). Studies with the electron microscope reveal that the chloroplast is composed of many layers of membranes. The centrally located **pyrenoid** is surrounded by plates of starch, the storage product of photosynthesis (Figure 1.3).

▶ Observe the blue pigment of *Stentor* (Figure 1.12A), called **stentorin**, and its location in longitudinal bands. *Blepharisma* (Figure 1.10) contains a pink pigment called **zoopurpurin,** which is enclosed in spheroidal double-membrane-bound bodies about 0.5 μm in diameter. The function of these two pigments is unknown, and colorless ciliates do exist in nature.

d. Photoreceptors

Protozoa respond to environmental stimuli such as chemicals, currents, food, touch, and light in the same way as other living cells. The ability to perceive and react to light may be a general property of the cell itself or of specialized organelles within the cell.

Some investigators believe that the photoreceptor in *Euglena* is a small red body near the gullet called the **stigma.** Others believe that a swelling at the base of the flagellum is the photoreceptor and that the stigma functions only as a light shield. In *Chlamydomonas* the stigma occupies a portion of the chloroplast.

▶ Examine wet mounts of *Euglena* (Figure 1.4) and attempt to locate the stigma. It is more easily seen in bleached strains without pigments.

▶ Phototaxis in *Euglena* can be demonstrated by placing a culture in a test tube and enclosing the entire tube in a cylinder made of black paper or aluminum foil. After allowing time for the organisms to take up position in the culture, remove the cover quickly and note the distribution of organisms in the medium. Now repeat the same procedure with a modification in the opaque cover so that it has a slit down its side or a hole ½ to 1 cm in diameter. Illuminate, then quickly remove the cover after allowing time for the organisms to take up their positions. What is the distribution of organisms relative to the area where light entered the tube? Using a culture that has been grown in diffuse light, illuminate the tube with a spot of light and note the migration of animals toward the light. Remember that light intensity as well as direction play a role in phototaxis, and intensities that are too great can produce a negative response.

e. Trichocysts

Trichocysts are structures that lie in the ectoplasm of some ciliates (Figures 1.8 and 1.22). Each trichocyst is a tiny vesicle containing a viscid solution, which can be squirted out through a pore in the pellicle to form a long thread. Trichocysts may be used as weapons of offense or defense or for attachment.

▶ Make wet mounts of *Paramecium* or other ciliates and put a drop of writing ink at the edge of the cover slip. Allow it to run underneath and observe the discharge of the trichocysts. (Dilute acetic acid or 0.5% aqueous methylene blue will also work, but not quite as well as ink.)

f. Contractile Vacuoles

Protozoa living in hypotonic media (fresh water) generally regulate the water content of the cell by expulsion of excess water via specialized organelles known as **contractile vacuoles.** Their principal function is osmoregulation.

▶ The contractile vacuole is particularly easy to see in *Amoeba proteus* (Figure 1.14A) and in

Paramecium (Figure 1.8A) when a drop of culture containing these organisms is mixed with a drop of 10% nigrosin, then covered with a cover slip and examined under high power.

▶ The osmoregulatory function of the contractile vacuole can easily be demonstrated by timing the frequency with which the contractile vacuole empties when a protozoan is in solutions of various osmotic concentration. Place drops from a culture of *Amoeba proteus* or *Paramecium* into a series of depression slides containing solutions of increasing salt concentration. Try not to dilute the salt solutions appreciably by adding large volumes of protozoan culture fluid. Remove a drop after a few minutes, make a wet mount, and examine under high power. Note the rate of emptying of the vacuole by counting the time between discharges. (Work in pairs, and have one person observe the contractile vacuole while the other reads a watch.) Count the rate of discharge in *Tetrahymena,* which have been grown in proteose peptone medium, and then transfer some organisms to a depression slide containing distilled water. What happens to the rate of discharge? (Try to avoid transferring too much proteose peptone to the water when adding the *Tetrahymena.*)

Does the contractile vacuole have a fixed position in the cell?

1–4. FEEDING

Nutrition in the Protozoa may be **holophytic (autotrophic), saprozoic,** and/or **holozoic.** The holophytic Protozoa are those containing the photosynthetic pigment **chlorophyll**; the organism manufactures its own food by photosynthesis, using carbon dioxide, nitrogen, and trace metals obtained from the environment, and the energy from sunlight. The presence of chloroplasts in these Protozoa demonstrates their plantlike nature, and this is reflected in their taxonomy (Phytomastigophorea). For many years, this group was of uncertain position, being claimed by zoologists and botanists alike. Today, their intermediate nature is recognized and the Protozoa form a natural link between the animal and plant kingdoms.

Saprozoic forms, e.g., colorless flagellates, do not ingest solid food, but absorb complex dissolved materials from the environment.

The majority of Protozoa are holozoic, ingesting solid, ready-made food such as bacteria, other Protozoa, or organic detritus. This particulate food has to be broken down by a process of enzymic digestion before it can be used by the organism for growth, energy, and repair. The presence of organelles, e.g., cytostome and tentacles, for the capture of food is correlated with the holozoic habit.

a. Phagocytosis (Particle Feeding)*

▶ Place a drop of *Paramecium* culture on a slide and mix it with a *small* drop of Congo-red-stained yeast suspension. The color of the mixture should be pink, not red. Cover with a cover slip, and observe the vortex of water produced by the ciliary activity and the way in which particles are trapped in the **oral groove** (Figures 1.8A and 1.17), which is situated on the ventral surface of the organism. At the posterior end of the groove is an invaginated pellicular region, the **vestibule,** which leads to the **cytostome** (cell mouth) and then to the **cytopharynx.** The cytopharynx is a complicated structure with numerous cilia on its inner surface, arranged in two bands; one band is the **peniculus,** the other is the **quadrulus** (Figure 1.17). The cytopharynx leads to an **esophageal sac** where the food vacuole is formed. As time passes note the

---

*If possible, this exercise should be started at the beginning of the laboratory period.

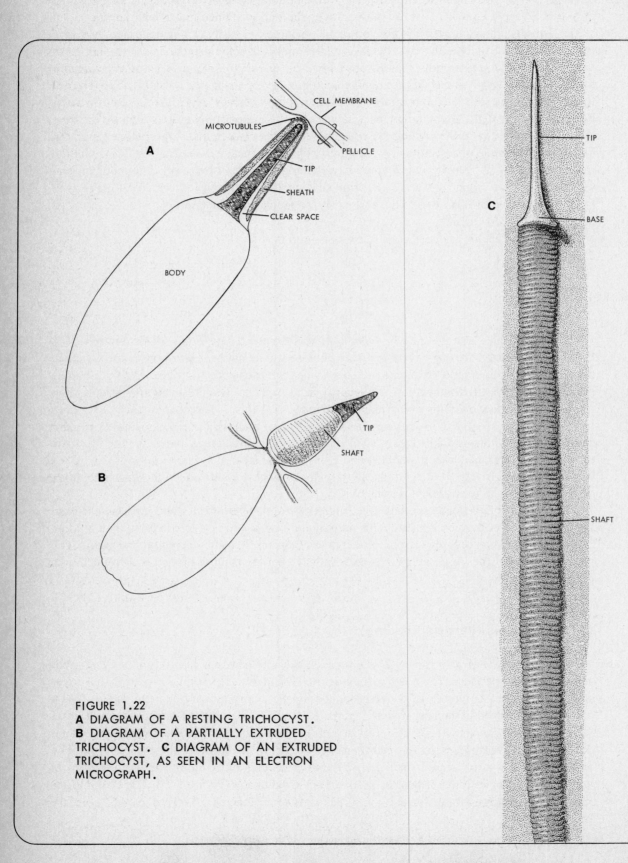

FIGURE 1.22
A DIAGRAM OF A RESTING TRICHOCYST.
B DIAGRAM OF A PARTIALLY EXTRUDED
TRICHOCYST. **C** DIAGRAM OF AN EXTRUDED
TRICHOCYST, AS SEEN IN AN ELECTRON
MICROGRAPH.

appearance of granules in the food vacuoles and the series of color changes that the granules subsequently undergo. What causes these color changes? What process is going on? Try similar feeding experiments with *Stentor* and *Vorticella*.

▶   Obtain a concentrated culture of starved amoebae. Add samples of the freshwater Protozoa *Chilomonas, Tetrahymena,* and/or the pigmented ciliate *Blepharisma* to the culture dish. Add one drop of 0.1% neutral red to 10 ml of medium containing amoebae and prey. After a while, make wet mounts and observe the movement of the amoebae, the formation of the "food cup," and the digestive process. Can you see changes of pH in the food vacuoles during digestion? How do you account for the fact that a slow-moving organism such as *Amoeba* is able to capture rapidly moving ciliates?

b. Pinocytosis (Cell Drinking)

▶   One of the simplest and most effective means of inducing pinocytosis, or "cell drinking," is by placing *Amoeba proteus* in a solution of 0.125 M NaCl in 0.01 M phosphate buffer at pH 6.5 to 7.0. At room temperature, channels will form in 2 or 3 minutes and continue to do so for up to 30 minutes. Apply a cover slip above a wet mount of a sample of the treated amoebae by ringing the edge of the drop with petroleum jelly. Notice the rounding of the amoebae, the formation of fine channels in the pseudopodia, and droplet formation. (See Chapman-Andresen for details and Figure 1.18.)

What is the difference between phagocytosis and pinocytosis?

c. Holophytic Versus Saprozoic Nutrition

Some flagellates can vary their mode of nutrition, behaving as autotrophs or saprozoites according to environmental conditions.

▶   Obtain a log-phase culture of *Euglena gracilis*. Observing aseptic conditions,* examine a loopful of the culture on a slide to be sure the cells are motile. If so, transfer a loopful of the culture aseptically to each of three test tubes containing 5 ml each of sterile peptone–acetate medium. Place one in a water bath in the light at 35°C, one in the dark at room temperature (20° to 30°C), and leave one in the light at room temperature. After 45 minutes, remove the culture from the water bath and return it to the light at room temperature. After 2 or 3 days, compare the cultures by examining a loopful of the cultures on a slide for greenness, observing aseptic conditions. Record the numbers of green, yellow-green, and white in each culture. Place the culture from the dark in the light and compare the cultures again after 2 or 3 more days.

▶   Record and write up your experimental results.

What implications can you draw from this experiment concerning the possible evolution of holozoic and saprozoic Protozoa?

1–5. GROWTH AND REPRODUCTION

a. Asexual Reproduction

Binary fission is the simplest and most common method of division in Protozoa. The cell divides mitotically into approximately equal halves, giving rise to two daughter cells of

---

*Flame the loop and the mouth of the test tube before inserting the loop into the medium (allow the loop to cool briefly after flaming). Flame the test tube mouth before replacing the plug.

identical genetic constitution. The plane of division is longitudinal in flagellates, transverse in ciliates, and without fixed orientation in many sarcodines.

▶ Examine demonstration slides of fission. Does longitudinal binary fission in flagellates begin at the anterior (flagellar) or posterior end?

b. Sexual Reproduction

Protozoa exhibit two kinds of sexual phenomena.

i. SYNGAMY, OR THE FUSION OF GAMETES. Gametes may be similar in form (**isogametes**) or different, as egg and sperm (**anisogametes**). The fusion of gametes (isogamy or anisogamy) gives rise to a **zygote,** and gametic nuclear fusion produces a **synkaryon.** *Chlamydomonas reinhardi* will be used to demonstrate isogamy.

▶ Using clean pipets for each operation, place relatively equal concentrations of two mating types of *Chlamydomonas reinhardi* in separate drops on a slide, about 1 cm apart. Do not add cover slips at this time. Examine the drops under the microscope and note the distribution of flagellates. Carefully mix the two drops with a pipet. Clumping of gametes should follow in a few minutes. The pairs swim attached by their anterior ends; cytoplasmic fusion or **plasmogamy** does not occur for 6 to 8 hours.
What adaptations for sexual recombination are found in this organism? Diagram its life cycle.

ii. CONJUGATION. Conjugation is a method of exchanging genetic material that is restricted to members of the subphylum Ciliophora. It involves the temporary adhesion of individuals of complementary mating types, breakdown of the macronucleus, meiotic division of the micronuclei, micronuclear exchange, and fusion of the migratory micronuclei with nonmigratory (stationary) micronuclei. The resultant **synkaryon** undergoes mitotic division and the conjugants separate. By appropriate numbers of mitotic cell divisions, the correct micronuclear number is restored; one micronucleus in each cell forms the new macronucleus (Figure 1.23).

▶ Obtain cultures of mating types of *Paramecium bursaria*. This ciliate has green **zoochlorellae** (symbiotic unicellular photosynthetic algae) in its cytoplasm and is quite distinct from the paramecia you have previously examined. Add 1 or 2 drops of each mating type to a series of depression slides and do not mix the mating types. Use a clean pipet each time you sample a mating type. Observe the movements of the paramecia. Is there any clumping? Now mix the mating types by adding a drop of one mating type to all the others. Observe the reactions that occur. Where do the animals attach? Do they attach by random contact? What is the meaning of the term **mating type**? What is the net effect of conjugation?

c. Growth of Protozoan Populations

Because of their small size the study of structure and function in individual Protozoa is difficult. Physiological investigations, however, can be simplified by the use of populations. Protozoa that reproduce asexually by binary fission yield individuals of identical genetic constitution. It is therefore possible to start a population from a single individual, to study the characteristics of all the progeny, and to relate these to the physiology of the original isolate. A population derived by asexual reproduction from a single individual is called a **clone.** This exercise is designed to acquaint you with the concept of **generation time** (time for doubling of the population) and with characteristics of population growth.

▶ *Tetrahymena* is a ciliate that is easily grown in proteose peptone medium. With a sterile pipet, inoculate tubes containing sterile medium with 1 or 2 drops of *Tetrahymena* culture. Flame the mouth of the tube and the pipet before and after inoculation. Incubate in subdued light at room temperature. The culture should become rich in about 24 to 48 hours. Once this has occurred, take up samples from each culture in the capillary ends of Pasteur pipets that have their wide ends plugged with cotton. Seal the organisms in the tip of each pipet by dipping in melted paraffin or in some plasticine. Count the number of individuals in the tube under a dissecting or compound microscope. If there are too many cells, dilute the culture with proteose peptone medium so that the number may be counted easily (e.g., five to ten). At intervals of 6, 12, and 24 hours, count the number of cells in the tubes. Record zero time and the numbers found at the time intervals. What is the generation time? Plot your data on a graph, with numbers of organisms on the ordinate and time on the abscissa.

What is meant by the terms **lag phase, log phase, stationary phase**?

1–6. TESTS, SHELLS, AND SKELETONS

Protozoa are capable of building protective shells around themselves by the secretion of materials or by cementing substances from the environment in such a way as to form a "house" or test. Great variety is found among protozoan skeletons, many of which have a complex architecture. Only a few examples will be studied.

▶ i. CELLULOSE. Among the flagellates, the armored dinoflagellates best show the arrangement of skeletal cellulose plates. Observe the form and architecture of the plates in prepared slides of *Ceratium* (Figure 1.24**B**). What is the purpose of the two grooves? What is the most obvious function of the skeleton in dinoflagellates?

▶ ii. CHITIN. Another organic material formed by Protozoa is chitin. The difference between cellulose and chitin is that the former is a polymer of glucose units, whereas the latter is a polymer of *N*-acetylglucosamine. Obtain a culture of *Arcella*. Make wet mounts and observe the form of the test and its texture (Figure 1.24**A**). What is the function of the opening? If this form divides by binary fission, what happens to the test?

▶ iii. SILICA. Radiolaria characteristically have a central capsule with radiating filaments composed of silica. Because the skeleton is extremely resistant to chemical degradation, the ocean floor is strewn with the remains of these marine creatures. The deposit thus formed is referred to as **radiolarian ooze.** Examine slides showing something of the varied and beautiful form of radiolarian skeletons (Figure 1.24**C**).

▶ iv. CALCIUM CARBONATE. Foraminifera, like their radiolarian cousins, also form a shell of complex beauty, but in this case it is composed largely of calcium carbonate. The form of these Protozoa can best be seen in dry shells. The shell is usually permeated by pores (*foram,* window; *fer,* bearer; L.) through which a bewildering array of filamentous pseudopods called reticulopodia pass in the living animal (Figure 1.24**E**). The shell is extremely resistant, consisting of 90% $CaCO_3$, which in some cases forms large deposits on the ocean floor called *Globigerina* ooze, after the predominant genus forming the ooze. The white cliffs of Dover

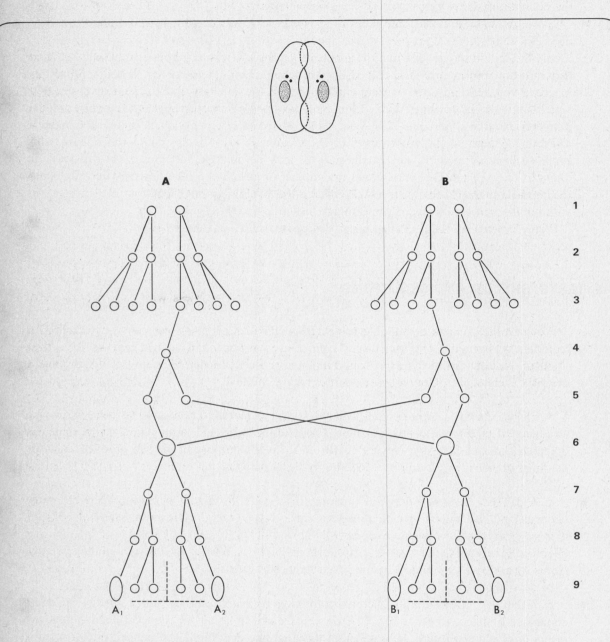

FIGURE 1.23 CONJUGATION IN *PARAMECIUM AURELIA*. TWO ANIMALS SHOWN IN POSITION PRIOR TO CONJUGATION. **A** AND **B** SHOW THE SEQUENCES OF EVENTS FOR THE RESPECTIVE MICRONUCLEI. EACH CONJUGANT HAS TWO MICRONUCLEI (1), WHICH DIVIDE TWICE (2 AND 3) TO YIELD EIGHT PRONUCLEI, SEVEN OF WHICH DEGENERATE. THE GONAL NUCLEUS (4) DIVIDES TO FORM TWO GAMETIC NUCLEI (5). ONE OF THESE MIGRATES INTO THE OTHER CONJUGANT AND FUSES WITH ITS STATIONARY GAMETIC NUCLEUS TO FORM A ZYGOTIC NUCLEUS (6). THE CONJUGANTS SEPARATE, AND THE SYNKARON DIVIDES TWICE (7 AND 8). A CELL DIVISION AND A FURTHER NUCLEAR DIVISION FOLLOW TO YIELD OFFSPRING WITH ONE MACRO- AND TWO MICRONUCLEI. THE TWO CONJUGANTS (A AND B) HAVE PRODUCED FOUR OFFSPRING (A_1, A_2, B_1, B_2) (COURTESY A. R. JONES).

were formed from such an ooze and were raised by orogenic movements at a later geological period from that in which they were deposited under water. Observe the varied form of the shell of different foraminiferan species; this is a diagnostic character in species identification.

▶ v. ARENACEOUS. Protozoa that cement sand grains together to build shells are described as being arenaceous. It is of interest to note that the form of the "house" and the size of the sand grains chosen are remarkably uniform within a species, so that these features can be used as diagnostic characters in taxonomy. Long after the living animal is gone, the species can be recognized by the structure of its shell. Observe demonstration slides of the amoeba *Difflugia* (Figure 1.24**D**) and/or the ciliate tintinnids (Figure 1.24**F**).

1–7. ECOLOGICAL SUCCESSION OF PROTOZOAN COMMUNITIES

Protozoa are restricted to a watery medium, but during adverse conditions, such as times of drought, many are able to resist desiccation or cold or lack of food by going into a dormant and protected state and forming **cysts** around themselves. We understand little about the physiological triggers that produce encystment, but we can take advantage of the process of excystment to produce a population in the laboratory and watch its progress with time and changing environment.

▶ Place some dry hay or grass in tap or distilled water in a jar and cover with a glass plate to prevent evaporation. At intervals during the next week or so, remove drops of the water from this **hay infusion** and examine them under the compound microscope. At the beginning, bacteria will predominate, followed by saprophytic flagellates, then herbivorous ciliates and amoebae that feed on bacteria, and eventually the carnivorous ciliates and amoebae. The appearance of each new species is related to factors such as light intensity and quality, ratio of gases present, pH, and concentration of organic compounds. From your observations and reference to Jahn and Jahn, try to identify some of the species in the infusion. Estimate the numbers present (e.g., number of organisms per high power field) and plot the distribution of each species or group of species on a graph.

▶ Alternatively, succession in a protozoan community may be followed in some pond water brought into the laboratory or in tap water to which one of the protozoan tablets supplied by commercial supply houses has been added.

The sequence of appearance of organisms forms what is known as a **food chain.** What are the earliest members of the food chain? In this context what are **producers** and **consumers**? If biological succession is the rule, why do all ponds not consist exclusively of predators?

References

Allen, R. D., and N. Kamiya (eds.). *Primitive Motile Systems in Cell Biology*. New York: Academic Press, 1964.

Barnes, R. D. *Invertebrate Zoology,* 3rd ed. Philadelphia: W. B. Saunders Co., 1974.

Beale, G. *The Genetics of Paramecium aurelia*. London: Cambridge University Press, 1954.

Borradaile, L. A., and F. A. Potts. *The Invertebrata*. London: Cambridge University Press, 1961.

Brown, F. A., Jr. (ed.). *Selected Invertebrate Types*. New York: John Wiley & Sons, 1950.

Buchsbaum, R. *Animals Without Backbones,* 2nd ed. Chicago: University of Chicago Press, 1972.

Bullough, W. S. *Practical Invertebrate Anatomy*. New York: St. Martin's Press, 1962.

Carter, G. S. *A General Zoology of the Invertebrates*. London: Sidgwick and Jackson, 1951.

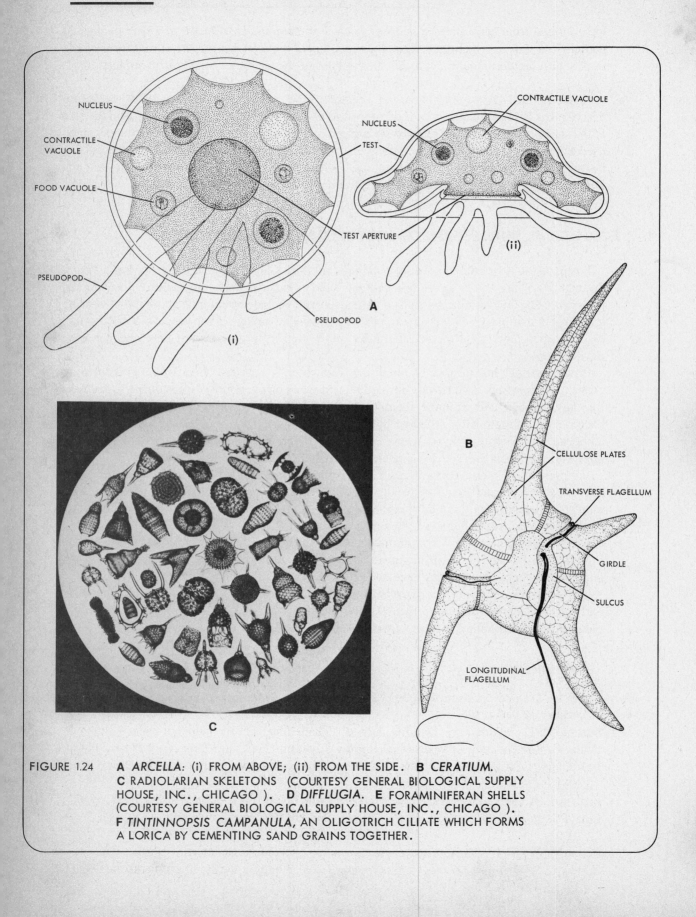

FIGURE 1.24   **A** *ARCELLA*: (i) FROM ABOVE; (ii) FROM THE SIDE. **B** *CERATIUM.*
C RADIOLARIAN SKELETONS (COURTESY GENERAL BIOLOGICAL SUPPLY
HOUSE, INC., CHICAGO). **D** *DIFFLUGIA.* **E** FORAMINIFERAN SHELLS
(COURTESY GENERAL BIOLOGICAL SUPPLY HOUSE, INC., CHICAGO).
F *TINTINNOPSIS CAMPANULA*, AN OLIGOTRICH CILIATE WHICH FORMS
A LORICA BY CEMENTING SAND GRAINS TOGETHER.

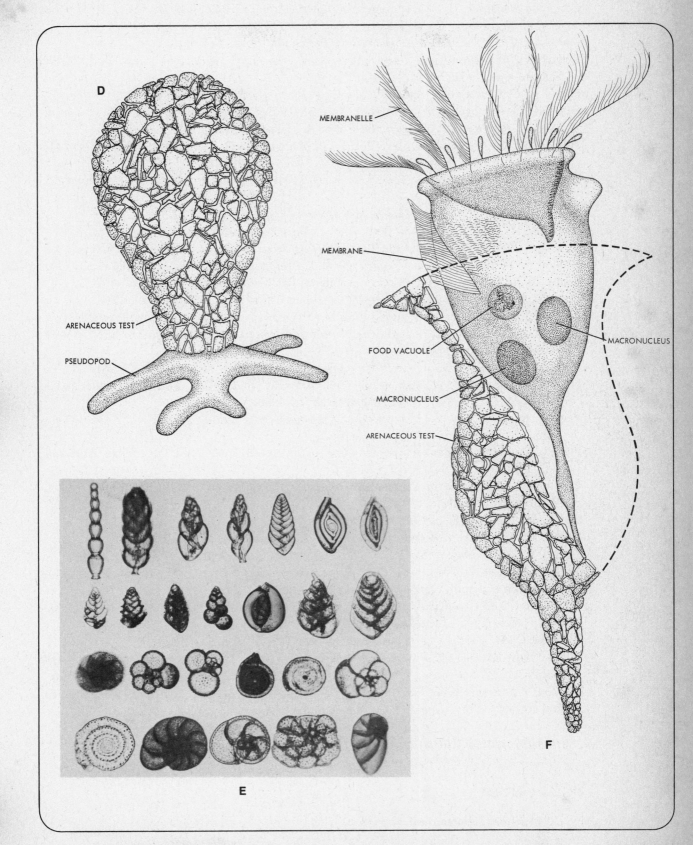

D

MEMBRANELLE

ARENACEOUS TEST

PSEUDOPOD

MEMBRANE

FOOD VACUOLE

MACRONUCLEUS

ARENACEOUS TEST

MACRONUCLEUS

E

F

Chapman-Andresen, C. "Measurement of Material Uptake by Cells: Pinocytosis," in *Methods in Cell Physiology,* D. M. Prescott (ed.). New York: Academic Press, 1964.

Corliss, J. *The Ciliated Protozoa.* Oxford: Pergamon Press, 1961.

Dobell, C. C. *Antony van Leeuwenhoek and His "Little Animals."* New York: Dover Publications, 1960.

Giese, A. C. *Blepharisma.* Stanford: Stanford University Press, 1973.

Grell, K. G. *Protozoology.* New York: Springer-Verlag, 1973.

Hall, R. P. *Protozoology.* Englewood Cliffs, N.J.: Prentice-Hall, 1953.

Hyman, L. H. *The Invertebrates,* Vol. I. *Protozoa Through Ctenophora.* New York: McGraw-Hill Book Co., 1940.

Jahn, T. L., and F. F. Jahn. *How to Know the Protozoa.* Dubuque, Iowa: W. C. Brown Co., 1949.

Jeon, K. W. *The Biology of Amoeba.* New York: Academic Press, 1973.

Jones, A. R. *The Ciliates.* New York: St. Martin's Press, 1974.

Jurand, A., and G. G. Selman. *The Anatomy of Paramecium aurelia.* London: Macmillan, 1969.

Kidder, G. W. (ed.). *Chemical Zoology,* Vol. I. *Protozoa.* New York: Academic Press, 1967.

Kudo, R. R. *Protozoology,* 5th ed., Springfield, Ill.: C. C. Thomas Publishers, 1971.

Leedale, G. F. *Euglenoid Flagellates.* Englewood Cliffs, N.J.: Prentice-Hall, 1967.

Lwoff, A. *Biochemistry and Physiology of the Protozoa.* New York: Academic Press, 1951.

Mackinnon, D. L., and R. S. J. Hawes, *An Introduction to the Study of Protozoa.* New York: Oxford University Press, 1961.

Manwell, R. D. *Introduction to Protozoology.* New York: St. Martin's Press, 1961.

Nanney, D. L., and M. A. Rudzinska. "The Protozoa," in *The Cell,* Vol. I, J. Brachet and A. E. Mirsky (eds.). New York: Academic Press, 1960.

Pennak, R. W. *Fresh Water Invertebrates of the United States.* New York: Ronald Press, 1953.

Pitelka, D. *Electron Microscopic Structure of Protozoa.* New York: Macmillan Publishing Co., 1963.

Pratt, H. S. *A Manual of Common Invertebrate Animals.* New York: Blakiston Division, McGraw-Hill Book Co., 1935.

Sleigh, M. *The Biology of Cilia and Flagella.* Oxford: Pergamon Press, 1962.

Sleigh, M. *The Biology of Protozoa.* London: Arnold, 1973.

Tartar, V. *The Biology of Stentor.* Oxford: Pergamon Press, 1961.

Taylor, D. L., et al. The contractile basis of ameboid movement, I. The chemical control of motility in isolated cytoplasm, *J. Cell Biol.* **59:**378–94, 1973.

Trager, W. "The Cytoplasm of Protozoa," in *The Cell,* Vol. VI, J. Brachet and A. E. Mirsky (eds.). New York: Academic Press, 1964.

Wenyon, C. M. *Protozoology.* London: Bailliere, Tindall and Cox, 1926.

Wichterman, R. *The Biology of Paramecium.* New York: Blakiston Division, McGraw-Hill Book Co., 1951.

Wolken, J. J. *Euglena: An Experimental Organism for Biochemical and Biophysical Studies,* 2nd ed. New York: Appleton-Century-Crofts, 1967.

An Abbreviated Classification of the Phylum Protozoa
(*J. Protozool.* 11:7–20, 1964.)

Phylum **Protozoa**

Subphylum I **Sarcomastigophora**
Having flagella, pseudopodia or both; single type of nucleus.

Superclass I **Mastigophora**

One or more flagella. Asexual reproduction by longitudinal binary fission.

Class 1 **Phytomastigophorea**

Chromatophores, and one or two flagella most commonly.

Order 1 **Chrysomonadida**

One to three flagella, yellow-brown chromatophores. Store leucosin and lipid.

Order 2 **Cryptomonadida**

Two flagella, body compressed, typically two chromatophores. Store starch, e.g., *Chilomonas*.

Order 3 **Dinoflagellida**

Two flagella, one transverse and one trailing, body with transverse and longitudinal grooves. Store starch and lipids, e.g., *Ceratium*.

Order 4 **Euglenida**

One or two flagella, green chromatophores, body plastic. Store paramylon, e.g., *Euglena, Astasia*.

Order 5 **Chloromonadida**

Two flagella, one trailing; body flattened. Store lipids and glycogen.

Order 6 **Volvocida**

Two to four flagella, green chromatophores. Store starch, e.g., *Chlamydomonas, Volvox*.

Class 2 **Zoomastigophorea**

Chromatophores absent; one to many flagella.

Order 1 **Choanoflagellida**

Anterior flagellum surrounded by a collar, e.g., *Codosiga, Proterospongia*.

Order 2 **Rhizomastigida**

Pseudopodia and one to four flagella, e.g., *Naegleria*.

Order 3 **Kinetoplastida**

One to four flagella; kinetoplast present, e.g., *Trypanosoma*.

Order 4 **Hypermastigida**

Numerous flagella, e.g., *Trichonympha*.

Order 5 **Trichomonadida**

Four to six flagella, with one recurrent, e.g., *Trichomonas*.

Superclass II **Opalinata**

Numerous cilia-like organelles, many similar nuclei, e.g., *Opalina*.

Superclass III **Sarcodina**

Pseudopodia; flagella, if present, restricted to developmental stages.

Class 1 **Rhizopodea**

Locomotion by lobopodia, filopodia or reticulopodia.

Subclass 1 **Lobosia**

Pseudopodia lobose, rarely filiform or anastomosing.

Order 1 **Amoebida**

Naked, typically uninucleate, e.g., *Amoeba, Pelomyxa, Entamoeba.*

Order 2 **Arcellinida**

Testate, e.g., *Arcella, Difflugia.*

Subclass 2 **Granuloreticulosia**

Fine, granular reticulopodia.

Order 1 **Foraminiferida**

Test with chambers; reticulopods protrude from aperture or perforations or both.

Class 2 **Actinopodea**

Spherical, axopodia, test of silica or strontium sulfate.

Subclass 1 **Radiolaria**

Siliceous skeleton with central capsule. Marine.

Subclass 2 **Heliozoa**

No central capsule, skeleton siliceous. Marine or freshwater.

Subphylum II **Sporozoa**

All species parasitic; spores typically present.

Subphylum III **Cnidosporidia**

All species parasitic; spores with polar filaments.

Subphylum IV **Ciliophora**

Cilia at some stage in the life cycle; two types of nuclei.

Subclass 1 **Holotrichia**

Uniform ciliation on body, e.g., *Paramecium, Tetrahymena.*

Subclass 2 **Peritrichia**

Body ciliation reduced or absent; oral cilia conspicuous, winding around apical pole counterclockwise toward cytostome. Body often attached, e.g., *Vorticella.*

Subclass 3 **Suctoria**

Adults lack cilia. Typically sessile with ingestion by sucking tentacles, e.g., *Tokophrya, Podophrya.*

Subclass 4 **Spirotrichia**

Body cilia sparse, adoral zone with membranelles winding clockwise toward cytostome.

Order 1 **Heterotrichida**

Somatic cilia when present uniform, body large, often pigmented, e.g., *Stentor, Blepharisma, Spirostomum.*

Order 2 **Oligotrichida**

Somatic cilia sparse.

Order 3 **Tintinnida**

Having lorica.

Order 4 **Hypotrichida**

Cirri on ventral surface, adoral zone of membranelles prominent. Organism flattened, e.g., *Euplotes.*

Equipment and Materials

Equipment

Microscope and lamp
Immersion oil
Xylene
Lens tissue
Facial tissue
Microscope slides
Cover slips
Blotting paper
Petroleum jelly
Toothpicks
Disposable pipets and rubber bulbs
Hay
Depression slides
Test tubes
Water bath
Loop (bacteriological)

Solutions and Chemicals

10% methocel
10% nigrosin
Congo-red-stained yeast suspension
0.125 M NaCl in 0.01 M PO_4 buffer
Acidified methyl green
Janus green B plus neutral red
Dilute acetic acid
0.5% aqueous methylene blue
1% proteose peptone
Lugol's iodine
Giemsa stain

Neutral red
Peptone–acetate medium

*Living Material**

Chlamydomonas reinhardi† (1)
Volvox (1)
Euglena (1)
E. gracilis (for nutrition experiment) (1)
Trypanosoma lewisi in rat blood (3)
Termite (for Protozoa of the gut) (6)
Chilomonas (11)
Peranema (1)
Astasia (1)
Paramecium caudatum (4,11)
Stentor (4,11)
Blepharisma (4,11)
Spirostomum (4,11)
Euplotes (4,11)
Vorticella (4,11)
Tetrahymena (4,11)
Amoeba proteus (4,11)
Pelomyxa (4,11)
Entamoeba terrapinae (3)
Paramecium bursaria, mating types (4,11)

Prepared Slides

Volvox
Chlamydomonas

*Numbers in parentheses identify sources listed in Appendix B.
†See directions for preparing mating types following.

Euglena	*Ceratium*
Trypanosoma sp.	*Arcella*
Flagellates of white ant (termite)	Radiolaria
Paramecium	*Difflugia*
Paramecium pellicle	Tintinnids
Stentor	Foraminifera
Blepharisma	*Volvox* cross section
Spirostomum	*Vorticella*
Euplotes	Silver line preparation of
Binary fission in *Paramecium*	*Tetrahymena*

Directions for Preparing Mating Types of Chlamydomonas reinhardi

The afternoon before demonstrations are desired, the cells from each strain should be washed from the agar on which they have been grown, with approximately 50 ml of sterile glass-distilled water; each cellular suspension should be placed in a separate 125 ml flask. Illuminate the flasks for several hours and then remove to a dark place until about 2 hours before they are needed the following day, at which time the flasks should be illuminated again.

Use of the Microscope: A Review

The compound microscope is an optical instrument equipped for double magnification of an object; it has accessory devices for gathering light and for holding the object to be magnified in a fixed position. The **lenses** are held in position in the body tube, with the **ocular** lens (or eyepiece) at the top and the **objective** lens at the bottom. The objective lens gives a primary enlarged image of the object; the ocular lens acts as a simple magnifier and further enlarges the image (Figure 1.25). Below the objective lens and parallel to the base of the microscope is the **stage.** The body tube, stage, and base are connected to each other by the **arm** of the microscope (Figure 1.26). The bottom of the tube contains the objective lenses, which are held in position by a revolving **nosepiece.** By rotation of the nosepiece, the lenses and thus the magnification may be changed. Your microscope is equipped with three objectives; the smallest objective is marked 10×, indicating a tenfold magnifying power; the next in size is marked 43×, and the third and longest objective, which carries a black band, magnifies about 90×. This last lens requires some special techniques for its use and is called the **oil-immersion objective.** Two adjustments are provided for focusing the lens system on the object: the **coarse adjustment** (large wheel) will move the lens system over a vertical distance of several inches to bring the object into approximate focus. The **fine adjustment** (small wheel) operates over a distance of a small fraction of an inch and is used for final focusing.

When not in use, the microscope should be stored in the cabinet provided, with the low-power objective locked into place and racked down to its lowest position. NEVER LEAVE THE MICROSCOPE WITH THE HIGH-POWER OR OIL-IMMERSION OBJECTIVE IN POSITION. The latter can be wound down fully and cracked into the stage or substage condenser.

▶ Carry your assigned microscope from its cabinet to your desk, holding it by the handle in one hand and supporting it by the horseshoe base with the palm of the other hand. Identify the coarse and fine adjustments, ocular, body tube, objective lenses, nosepiece, and stage. Note also the mirror, between the legs of the horseshoe base. With your left hand hold the base firmly, then grasp the handle in your right hand and move the handle back and down so as to orient the long axis of the microscope from a vertical to a horizontal position. Beneath the stage you will see the **substage condenser,** which can be moved up and down by means of the small wheel at its side, and the **substage iris diaphragm,** which can be opened and closed by means of

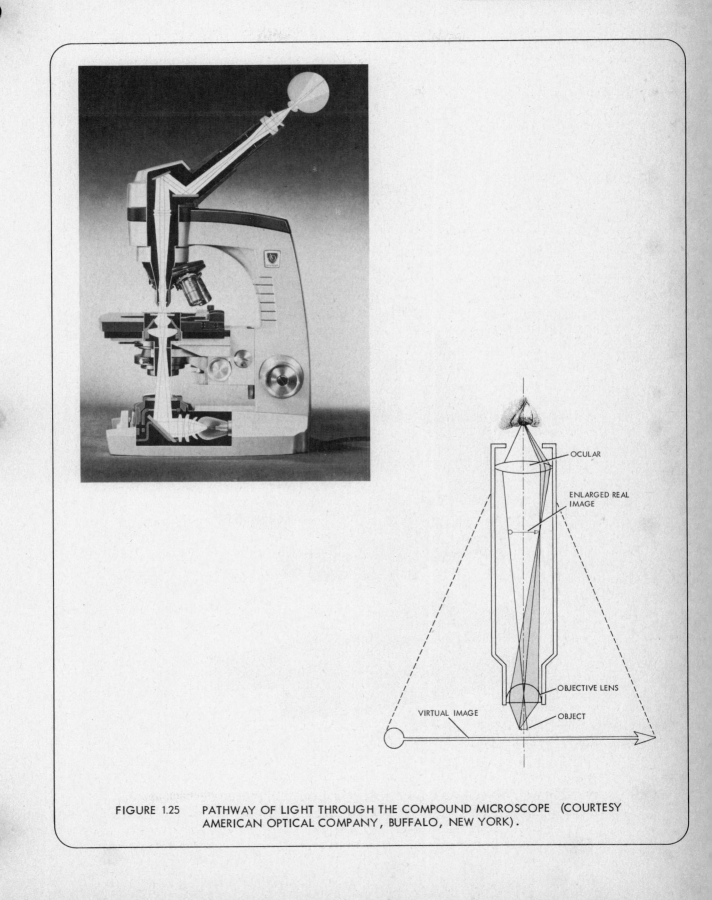

FIGURE 1.25   PATHWAY OF LIGHT THROUGH THE COMPOUND MICROSCOPE   (COURTESY AMERICAN OPTICAL COMPANY, BUFFALO, NEW YORK).

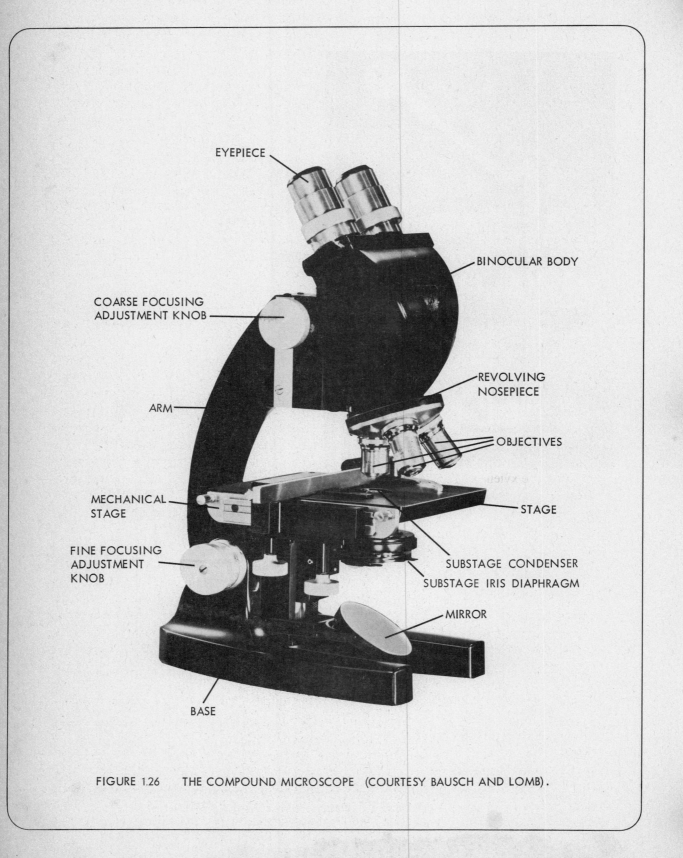

EYEPIECE

BINOCULAR BODY

COARSE FOCUSING
ADJUSTMENT KNOB

REVOLVING
NOSEPIECE

ARM

OBJECTIVES

MECHANICAL
STAGE

STAGE

FINE FOCUSING
ADJUSTMENT
KNOB

SUBSTAGE CONDENSER
SUBSTAGE IRIS DIAPHRAGM

MIRROR

BASE

FIGURE 1.26   THE COMPOUND MICROSCOPE (COURTESY BAUSCH AND LOMB).

a lever. Note the position of the lever when the diaphragm is opened and when closed. Now return the microscope to its original position, i.e., body tube vertical and the stage exactly parallel to the table top. THE MICROSCOPE SHOULD NEVER BE USED WITH THE STAGE INCLINED FROM THE HORIZONTAL. Rack the substage condenser all the way up, open the substage iris diaphragm to its widest position, rotate the nosepiece so that the 10× objective is in place, and make sure the latter clicks into position. Make sure all lenses are free of dust, oil, etc. Use only lens tissue and nothing else, for your handkerchief or facial tissues may have fine dust particles adhering to them which will scratch the lenses. Wipe clean the ocular and objective lenses, the top of the substage condenser, and the mirror. Tilt the mirror so that the flat side is directed away from the microscope and faces the microscope lamp.

One of the great faults of beginning microscopists is failing to let enough light enter the microscope. **Critical** or **Köhler** illumination is obtained when the condenser focuses light onto the object and the light transmitted from the object to the objective lens almost completely fills this lens. Illumination by artificial light is far superior to daylight.

▶ Place the microscope lamp about 30 cm from the mirror, remove the ground glass of the lamp, turn the lamp on, and direct the light so that it shines on the center of the mirror. Adjust the position of the bulb and the iris diaphragm of the lamp so that the coils of the bulb filament are clearly seen in the face of the mirror. (Lens tissue placed over the surface of the mirror may help to give better definition of the filament coils.) Now look through the eyepiece and rotate the mirror until the beam of light passes through the hole in the stage and reaches your eye. Make sure that you have the flat side of the mirror facing the lamp, that the light intensity is maximum, and that the substage iris diaphragm is wide open. Looking through the eyepiece, focus the substage condenser by moving it up and down until you see a sharp image of the iris diaphragm of the lamp. Replace the ground glass in the lamp; then open the lamp's iris diaphragm so that light fills the entire field of view. To assure that light fills the objective lens, remove the eyepiece and look down the tube of the microscope. Replace the ocular and place a clean slide (e.g., a diatom) in position on the stage. A useful technique for removing oil from the surface of a dirty slide is to place a drop of xylene on the oily area, lay a piece of lens tissue on top of the xylene and drag the lens tissue across the slide. No rubbing is necessary and only the weight of the lens tissue should be applied. If the slide is not free of oil, repeat the procedure.

▶ Move the object on the slide to the center of the light beam. With the 10× objective in position, focus the object by using the coarse and fine adjustments. You may have to reduce the cone of light by using the substage iris diaphragm. The iris diaphragm reduces the cone of light so that after passing through the object, the light does not diverge beyond the diameter of the objective lens. Move the specimen from left to right and notice the direction in which the object moves as you view it. Is the image you see reversed? The total magnification of the object is the magnification of the objective lens multiplied by the ocular magnification. If the ocular magnifies 10×, what is the total magnification of the object under each of the three objective lenses?

▶ Switch from the 10× to the 43× objective. The center of the field of the low-power objective should correspond with the center of the field of the high-power objective. Because the lenses are **parfocal,** only a minor adjustment in focus should be necessary. Carefully move the fine adjustment wheel; if this does not bring the specimen into view, use the coarse adjustment and rotate the wheel toward you and then toward the microscope slide. If the object appears too dark, open the diaphragm. How does the size of the field (visible area) compare under 10× and 43×? How does magnification of the object relate to field size? Is more or less light required with higher magnification? Which lens has the greatest working distance (i.e., distance between objective and object)? Which has the shortest working distance?

The oil-immersion lens requires some special attention. The **resolving power** of a lens (i.e., the ability to separate adjacent objects as distinct entities) depends on the **refractive index** of the medium (n) through which the light passes between objective and object, as well as the angle through which the objective lens gathers light (A). The relationship is expressed by a term known as numerical aperture (NA):$NA = n \times \sin \frac{1}{2}A$. The greater the numerical aperture, the better the resolution. Because the objectives are physically limited by the angle through which the light can be gathered, the numerical aperture becomes limited by the refractive index of the least refractive medium in the light path. Increasing the refractive index of the medium between lens and object increases resolution. Immersion oil has a refractive index of 1.5 and the lens has a refractive index of about 1.4, whereas that of air is 1.0. Therefore, immersion oil is placed on the surface of the slide and in contact with the oil-immersion lens in order to increase the numerical aperture and the resolution.

► The oil-immersion objective has a very short working distance; it can therefore be moved too close to the slide or ground into the slide, causing the slide to fracture and the lens to be scratched. Because field size diminishes with increasing magnification, it is imperative that you first locate the object you are interested in under high power (43×), rack the nosepiece up by using the coarse adjustment, and then move the oil-immersion lens into position. Add a drop of immersion oil to the top surface of the slide in the illuminated area. Avoid air bubbles in the oil droplet. Make sure the substage iris diaphragm is open wide and the substage condenser is properly focused.) Carefully lower the nosepiece with the coarse adjustment while watching its downward progress toward the slide from the side. NEVER FOCUS DOWNWARD WITH THE COARSE ADJUSTMENT WHILE LOOKING INTO THE MICROSCOPE. As the oil-immersion objective touches the oil droplet, look through the ocular and focus UP and AWAY from the slide by rotating the fine adjustment toward you. This usually brings the object safely into view. If a minimum amount of effort does not produce an image, focus downward very carefully. If the image is blurred, begin again, making sure all details have been strictly observed.

► When for any reason the image of an object cannot be resolved or brought into focus clearly, call this to the attention of the instructor. This condition arises most often from dirty oculars or objectives, air bubbles in the oil droplet, or objectives that are not locked into position on the nosepiece or screwed in place tightly. Check for these and check for dirt on the ocular by rotating it and noting whether there is a corresponding rotation of foreign elements. What is the total magnification of the object under the oil-immersion objective? What is the diameter of the smallest resolvable object? What sets the limit on resolving power (i.e., the ability to distinguish the minimum distance between two points, thus enabling them to be seen as two and not one point)?

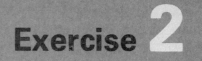

PORIFERA

Contents

The potential for cellular aggregation and division of labor is well demonstrated by some Protozoa, for example, colonial forms. This potential is more fully realized by a group of sessile aquatic organisms called sponges.

Sponges have no mouth, digestive tract, nervous system, or any true tissues in the sense that other multicellular organisms (Metazoa) have tissues, but instead are composed of layers of cells functioning in loose concert. Different cells have become specialized for different functions, yet tend to act somewhat independently of each other, as demonstrated by the dramatic ability of a sponge to reaggregate after breakup and resuspension.

Sponges have a saclike form and, alive or dead, are able to take up fluids in relatively large quantities. This ability is related to the fact that the body is riddled with pores, channels, and chambers through which, in life, a constant current is kept moving. For this reason sponges have been given the phylum name Porifera (*porus,* hole; *fer,* bearer; L.). The microscopic pores through which water enters are called **ostia** and the current exits through larger openings called **oscula** (singular: **osculum**). Externally, sponges are covered by a layer of syncytial epithelium; in some sponges this may be made up of distinct epithelial cells called **pinacocytes.** Internally, the body cavity (**spongocoel**) is lined completely or partially with specialized flagellated collar cells (**choanocytes**) that produce the water current and are important in

*The exercises on reassociation and regeneration should be started at the beginning of the laboratory period so that observations can be made over several hours.

43

feeding. Between these two layers is found a loosely organized **mesenchyme** in which may be found the supporting fibers and spicules of the skeleton and a variety of amoeboid cells.

The sponge thus consists of a mass of cells specially modified for the production and maintenance of a water current through the body. The flagellated choanocytes may be enclosed in chambers, and various passages lead the inhalent current from the ostia through these chambers and the exhalent current out via the oscula.

Because sponges have evolved along rather specialized multicellular lines, they are often put in a subkingdom of their own called the **Parazoa,** alongside the mainstream of metazoan evolution.

2–1. BODY ORGANIZATION

The evolution of the sponges demonstrates the transition from small, vaselike forms with the ability to move relatively small volumes of water through their bodies, to larger, more complex organisms with complicated, highly efficient pumping chambers.

a. Asconoid Type

The simplest form is seen in the **asconoid** sponges (Figure 2.1A). The radially symmetrical, tubelike body consists of a cellular wall that encloses a large central **spongocoel.** The spongocoel opens to the exterior via the **osculum,** situated at the free end of the body tube, which is closed and attached at the base.

The body wall is composed of an outer and an inner epithelium, with a mesenchyme between. The outer epithelium or **epidermis** consists of a single layer of close-fitting polygonal **pinacocytes,** which are interrupted at frequent intervals by pore cells or **porocytes.** Each porocyte is a tubular cell with an intracellular pore. These incurrent pores or **ostia** extend from the external surface to the spongocoel. In life, a water current passes through the ostia to the spongocoel and out through the osculum. The mesenchyme contains several types of amoeboid cells and skeletal elements, all embedded in a gelatinous matrix. The inner epithelium lining the spongocoel consists of a single layer of flagellated **choanocytes** (Figures 2.1A and 2.2). These produce the water current, collect food, and in some cases are involved in the formation of gametes. Unfortunately, they are difficult to observe in the living animal. Electron micrographs of the collar cell show that the collar is composed of 30 to 40 filaments that resemble microvilli (Figure 2.2). The transparent, contractile collar surrounds a single flagellum. The movement of the flagellum produces a current of water such that food particles are caught on the outside of the mucus-lined collar; the mucus may be passed downward and then ingested at the base of the collar. The choanocyte cell type is unique to sponges, and therefore diagnostic of them, although the protozoan protomonads closely resemble the choanocyte in their structure. Some workers believe that a colonial protozoan much like *Proterospongia* (Figure 2.3) gave rise to sponges.

▶   The asconoid form can best be seen in *Leucosolenia* (Figure 2.4). If living specimens are available, take a single sponge tube and place it in a wax pan. Make a longitudinal cut (from osculum to base) and place the two halves of the sponge tube on a slide so that half the tube shows the inner surface and the other half shows the outer surface. Cover gently with a cover slip, making sure that a thin layer of water covers the sponge. Examine under the microscope. Porocytes should be visible on each side; they are clear cells with a canal that is frequently

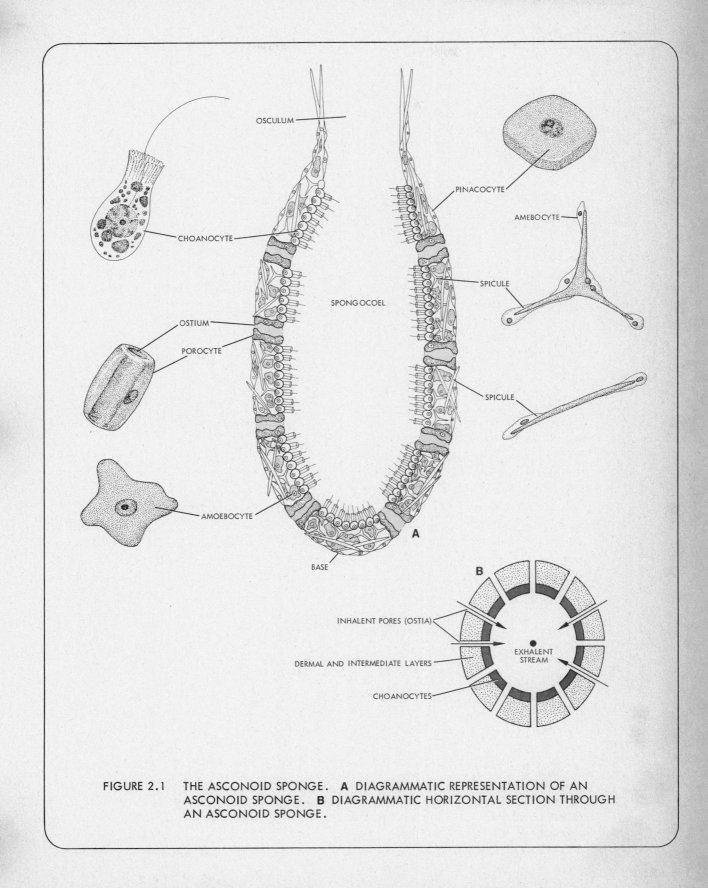

FIGURE 2.1   THE ASCONOID SPONGE.   **A** DIAGRAMMATIC REPRESENTATION OF AN
ASCONOID SPONGE.   **B** DIAGRAMMATIC HORIZONTAL SECTION THROUGH
AN ASCONOID SPONGE.

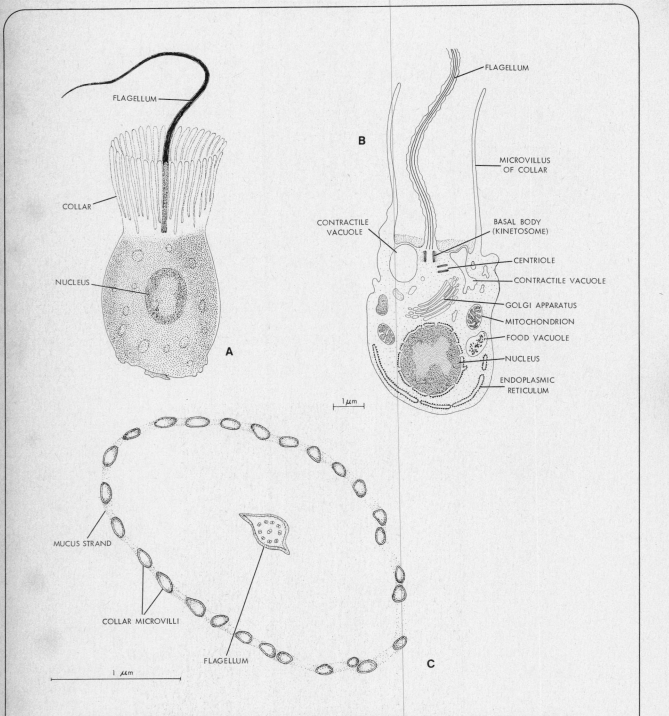

FLAGELLUM

COLLAR

NUCLEUS

A

B

FLAGELLUM

MICROVILLUS
OF COLLAR

CONTRACTILE
VACUOLE

BASAL BODY
(KINETOSOME)

CENTRIOLE

CONTRACTILE VACUOLE

GOLGI APPARATUS

MITOCHONDRION

FOOD VACUOLE

NUCLEUS

ENDOPLASMIC
RETICULUM

1 μm

MUCUS STRAND

COLLAR MICROVILLI

FLAGELLUM

C

1 μm

FIGURE 2.2 THE SPONGE CHOANOCYTE. **A** ENTIRE CELL AS SEEN WITH THE ELECTRON MICROSCOPE (AFTER RASMONT, 1959). **B** SCHEMATIC LONGITUDINAL SECTION OF THE CHOANOCYTE OF THE FRESHWATER SPONGE *EPHYDATIA* AS SEEN WITH THE ELECTRON MICROSCOPE (AFTER BRILL, 1973). **C** A CROSS SECTION THROUGH THE COLLAR OF A CHOANOCYTE AS SEEN WITH THE ELECTRON MICROSCOPE, SHOWING THE COLLAR MICROVILLI AND THE FLAGELLUM.

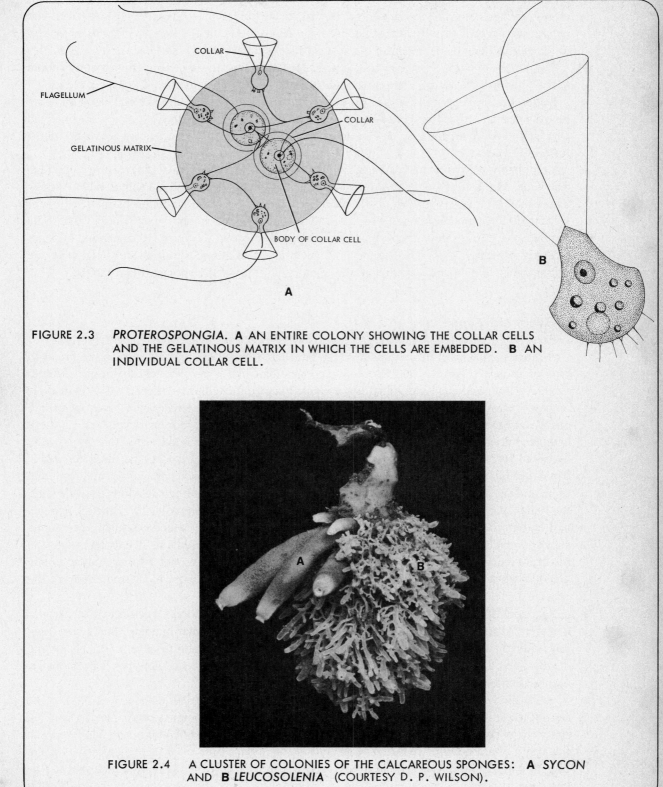

FIGURE 2.3   *PROTEROSPONGIA.* **A** AN ENTIRE COLONY SHOWING THE COLLAR CELLS AND THE GELATINOUS MATRIX IN WHICH THE CELLS ARE EMBEDDED. **B** AN INDIVIDUAL COLLAR CELL.

FIGURE 2.4   A CLUSTER OF COLONIES OF THE CALCAREOUS SPONGES: **A** *SYCON* AND **B** *LEUCOSOLENIA* (COURTESY D. P. WILSON).

crooked. Sometimes a membrane can be seen, leaving only a tiny central hole. Pinacocytes form a pavement on the exterior that is not always easy to see. The choanocytes are small, closely packed cells; the collars will not be visible. As you rack down on the inner surface, you should see flagellar activity before the cells come into focus. In all cases, avoid the temptation to focus on a deep plane, because the very surface will best show what you are looking for.

▶ Cut a piece of living sponge into O- or C-shaped sections; phase-contrast microscopy should reveal choanocytes. Cross sections should be as thin as possible, so use a sharp razor blade.

▶ Better specimens of cell types (e.g., choanocytes, amoebocytes, and skeletal elements) may be obtained by teasing a sponge apart with microforceps or tiny needles. Spicules dissected carefully from living material may in rare instances show the amoebocytes (**calcoblasts**) which form the spicules.

▶ Completely immerse a cluster of living *Leucosolenia* tubes in a finger bowl of clean seawater. Mix a dilute suspension of carmine and seawater and fill a Pasteur pipet. Gently squirt the carmine suspension in the vicinity of the sponge, and trace the pathway of water movement through the sponge. Where do the carmine particles enter? Where do they exit? The absence of a current in your specimen probably indicates that the sponge is moribund or dead.

▶ Study prepared cross and longitudinal sections of asconoid sponges and determine the arrangement of elements comprising the pumping system of the sponge (Figure 2.1**B**).

b. Syconoid Type

A more complicated development of the flagellated chambers may be seen in the **syconoid** type of sponge (Figure 2.5**A**). Syconoid sponges look like large asconoid sponges, having a tubular shape, and each individual has a single excurrent osculum. The body wall is thicker, however, and the spongocoel is lined with pinacocytes. The choanocytes line finger-like chambers (**radial canals**), which radiate from the spongocoel; during development, these chambers arise from the spongocoel by simple evagination of the body wall, showing how the asconoid stock could have given rise to syconoid sponges. The spaces between the radial canals are filled with **incurrent canals,** and tiny openings, **prosopyles,** connect the incurrent and radial canals. Water flows through external ostia via the incurrent canals and prosopyles into the radial canals. The flagellar current forces the water through internal ostia to the spongocoel and on out via the osculum. In some syconoid sponges, the body wall has a highly developed **cortex;** water flows through **dermal pores** in the cortex, along branching canals and so to the incurrent canals. Syconoid mesenchyme resembles that of asconoid sponges, containing amebocytes and spicules of various kinds arranged in specific patterns characteristic of the species.

▶ Syconoid sponges usually consist of unbranched colonies. Obtain a specimen of *Scypha* (Figure 2.4) and examine the arrangement of spicules at the oscular opening. Make cross and longitudinal sections and attempt to see the structural organization (Figure 2.5**B**). Using fine needles, tear apart speciments of *Scypha* and observe the cell and spicule types. Occasionally, eggs and embryos may be found in the body wall.

▶ Trace the water current in living specimens with carmine-seawater suspensions. How is the flow of water in a sponge regulated? Can the size of the oscular openings be changed? How? To observe the reaction of the osculum to stimulation, take a cluster of sponges and allow them to relax in a finger bowl of seawater. Carefully place a glass tube of narrow diameter over the oscular rim of one of the sponge tubes. Note the reaction of this individual and contrast its behavior with that of other individuals in the sponge colony. What is the explanation for the observed effect? Does the osculum respond to mechanical stimulation, such as touch? Cut a cross section just below the oscular opening and look for the irislike **oscular membrane.** What is its function? If sponges have no nervous system, how is coordination obtained?

c. Leuconoid Type

The great majority of sponges are of the **leuconoid** type. Most leuconoids are colonial; although individual oscula can be distinguished, it is difficult to separate members of the colony. Leuconoid sponges resemble more elaborate syconoid types in which the cortex is well developed. Water enters the sponge via dermal pores and branching incurrent canals (Figure 2.5C). The choanocytes are restricted to flagellated chambers surrounding excurrent canals, which merge to form a main excurrent canal. There are three major types of leuconoid sponges, depending on how the choanocyte chambers join up with the incurrent and excurrent canals. *Rhabdodermella* has flagellated chambers but retains the spongocoel and shows a simple leuconoid body form; structurally, it is intermediate between leucon and sycon types.

▶   Using a dilute suspension of carmine in seawater, trace the pathway of water movement through *Rhabdodermella, Dysidea* or *Hymeniacodon. Microciona, Haliclona, Halichondria,* and *Hippospongia* show more complex arrangements of the flagellated chambers. Cut up at least one of these sponges and examine the structure as previously directed. When cutting up *Rhabdodermella,* you may find **amphiblastula larvae** in the spongocoel (see Section 2–3a).

2–2. THE SKELETON

The taxonomy of the Porifera is based on the structural variety of the spicules of the skeletal system (Figure 2.6); this is the most easily recognized and stable characteristic of each group. Spicules may be composed of calcium carbonate or silicon dioxide, or the skeleton may consist entirely of collagenous fibers **(spongin)** or a combination of spicules and spongin. (See Classification of Porifera, p. 57.)

▶   You may have observed spicules during your studies of body form. Recall which spicule type lined the osculum of *Scypha* and which formed the main skeletal element in the rest of the body. Prepare isolated spicules by boiling a piece of sponge in 5% sodium hypochlorite solution until the organic materials disintegrate. After the preparation has cooled, place a drop of the resuspended sediment on a clean slide, cover with a cover slip, and observe the form of the spicules. Study and sketch the types that you see. The inorganic chemical nature of spicules may be determined by drawing some acetic acid or dilute hydrochloric acid under the cover slip of a wet mount of some spicules. What is the effect on spicules made of calcium carbonate? Of silicon dioxide?

A broad range of spicule types from a variety of sponges will be provided. Examine the structure of the spicules, and with reference to Light's manual or other available source, attempt to classify the sponges from which the spicules were obtained. Check your identifications with the classification scheme given at the end of this exercise. Which common forms belong to the Calcarea? Hexactinellida? Demospongiae? How are spicules formed?

▶   The proteinaceous skeletal material of sponges is called **spongin** and can be seen in macerated specimens. Why was it not seen in the 5% sodium hypochlorite preparation? Some sponges have no spicules, e.g., the bath sponges *Spongia* and *Hippospongia,* and contain only spongin fibers. Feel the form of specimens of these sponges, remove a small piece and observe under the microscope. Is there a fixed orientation of the fibers?

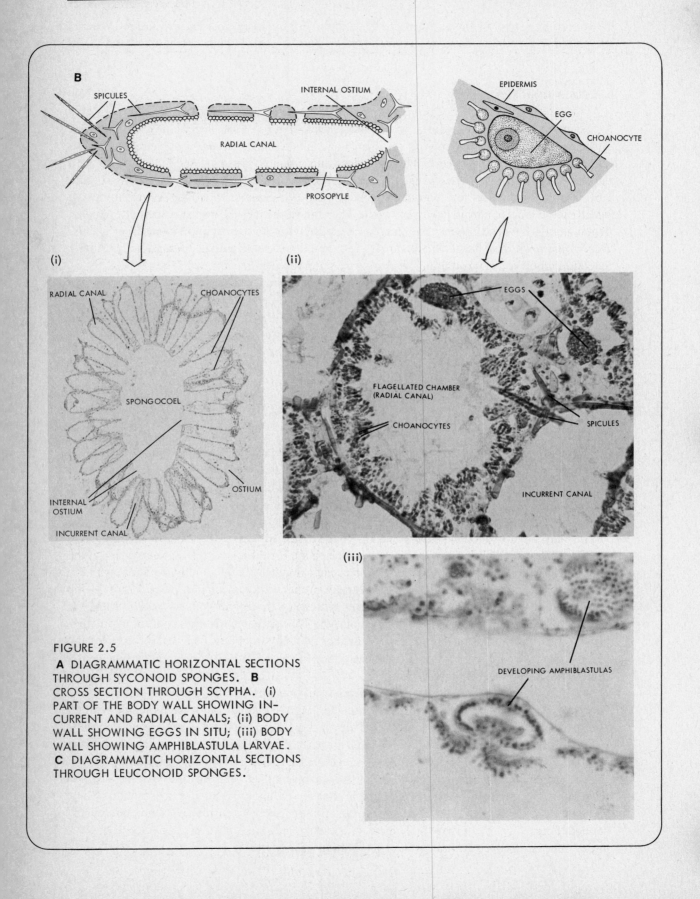

B

SPICULES

INTERNAL OSTIUM

RADIAL CANAL

PROSOPYLE

EPIDERMIS

EGG

CHOANOCYTE

(i)

RADIAL CANAL

CHOANOCYTES

SPONGOCOEL

INTERNAL OSTIUM

OSTIUM

INCURRENT CANAL

(ii)

EGGS

FLAGELLATED CHAMBER (RADIAL CANAL)

CHOANOCYTES

SPICULES

INCURRENT CANAL

(iii)

DEVELOPING AMPHIBLASTULAS

FIGURE 2.5
 A DIAGRAMMATIC HORIZONTAL SECTIONS THROUGH SYCONOID SPONGES. **B** CROSS SECTION THROUGH SCYPHA. (i) PART OF THE BODY WALL SHOWING IN-CURRENT AND RADIAL CANALS; (ii) BODY WALL SHOWING EGGS IN SITU; (iii) BODY WALL SHOWING AMPHIBLASTULA LARVAE. **C** DIAGRAMMATIC HORIZONTAL SECTIONS THROUGH LEUCONOID SPONGES.

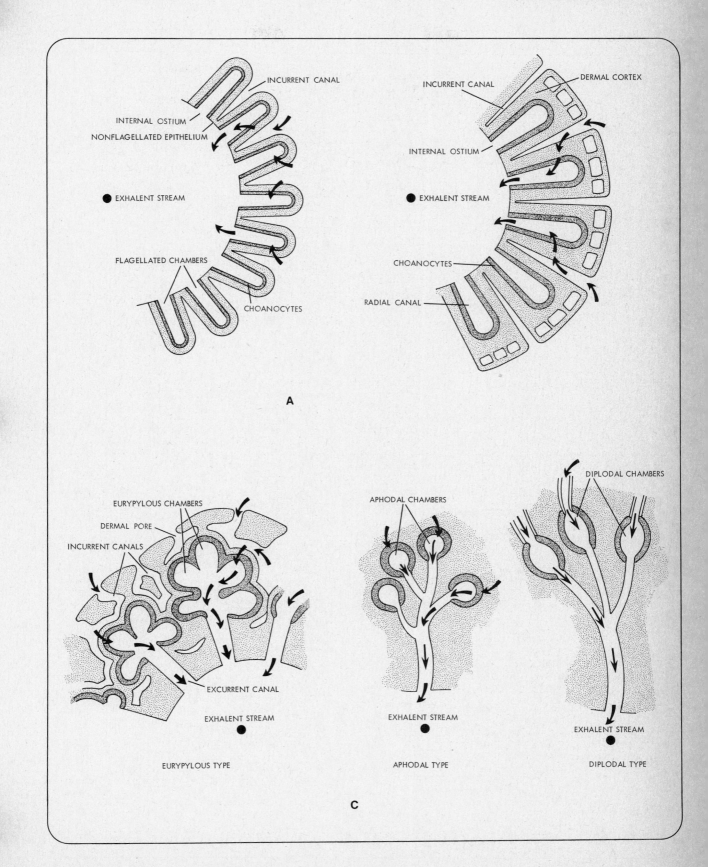

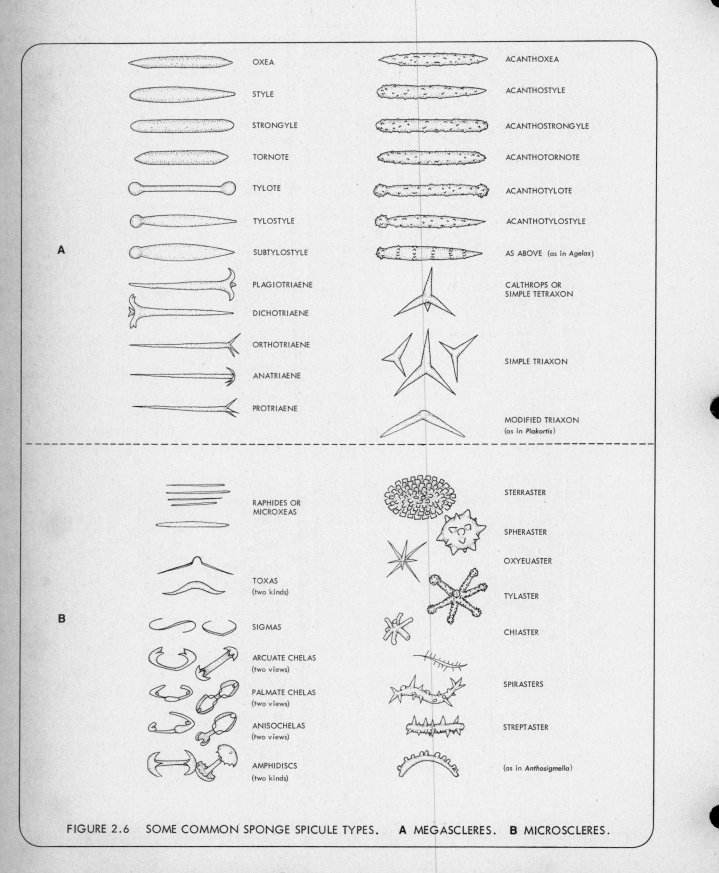

FIGURE 2.6   SOME COMMON SPONGE SPICULE TYPES.   **A** MEGASCLERES.   **B** MICROSCLERES.

▶ Some of the siliceous sponges reach a highly complex degree of skeletal development. One of the most beautiful and spectacular of all sponge skeletons is found in the glass sponge *Euplectella* (Venus's flower basket), which inhabits the abyssal regions of the ocean. Examine a demonstration specimen. The skeleton is cylindrical with longitudinal struts running along the length of the cylinder and siliceous hoops running at right angles to the longitudinal. There are also oblique bands. Of what advantage is this arrangement to *Euplectella* and how is it related to the habitat of the animal? How does water flow through *Euplectella?*

▶ Observe the long, filamentous skeletal stalk of a demonstration specimen of *Hyalonema* (the glass rope sponge), in which the spicules fuse to form long glassy fibers. What is the functional significance of this form?

2–3. REPRODUCTION AND REASSOCIATION*

Sponges can reproduce both sexually and asexually and also have great powers of regeneration and reassociation.

a. Sexual Reproduction

Some authors state that all gametes are the product of specialized amoeboid cells, but others consider that sperm arise from choanocytes and eggs from amoebocytes with stored food (**archeocytes**). The resolution of the problem and demonstrations of gametogenesis are beyond the scope of the present course; however, cross sections of *Scypha* will show the position and form of the egg (Figure 2.5**B**).

In the living sponge, sperm are carried from one organism to another in the water current. They leave via the osculum and enter the ostia of the recipient sponge. Transfer of sperm to the egg occurs via choanocytes or amoebocytes. The fertilized egg undergoes cleavage and develops into a blastula *in situ*. The blastula becomes flagellated, works its way through the parent body until it gets into the excurrent stream, and emerges as a flagellated larva.

▶ Two types of larvae are found: a uniformly ciliated **parenchymula** and the **amphiblastula** larva, which has flagellated cells at one pole. Most sponges, including *Leucosolenia* and the freshwater *Spongilla,* have a parenchymula larva, whereas *Scypha* is one of the more unusual forms with amphiblastula larvae. Depending on the availability of living material, look for these larval types (Figure 2.7). Snip apart adult specimens in a finger bowl of water. If present, larvae should swim up to the surface layer, where they can be easily collected by drawing the surface layer up into a fine pipet and transferred to a syracuse watch glass. Some may be examined on a slide under a cover slip. Larvae may settle after about 24 hours and metamorphose into tiny sponges in a few days.

What is the significance of a motile larval form such as the amphiblastula to a sponge like *Scypha?*

b. Asexual Reproduction

i. BUDDING.   One of the most common methods of asexual reproduction in sponges is by budding. You have seen examples of this phenomenon in the cluster of tubes of *Leucosolenia* and *Scypha* and in the encrusting growth of members of the Demospongiae.

---

*The exercises on reassociation and regeneration should be started at the beginning of the laboratory period so that observations can be made over several hours.

ii. REASSOCIATION AND REGENERATION.* As has been noted, sponges exhibit little coordination between cells. Related to this is the fact that if broken up, they are able to reassociate again with relative ease; they also have considerable powers of regeneration. The classic experiment of H. V. Wilson will be attempted to demonstrate powers of reassociation in sponges.

▶ Press 1 g of fragments of fresh *Microciona* through number 21 standard quality silk bolting cloth into 100 ml of cold (approximately 4°C) seawater in a beaker. One original suspension will serve a whole class. Prepare several known dilutions in seawater. If time permits, a hemocyto-meter may be used to count the number of cells in the suspension. Fill small (16 cm) finger bowls two thirds full of cool seawater; place a syracuse watch glass on the bottom of each finger bowl and put two slides on each syracuse dish. Using a Pasteur pipet, gently dispense equal aliquots of the original suspension and each dilution onto a series of slides. With two or three setups on each slide, you can insure against accidents. (Note that large masses do not aggregate well.) Cover the finger bowls and allow one group to stand undisturbed for 24 hours at 4°C and another at room temperature (25°C). Gently remove the slides, cover with a cover slip, and examine under the microscope. Count the aggregates and note their size distribution. How do cell density and temperature relate to aggregation? What form do the aggregates take?

▶ Prepare a fresh suspension as just described, place a drop on a slide, and cover with a cover slip. At intervals during the laboratory period, examine under the microscope at 100× and observe any tendency of the cells to aggregate. Watch for pseudopodia (**filopodia**) to be extruded from groups of cells. Can you relate filopodium size to cell group size? Can you recognize different cell types?

▶ Aggregation may also be obtained by rotation. Prepare a suspension of *Microciona* (1×10^7 cells per milliliter seawater gives the best results), distribute 3-ml quantities in 25-ml Erlen-meyer flasks, and seal with parafilm. Place the flasks on a gyratory shaker at 25°C and rotate at 80 rpm. Sample at intervals of 15 minutes, 30 minutes, 1 hour, and 12 or 24 hours; examine the samples under the dissecting microscope to determine the degree of aggregation. To observe later stages of development in a healthy condition, the seawater should be changed daily. What is the advantage of studying aggregation by rotation of the suspensions?

▶ It is possible to dissociate sponges chemically, when "cleaner" suspensions of cells are obtained (e.g., individual cells rather than clumps of varying size) by placing them in seawater free of Ca^{++} and Mg^{++}. The absence of these ions, which are required for the maintenance of cell adhesion, causes the small fragments to break up into isolated cells. (See Humphreys for further details.)

Is it possible to get a graft of one sponge to take in another species? Can you mix two dissociated species and get chimerical aggregates?

iii. GEMMULES.   All freshwater sponges and some marine forms are capable of producing resistant overwintering bodies which reproduce asexually and are called **gemmules.**

▶ Obtain living or preserved samples of the freshwater Demospongiae *Spongilla* and/or *Ephydatia*. In specimens collected in the autumn, gemmules are often found. Extract gem-mules from the parent sponge by maceration or expose a fragment of sponge in a petri dish to a jet of water so that the gemmules are freed. Collect gemmules with a Pasteur pipet, make wet mounts, and examine their structure (Figures 2.8**A** and **B**). Each gemmule is pierced by an outlet, the **micropyle,** which may be at the end of a short neck, depending on the species. Those with the longest necks are easiest to use in the following manipulations.

▶ Place some gemmules obtained from a living sponge on a slide and clean them of debris with

*The exercises on reassociation and regeneration should be started at the beginning of the laboratory period so that observations can be made over several hours.

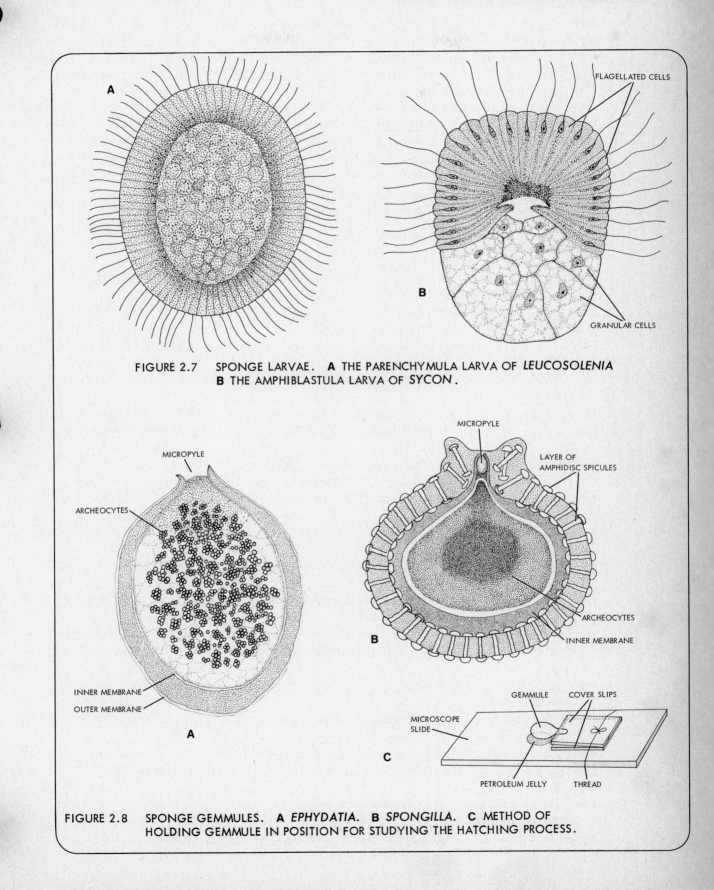

FIGURE 2.7   SPONGE LARVAE.   **A** THE PARENCHYMULA LARVA OF *LEUCOSOLENIA*
B THE AMPHIBLASTULA LARVA OF *SYCON*.

FIGURE 2.8   SPONGE GEMMULES.   **A** *EPHYDATIA*.   **B** *SPONGILLA*.   **C** METHOD OF
HOLDING GEMMULE IN POSITION FOR STUDYING THE HATCHING PROCESS.

microforceps. Immerse them in a 1% hydrogen peroxide solution in a syracuse watch glass for about 1 minute to kill any organisms on the outside of the specimens. Replace the gemmules on slides, and manipulate them with forceps so that the micropyle is held between two cover slips (Figure 2.8**B**). It may be helpful to hold the gemmule in position with a spot of petroleum jelly. Place the slides in petri dishes containing boiled, filtered, or aerated pond water at a depth of about 5 mm so that oxygen can diffuse in easily, cover the dishes, and leave at room temperature. Examine under the dissecting microscope after about 18 hours and at periodic intervals thereafter. With some luck the gemmules will hatch; the cells migrate outward and begin forming spicules and canals within about 4 to 5 days, depending on conditions. After 1 week's growth, slides with young sponges attached may be transferred to a freshwater aquarium.

Why do you think gemmules are characteristic of freshwater and not marine species?

References

Ankel, W. E., and H. Eigenbrodt. Über die Wuchsform von *Spongilla* in sehr flachen Räumen, *Zool. Anz.* **145**:195–204, 1950.

Barrington, E. J. W. *Invertebrate Structure and Function*. Boston: Houghton Mifflin Co., 1967.

Bidder, G. P. The relation of the form of a sponge to its currents, *Quart. J. Microsc. Sci.* **69**:293–323, 1923.

Brill, B. The ultrastructure of choanocytes in *Ephydatia fluviatilis* L., *Zeitschr. f. Zellforsch.* **144**:231–46, 1973.

Brown, F. A., Jr. (ed.). *Selected Invertebrate Types*. New York: John Wiley & Sons, 1950.

Buchsbaum, R. *Animals Without Backbones,* 2nd ed. Chicago: University of Chicago Press, 1972.

Bullough, W. S. *Practical Invertebrate Anatomy*. New York: St. Martin's Press, 1962.

Carter, G. S. *A General Zoology of the Invertebrates*. London: Sidgwick and Jackson, 1951.

De Laubenfels, M. W. The marine and freshwater sponges of California, *Proc. U.S. Nat. Mus.* **81**:1–40, 1932.

Fry, W. G. (ed.). *The Biology of the Porifera. Symp. Zool. Soc. London,* No. 25. New York: Academic Press, 1970.

Florkin, M., and B. Scheer (eds.). *Chemical Zoology,* Vol. II. *Porifera, Coelenterata and Platyhelminthes*. New York: Academic Press, 1968.

Henkart, P., et al. Characterization of sponge aggregation factor. A unique proteoglycan complex, *Biochemistry* **12**:3045–50, 1973.

Humphreys, T. Chemical dissolution and "in vitro" reconstruction of sponge cell adhesions, I. Isolation and functional demonstration of the components involved," *Dev. Biol.* **8**:27–47, 1963.

Hyman, L. H. *The Invertebrates*, Vol. I. *Protozoa Through Ctenophora.* New York: Mc-Graw-Hill Book Co., 1940.

Jewell, M. "Porifera," in Ward and Whipple, *Freshwater Biology,* W. T. Edmonson (ed.). New York: John Wiley & Sons, 1959.

Moscona, A. How cells associate, *Sci. Amer.* **205**(3):143–66, 1961.

Pennak, R. W. *Fresh Water Invertebrates of the United States*. New York: Ronald Press, 1953.

Pratt, H. W. *A Manual of the Common Invertebrate Animals*. New York: Blakiston Division, McGraw-Hill Book Co., 1935.

Smith, R. I., and J. T. Carlton (eds.). *Light's Manual: Intertidal Invertebrates of the Central California Coast,* 3rd ed. Berkeley: University of California Press, 1975.

Vogel, S. Current-induced flow through the sponge *Halichondria, Biol. Bull.* **147**(2):443–56, 1974.

Wilson, H. V., and J. T. Penny. The regeneration of sponges from dissociated cells, *J. Exp. Zool.* **56**:73–148, 1930.

An Abbreviated Classification of the Phylum Porifera

Phylum Porifera

Class I Calcarea

Spicules of calcareous nature, one- or four-rayed.

Order 1 Homocoela

Asconoid, e.g., *Leucosolenia.*

Order 2 Heterocoela

Syconoid or leuconoid, e.g., *Scypha, Rhabdodermella.*

Class II Hexactinellida (Triaxonida or Hyalospongiae)

Spicules six-rayed, siliceous or modified triaxon.

Order 1 Hexasterophora

Hexasters, no amphidiscs, e.g., *Euplectella.*

Order 2 Amphidiscophora

Amphidiscs, no hexasters, e.g., *Hyalonema.*

Class III Demospongiae

Spicules of siliceous nature, but not six-rayed, or skeleton may have horny spongin fibers and spicules. Leuconoid.

Subclass 1 Tetractinellida

Tetraxon spicules, no spongin.

Subclass 2 Monaxonida

Monaxon spicules, with or without spongin, e.g., *Poterion, Halichondria, Cliona, Microciona, Haliclona, Spongilla, Dysidea, Hymeniacodon.*

Subclass 3 Keratosa

Skeleton of spongin and no siliceous spicules, e.g., *Hircinia, Hippospongia, Euspongia.*

Class IV Sclerospongiae

Coralline sponges with a calcium carbonate skeleton and siliceous spicules.

Equipment and Materials

Equipment

Microscope and lamp
Dissecting instruments

Immersion oil
Fine mesh silk bolting cloth
Bunsen burner

Beakers
Microscope slides and cover slips
Petri dishes
25-ml Erlenmeyer flasks
Gyratory shaker
Pasteur pipets and rubber bulbs
Parafilm

Living Material*

Scypha or *Rhabdodermella* (9)
Microciona (7,9,12)
Ephydatia or *Spongilla* (5)
Ephydatia or *Spongilla* gemmules
  (5,3)
Leucosolenia (9,12)
Dysidea or *Hymeniacodon* (9)

Solutions and Chemicals

Carmine powder
5% sodium hypochlorite

Dilute HCl
1% H_2O_2 solution

Preserved Material

Leucosolenia
Scypha
Haliclona
Hippospongia or *Spongia*
Halichondria
Spongilla
Euplectella
Hyalonema

Slides

Leucosolenia, cross section
Leucosolenia, longitudinal section
Scypha, cross section
Scypha, longitudinal section
Gemmules

*Numbers in parentheses identify sources listed in Appendix B.

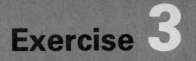

CNIDARIA

Contents

The cnidarians (*cnidos,* thread; Gk.) are an aquatic group of organisms, commonly known as hydroids, jellyfish, sea anemones, and corals. They are tentacle-bearing, radial or biradial animals with a saclike body composed of two basic cell layers. The inner layer of cells, the **gastrodermis,** lines a hollow space, the **coelenteron,** which functions in digestion and transport, and into which there is a single opening, the **mouth.** The outer cell layer consists of an

*Parts of the exercises on feeding and digestion should be started at the beginning of the laboratory period to allow time for completion: see pages 76 and 78.

epidermis. The gastrodermis and epidermis are separated by a thin, acellular matrix or **mesolamella** or by a jelly-like material containing cells and/or fibers called the **mesogloea** or **mesenchyme.**

All Cnidaria are constructed from only eight basic cell types: epithelial, muscular, nervous, glandular, male and female reproductive, interstitial, mesenchyme, and the diagnostic cell type, the **cnidoblast.** Cnidoblasts are most frequently located on the tentacles and contain stinging structures called **nematocysts** which function in defense and in the capture of food; all cnidarians are carnivorous in habit.

The Cnidaria are at the **tissue** level of organization, having epidermal, nervous, digestive, and muscular tissues, but lacking specialized organ systems. Fundamentally, two body forms are represented: the **polyp** and the **medusa.**

3–1. THE CNIDARIAN LIFE CYCLE—DIVERSITY IN FORM

The two basic body forms found in this phylum are variously modified in its three classes. The **polyp** is usually a sedentary form, adapted to a nonmotile existence (Figure 3.1A), and is dominant in the classes Hydrozoa and Anthozoa. In the class Scyphozoa the usually free-swimming **medusa** (Figure 3.1B) is most apparent and gives the group the name **jellyfish.** The classification of the Cnidaria (see p. 88) is primarily based on the degree of dominance or suppression of one of these body types within the life cycle, and secondarily upon other characteristics.

The cylindrical polyp and bell-shaped medusa are essentially similar, except for a change in the length of the oral-aboral axis and the amount of gelatinous material separating epidermis from gastrodermis (Figure 3.1C). The presence of these two body plans has caused considerable confusion, especially when they occur within the life history of the same species. The phenomenon has been called **metagenesis,** or alternation of generations. In fact, there is no alternation of a haploid individual with a diploid individual in the Cnidaria. (Contrast this with haplophase and diplophase in plants.) A juvenile, asexually reproducing polyp form gives rise to the sexually reproducing medusoid stage. Both individuals are diploid, and only the gametes are haploid. This change of form within the life cycle of any one species (Figure 3.2A) is **polymorphism** (*poly,* many; *morph,* form; Gk.). The nature of polymorphism will be more readily understood as you survey the phylum.

▶   The marine scyphozoan *Aurelia* will be used to illustrate the cnidarian life cycle (Figure 3.2B). Examine plastic mounts of the adult *Aurelia* (Figure 3.3A), preserved specimens, and prepared slides of the developmental stages, and study the development from polyp to medusa. The minute polyp (**scyphistoma**) of *Aurelia* has a hollow coelenteron and a circlet of tentacles that surrounds the mouth (Figure 3.3C). This polyp may give rise to other scyphistomae, or it undergoes a series of transverse "cuts" (**strobilation,** Figure 3.3D), resulting in the formation of a stack of saucer-like medusae called **ephyrae** (Figure 3.3E). These are eventually set free to grow into the adult medusoid (jellyfish) form. The sexes are separate and the medusa develops gonads that produce eggs or sperm. These are shed into the coelenteron; the spermatozoa escape to enter the female via the mouth and fertilize the eggs. After fertilization, the zygote lodges in the folds of the oral arms, where it develops into a small, ciliated larval form called a **planula** (Figure 3.3B). The planula swims free and eventually settles on rocks or weed to grow into a scyphistoma.

Variations on this basic life cycle will now be examined, together with body organization.

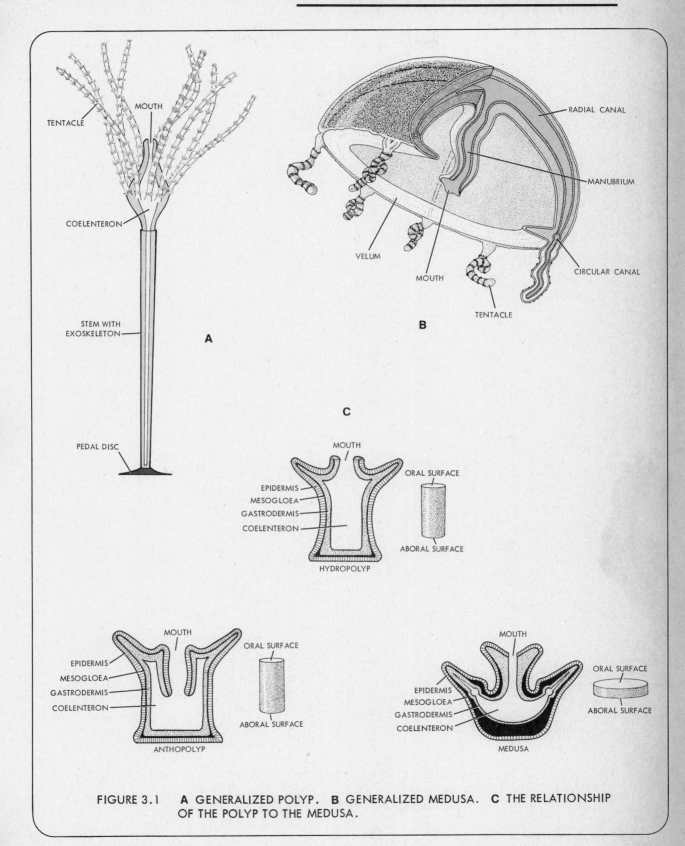

FIGURE 3.1 **A** GENERALIZED POLYP. **B** GENERALIZED MEDUSA. **C** THE RELATIONSHIP OF THE POLYP TO THE MEDUSA.

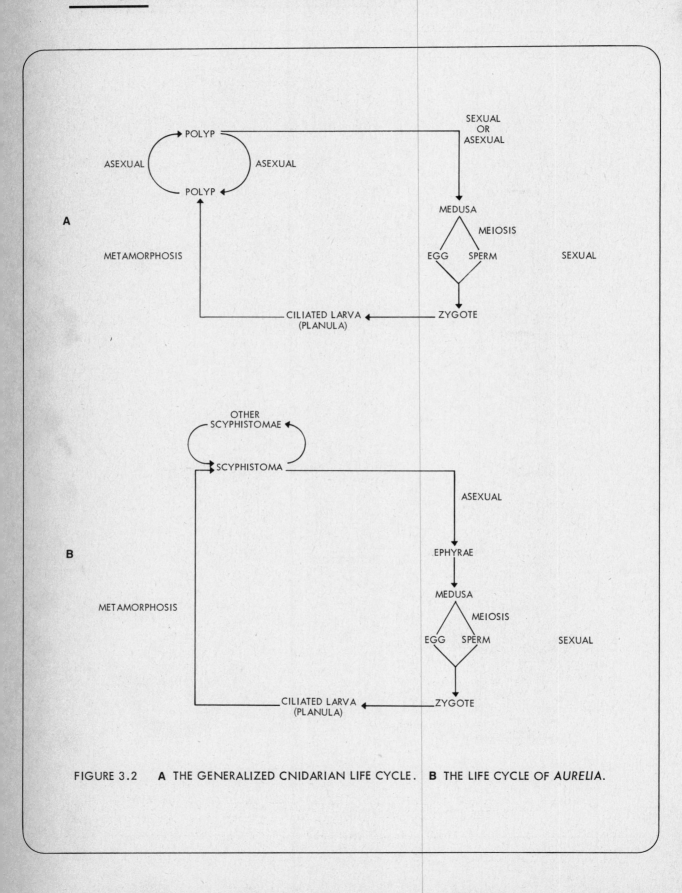

FIGURE 3.2   **A** THE GENERALIZED CNIDARIAN LIFE CYCLE.   **B** THE LIFE CYCLE OF *AURELIA*.

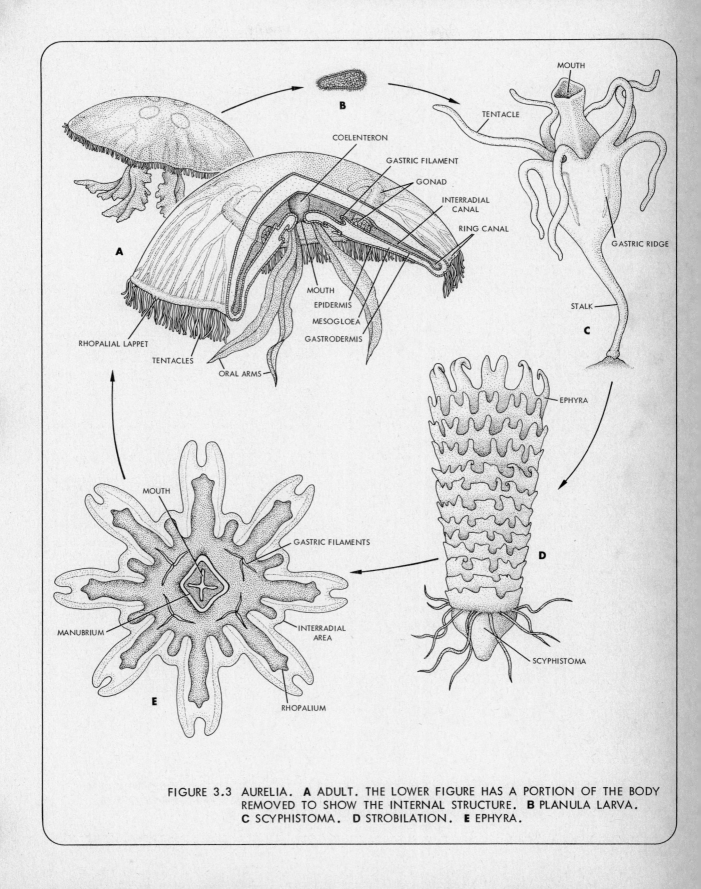

FIGURE 3.3 AURELIA. **A** ADULT. THE LOWER FIGURE HAS A PORTION OF THE BODY REMOVED TO SHOW THE INTERNAL STRUCTURE. **B** PLANULA LARVA. **C** SCYPHISTOMA. **D** STROBILATION. **E** EPHYRA.

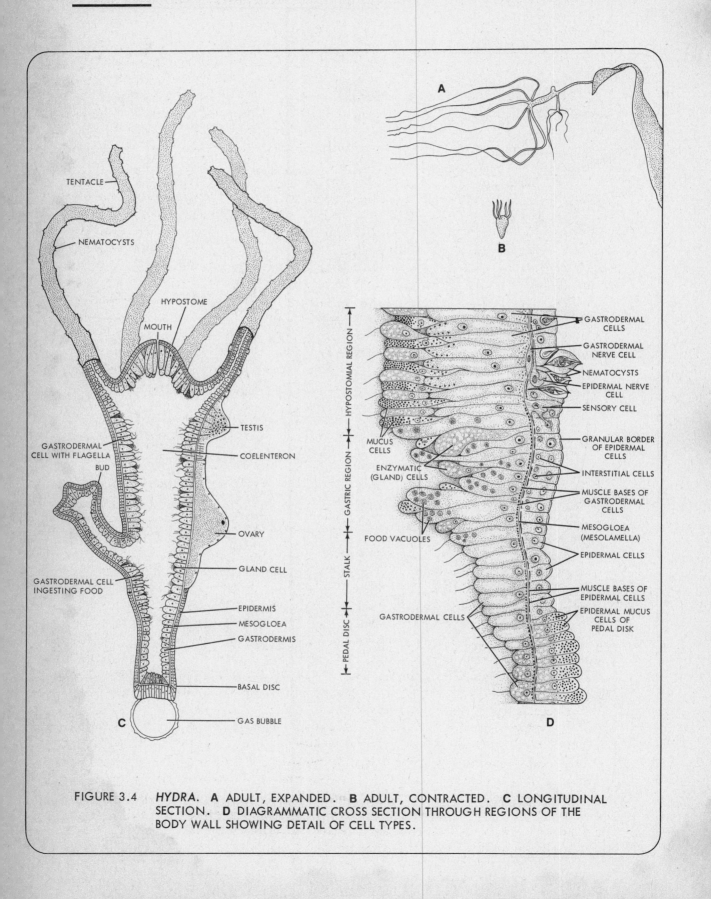

FIGURE 3.4   *HYDRA.* **A** ADULT, EXPANDED. **B** ADULT, CONTRACTED. **C** LONGITUDINAL
SECTION. **D** DIAGRAMMATIC CROSS SECTION THROUGH REGIONS OF THE
BODY WALL SHOWING DETAIL OF CELL TYPES.

a. Polyp

i. HYDROPOLYP.   In the class Hydrozoa, sessile solitary or colonial polyps usually give rise to medusae by asexual means; these medusae may or may not be released from the parent polyp before shedding their sexual products. Typically, both polyp and medusa stages are present in the life cycle, although in atypical forms such as the freshwater *Hydra* the medusoid phase may be entirely absent; the polyp itself gives rise to eggs and sperm, and the zygote develops in a protective egg capsule to form a miniature adult. Of what adaptive advantage do you think this mode of development is to *Hydra?*

▶   Obtain some living *Hydra* in a syracuse watch glass containing water and specimens mounted in plastic (Figures 3.4**A** and **B**). Observe under the dissecting microscope and note the arrangement of the tentacles and their hollow nature. Is the number of tentacles constant from individual to individual? The body of *Hydra* can be thought of as a cylinder composed of two cell layers and differentiated at both ends. Basally, the body column forms an adhesive **disc** by which the animal attaches itself. The free end has a circlet of tentacles surrounding an elevation called the **hypostome.** The mouth is situated in the center of the hypostome.

▶   The histology of *Hydra* can best be seen in prepared slides of longitudinal and cross sections of the animal (Figures 3.4**C** and **D**). The outer layer or **epidermis** consists of epitheliomuscular cells, interstitial cells, cnidoblasts containing nematocysts, nerve cells, and, if the animal is sexual, germ cells. The inner layer or **gastrodermis,** which lines the coelenteron, consists of flagellated gastrodermal cells, gland cells, and mucous cells. The two layers are separated by a thin acellular matrix, the **mesolamella.** Identify as many cell types as possible, and note their distribution in the body. (For example, where are nematocysts concentrated?) The cross section will clearly show the two-layered condition of the organism and the relation of these layers to the mesolamella. Relate the body organization of *Hydra* to that of the scyphistoma (scyphopolyp) of *Aurelia*. What are the similarities? What are the differences?

ii. ANTHOPOLYP.   In anthozoan life cycles, as in that of *Hydra,* the medusoid stage is absent, but the zygote develops into a ciliated **planula larva** (see Section 3–6b). The anthozoan polyp is more complex in structure than the hydrozoan polyp, and the members of this class may form complex colonies.

▶   Obtain a specimen of a sea anemone *(Metridium, Sagartia)* and observe that the anthopolyp is a cylinder that has the free end differentiated into an **oral disc** with a centrally located mouth and marginal tentacles (Figure 3.5A). The mouth opens into the coelenteron via a tubular **actinopharynx.** The mouth may be differentiated at one, two, or three places to form a ciliated groove or **siphonoglyph,** which runs down the length of the actinopharynx. The position of the mouth and siphonoglyph(s) superimposes a bilateral symmetry onto the radially symmetrical animal because only one plane divides the animal into mirror images. If only one siphonoglyph is present, the symmetry is bilateral. How many siphonoglyphs are present in your specimen? The attached end of the body cylinder forms the base or **pedal disc.**

▶   Using a preserved specimen, make two cross sections through the body, one about midway down the body and the other nearer to the pedal disc. In another preserved anemone, make a longitudinal cut from mouth to base so that the animal is cleanly cut in half. These sections of whole animals should be supplemented by prepared slides of cross sections and longitudinal sections. In the cross section (Figures 3.5**B** and **C**) note that the coelenteron is divided by a series of longitudinal partitions called **septa** or **mesenteries.** These septa are folds of mesogloea covered by gastrodermis, which arise from the column wall and run to the center of the body. Septa which reach and join the actinopharynx are called primary (**complete**) septa. How many pairs of primary septa are present in your specimen? Between the pairs of primary septa are

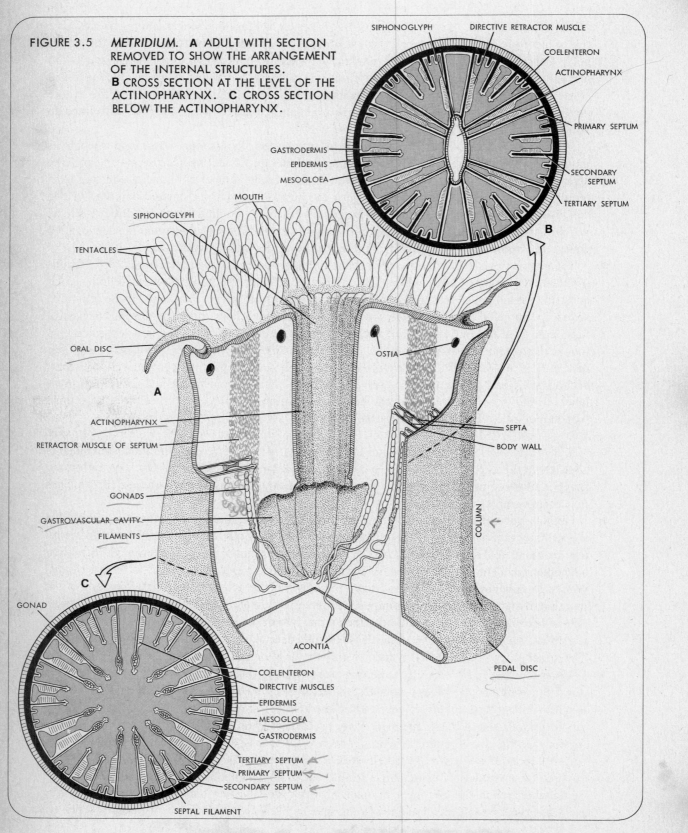

FIGURE 3.5   *METRIDIUM.* **A** ADULT WITH SECTION
REMOVED TO SHOW THE ARRANGEMENT
OF THE INTERNAL STRUCTURES.
B CROSS SECTION AT THE LEVEL OF THE
ACTINOPHARYNX. **C** CROSS SECTION
BELOW THE ACTINOPHARYNX.

shorter (**incomplete**) septa which do not reach the actinopharynx; they are called secondary, tertiary, and quaternary septa, depending on their length. Cross sections below the level of the actinopharynx, and longitudinal sections, will show that the free edges of the septa bear gonads or are drawn out to form threadlike **acontia.** Acontia bear nematocysts and may be extruded through small openings in the body wall called **cinclides** when the animal is disturbed. Notice the well-developed muscle bands in the body column and the mesenteries. Keep these cross sections and longitudinal sections under wet toweling for future study.

What are the major similarities and differences between anthopolyps and hydropolyps?

b. Medusa

The medusa is usually bell- or umbrella-shaped and the margin is fringed with tentacles (Figure 3.1**B**). Extending inward from the edge of the bell there may or may not be a shelf of epidermal tissue known as the **velum.** This structure is important in classification (see p. 88). Medusae with a velum are described as being **craspedote,** whereas those without a velum are **acraspedote.** The outer surface of the medusa is called the **exumbrellar** surface, and the inner or **subumbrellar** surface bears the mouth at its center. The coelenteron consists of a **gastric cavity** into which the mouth opens and four **radial canals,** which join a **ring canal** running around the margin of the bell.

i. SCYPHOZOAN MEDUSAE. In the class Scyphozoa the polyp or **scyphistoma** is reduced in size, and it fragments or **strobilates** to produce medusae asexually (e.g., *Aurelia,* Figure 3.3); the medusoid phase is more conspicuous than the polyp phase. Scyphozoan medusae are the typical jellyfish, in which the mesogloea is well developed.

▶ Take a plastic mount or a preserved specimen of *Aurelia* and study its body organization (Figure 3.3**A**). The body is umbrella-shaped and scalloped at the margin, and it bears short marginal tentacles. The subumbrellar surface bears four frilled tentacular **oral arms,** in the center of which is the square mouth. Running from the mouth are four **gastric pouches,** which communicate with the branched and unbranched radial canals; these in turn run to the marginal ring canal. The horseshoe-shaped **gonads** are located in the gastric pouches.

ii. HYDROZOAN MEDUSAE. Hydrozoan medusae may or may not be retained on the parent polyp before producing and/or releasing the gametes. Hydrozoan life cycles range from those in which the medusa phase is entirely absent (e.g., *Hydra*), through those in which it is reduced (e.g., *Obelia*), to those in which it is dominant (e.g., *Gonionemus, Polyorchis*). The medusa is dominant in a single order of Hydrozoa, the Trachylina. A planula larva gives rise to an **actinula larva** (a microscopic polyp-like form), which in turn gives rise to the medusa.

▶ Study the medusa of *Gonionemus* (Figure 3.6**A**) and/or *Polyorchis* (Figure 3.6**B**) and examine mounts of the medusa of the hydrozoan *Obelia* (Figure 3.6**C**). Note the shape of the bell and the number of tentacles on the margin. Is the margin scalloped? Is a velum present? Turn the specimen so that the subumbrellar surface faces you. In the center of the subumbrellar surface there is a tubular structure, the **manubrium,** which bears the mouth at its free end. How many radial canals are present? The gonads are borne on the radial canals. How many gonads are there?

▶ Compare hydrozoan to scyphozoan medusae, paying particular attention to the location of gonads, the presence or absence of a velum, and the nature of the polyp stage, if it is known. If sessile or so-called attached medusae (**stauromedusae**) are available, such as those of *Haliclystus* or *Lucernaria* (Figure 3.6**D**), note their similarity to both the medusoid and the polyp form.

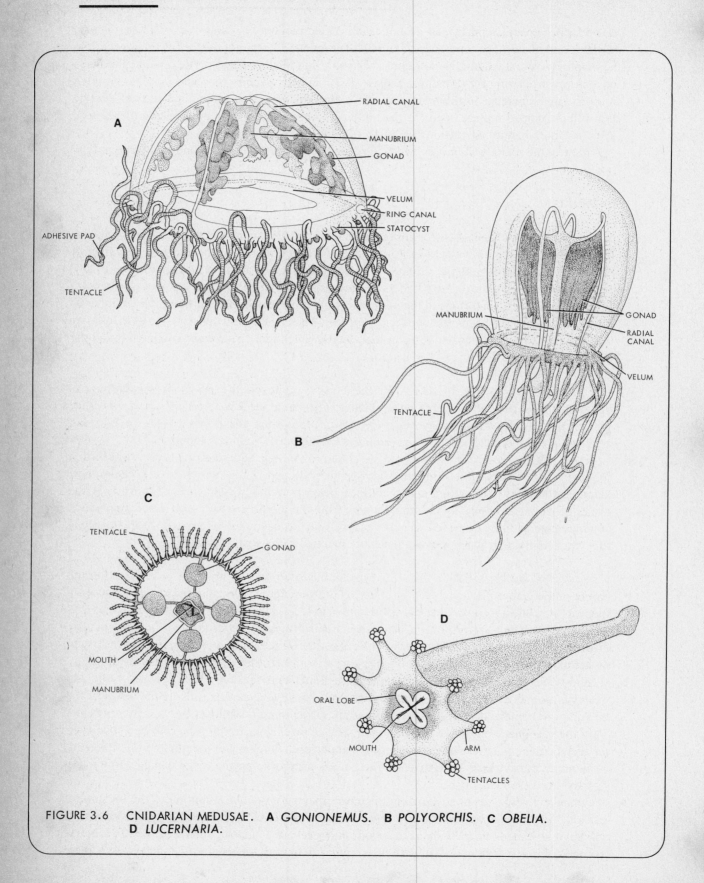

FIGURE 3.6   CNIDARIAN MEDUSAE.   **A** *GONIONEMUS*.   **B** *POLYORCHIS*.   **C** *OBELIA*.
D *LUCERNARIA*.

c. Polymorphic Colonies

There has been a recurrent tendency in cnidarian evolution for the polyp stage to become colonial, and many hydrozoan and anthozoan species form complex colonies. The colony is formed by asexual budding of polyps with incomplete separation of the buds.

i. HYDROZOA. In typical hydrozoan colonies both polyp and medusa are present. Colonial hydrozoans consist of tubes composed of gastrodermis and epidermis, collectively called the **coenosarc.** Surrounding the coenosarc, there may be an acellular, nonliving protective covering, the **perisarc,** or cuticle, secreted by the epidermis. At the free ends of the cylindrical tubes the polyps or **hydranths** arise. There is usually subdivision of function in the colonial hydrozoa; some polyps are specialized for feeding (**gastrozooids** or **trophozooids**), some for protection (**dactylozooids** or **tentaculozooids**), and others for reproduction (**gonozooids** or **blastostyles**). When present, the perisarc that surrounds the blastostyle is called a **gonotheca;** the blastostyle and gonotheca are together referred to as a **gonangium.** The reproductive polyps bear medusoid individuals called **gonophores,** which may or may not develop into fully fledged free-swimming medusae before developing and releasing eggs and sperm. Medusoid individuals may also develop from almost any other part of the colony.

▶ The colonial hydrozoans *Tubularia, Clava,* and *Campanularia* produce abortive medusae (gonophores), which are not set free into the surrounding water before releasing their eggs and sperm (Figure 3.7). Examine prepared slides of representatives of these species; note the arrangement of the polyps into a colony, the mode of branching, and the locations of tentacle-bearing individuals and those individuals lacking tentacles. Observe the form and extent of the perisarc. The gonophores may arise from the hydranths themselves, as in *Tubularia,* or from the hydranth stalk, as in *Clava,* or from modified hydranths, the blastostyles, as in *Campanularia*. In *Tubularia,* planula larvae develop in the female gonophores and subsequently actinulae are released. These settle and develop into new polyps (see p. 85). The female gonophores of *Clava* and *Campanularia* form chambers called **sporosacs,** and these retain the developing eggs until they are released as free-swimming planulae. Remember that all individuals are connected by a common coelenteron and that the reproductive individuals depend on the feeding polyps for their maintenance. These colonies grow in a fashion much like that of a plant; try to determine the pattern of branching, the zone of growth, and the locations of the oldest and youngest members of the colony.

▶ The other group of colonial Hydrozoa to be examined contains species in which the medusae are set free from the reproductive polyps. Obtain specimens of this group, e.g., *Obelia, Pennaria* or *Bougainvillea* (Figure 3.8), and note the organization of the colony, the locations of reproductive and feeding individuals, and the presence or absence of the perisarc.

▶ All hydrozoan colonies are at least dimorphic, but in certain species the colony displays a remarkable degree of polymorphism. Examine a specimen of *Hydractinia* (Figure 3.9). The colony grows as an encrusting mat on the shells of hermit crabs; the polyps spring up singly and irregularly from the **stolon,** which consists of roots or runners collectively called the **hydrorhiza.** The polyps of this species may be gastrozooids, and spiral dactylozooids or tentaculozooids. Are the different kinds of individuals distributed randomly over the shell?

▶ The order Siphonophora contains colonial species that are extreme in the number and kinds of polyps and medusae comprising them. Examine a plastic-enclosed specimen of the Portuguese man o' war *Physalia* (Figure 3.10). *Physalia* is not usually considered a single animal, but a hydrozoan colony of separate organisms that are inextricably linked. Presumably, the individual members arise by budding from a central stem, which is medusoid in form. There may be up to 1000 individuals in a single colony. Each group of individuals is specialized for a

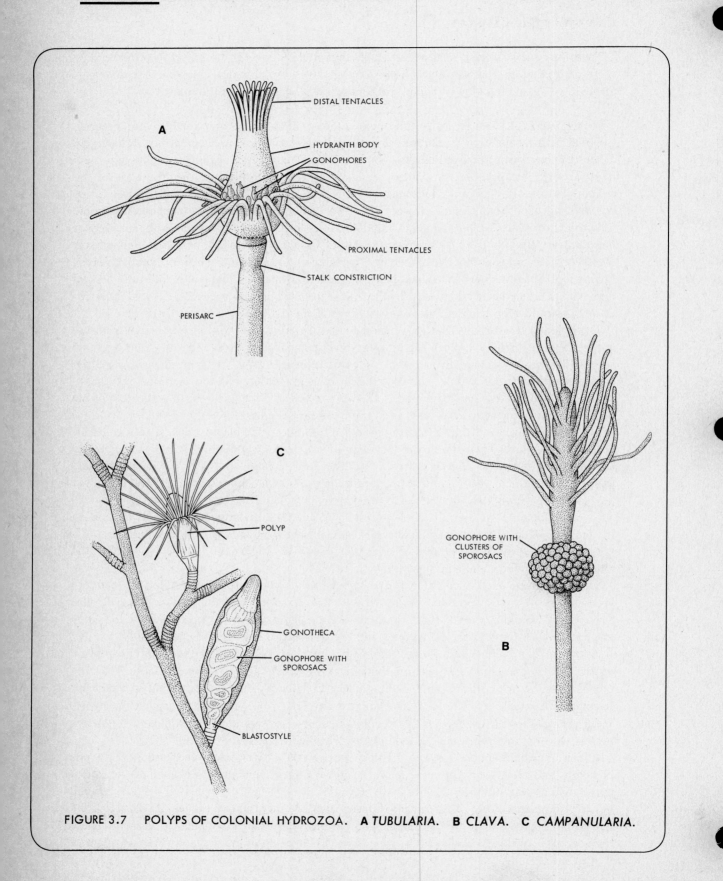

FIGURE 3.7   POLYPS OF COLONIAL HYDROZOA.   **A** *TUBULARIA.*   **B** *CLAVA.*   **C** *CAMPANULARIA.*

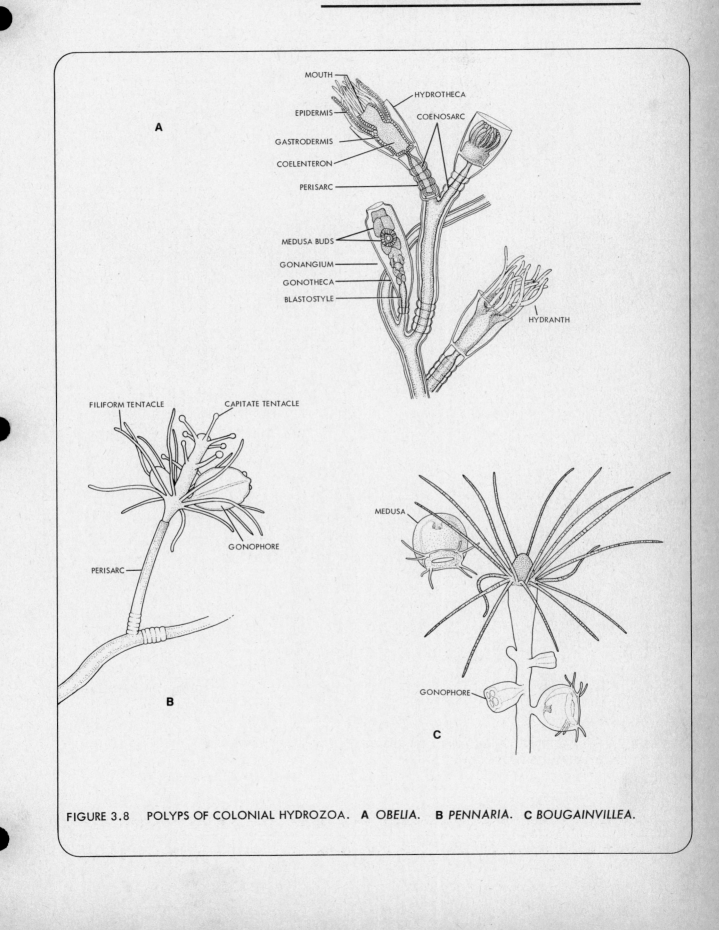

FIGURE 3.8   POLYPS OF COLONIAL HYDROZOA.   **A** *OBELIA.*   **B** *PENNARIA.*   **C** *BOUGAINVILLEA.*

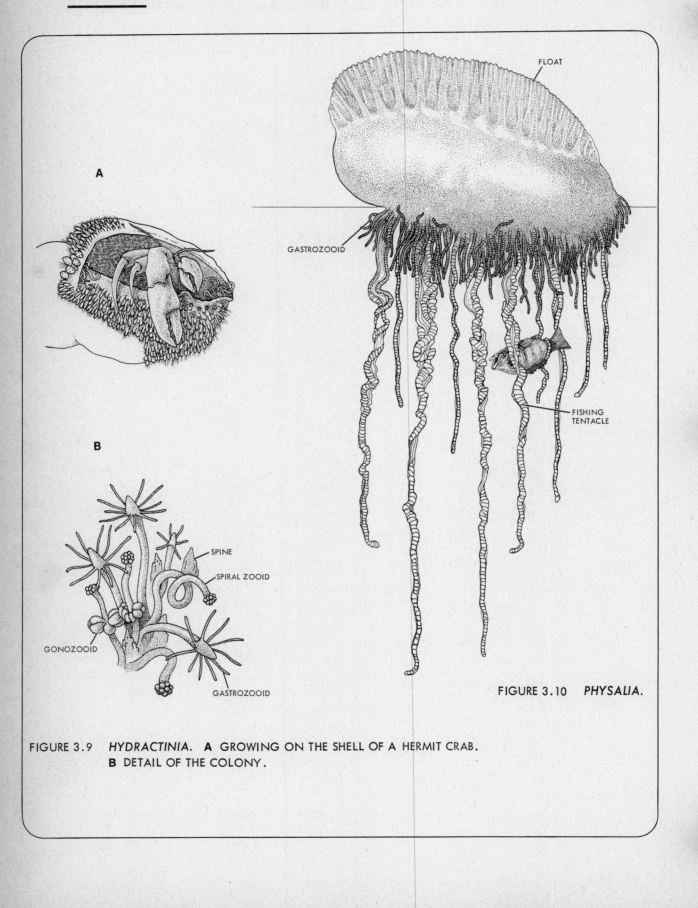

A

B

FLOAT

GASTROZOOID

FISHING
TENTACLE

SPINE

SPIRAL ZOOID

GONOZOOID

GASTROZOOID

FIGURE 3.10   *PHYSALIA.*

FIGURE 3.9   *HYDRACTINIA.* **A** GROWING ON THE SHELL OF A HERMIT CRAB.
B DETAIL OF THE COLONY.

particular task: fishing tentacles for the capture of prey, gastrozooids for digestion of food, reproductive individuals for production of gametes, and the float, which acts as a sail in locomotion. Attempt to identify the different components of the colony.

ii. ANTHOZOA.   Based on the structure of the polyp, members of the Anthozoa are divided into two subclasses: Alcyonaria and Zoantharia. (See Classification of Cnidaria, p. 88.) You have examined a typical zoantharian polyp, *Metridium.* Alcyonarian polyps are distinguished from zoantharian polyps by their octaradiate symmetry, pinnate (branching) tentacles, and specialized colonial habits. Both subclasses contain colonial species, but only one order of the Alcyonaria, the Pennatulacea, has species with dimorphic colonies.

Alcyonarian colonies arise by branching. However, the polyps are not directly connected to one another, but communicate by means of gastrodermal tubes called **solenia,** which are continuous with the coelenteron. In the dimorphic pennatulaceans **(sea pens)** one main axial polyp gives rise to the other polyps by way of lateral solenia, and the colonies have a feathery appearance. The colonies are supported by a skeletal material (see Section 3–5c) secreted by the mesogloeal cells.

▶   Study the form of the sea pansy *Renilla* (Figure 3.11). The body consists of a stemlike axial polyp and flat, fleshy lobes bearing two types of polyp: **autozooids** and **siphonozooids.** The autozooids are typical alcyonarian polyps, but the siphonozooids are small and wartlike with rudimentary tentacles. They do not feed, but serve to drive a water current through the colony (Figure 3.12).

3–2. FEEDING

Cnidarians are predaceous carnivores and employ nematocysts (Figure 3.13) to subdue and kill their prey.

a. Polyp

Polyps are sessile animals that show little obvious movement except for body contractions and waving of tentacles. They do not actively move from place to place in the search for their food, but wait until food organisms come close to them.

▶   i. HYDROPOLYP.   Take a specimen of *Pelmatohydra oligactis* or *Chlorohydra viridis* that has been starved for 48 hours and transfer it to a syracuse watch glass containing pond water. Allow the animal to relax and note how the body column elongates and the tentacles extend. With a Pasteur pipet transfer some *Artemia* larvae that have been washed in fresh water, or some small *Daphnia,* to the dish with the hydra. Note the movements of the tentacles and the reactions of the hypostome. What are the effects on the prey? Do the tentacles hold the prey by wrapping themselves around it or is the victim held by some other means?

▶   Remove the prey from the tentacles by means of watchmaker's forceps, make a wet mount, and examine under the compound microscope, first under high power and then under the oil-immersion lens. Usually the body of the prey is pierced by a variety of nematocyst types (stenotele penetrants); volvent (desmoneme)-type nematocysts are coiled around the bristles of the victim (Figure 3.13A).

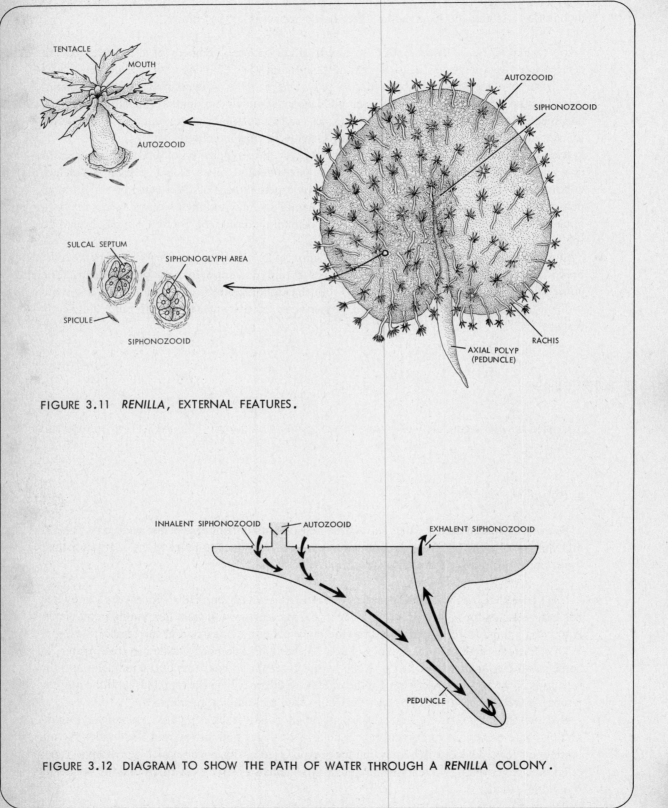

TENTACLE

MOUTH

AUTOZOOID

SULCAL SEPTUM

SIPHONOGLYPH AREA

SPICULE

SIPHONOZOOID

AUTOZOOID

SIPHONOZOOID

RACHIS

AXIAL POLYP
(PEDUNCLE)

FIGURE 3.11 *RENILLA*, EXTERNAL FEATURES.

INHALENT SIPHONOZOOID

AUTOZOOID

EXHALENT SIPHONOZOOID

PEDUNCLE

FIGURE 3.12 DIAGRAM TO SHOW THE PATH OF WATER THROUGH A *RENILLA* COLONY.

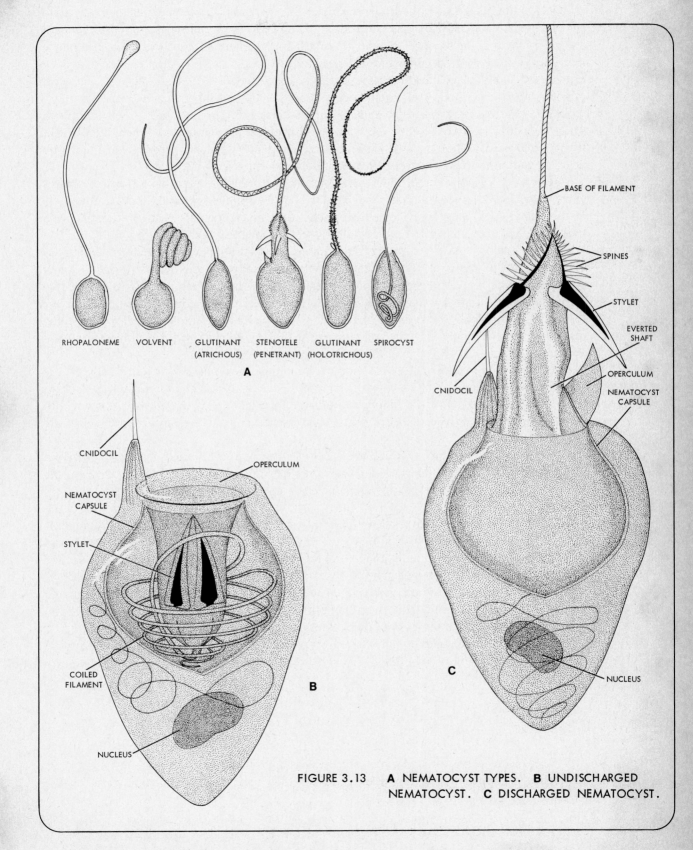

RHOPALONEME   VOLVENT   GLUTINANT   STENOTELE   GLUTINANT   SPIROCYST
(ATRICHOUS)   (PENETRANT)   (HOLOTRICHOUS)

A

BASE OF FILAMENT

SPINES

STYLET

EVERTED
SHAFT

OPERCULUM

NEMATOCYST
CAPSULE

CNIDOCIL

CNIDOCIL

OPERCULUM

NEMATOCYST
CAPSULE

STYLET

COILED
FILAMENT

B

C

NUCLEUS

NUCLEUS

FIGURE 3.13   **A** NEMATOCYST TYPES.   **B** UNDISCHARGED
NEMATOCYST.   **C** DISCHARGED NEMATOCYST.

▶   Remove a tentacle from a hydra by means of forceps and place it on a slide. Cover with a cover slip and examine under the compound microscope. What cell types are most abundant along the length of the tentacle? Do the nematocysts in the tentacle have the same appearance as those in the prey? The undischarged nematocysts in the tentacle may be discharged by drawing 0.5% methylene blue or 5% acetic acid under the cover slip. What happens? How is it possible for the fine threads of a nematocyst to subdue prey as large as *Artemia?*

▶   Shortly after the prey has been subdued, the "feeding reaction" will begin. This involves a series of coordinated movements: the tentacles bearing the prey move toward the hypostome, the mouth opens, and the impaled victim passes slowly into the coelenteron. Do the tentacles push the prey into the coelenteron or does the hypostome engulf it?

The stimulus responsible for inducing the feeding reaction has been the subject of controversy. Some workers suggest that it is related to chemicals in the prey and others to chemicals in the nematocyst capsule. Among the former class of compounds **reduced glutathione,** a tripeptide, has been shown by Loomis (1955) to elicit a feeding response.

▶   Transfer hydras* that have been starved for at least 48 hours to a syracuse watch glass containing freshly prepared 10^{-5} M reduced glutathione. How do the hydras respond? Allow some animals to remain in the glutathione solution for 1 hour. What happens to the feeding response of these animals? Can *Artemia* induce the feeding response in animals that no longer respond to glutathione? What does this tell you about the nature of the inducer for the feeding response? Can you induce the feeding response in starved animals with inert objects or dried *Artemia?*

▶   Attempt to feed hydras on *Artemia* until they no longer respond to food. How many larvae are eaten by one hydra? What causes the hydra to stop feeding? Is it related to lack of nematocysts? Do nematocysts no longer discharge from satiated hydra when in contact with prey? What factors do you think are responsible for preventing the mouth from opening in a fully fed hydra?

▶   Similar feeding reactions may be obtained with the marine hydroid *Tubularia*. *Artemia* larvae may be fed directly to marine hydroids without rinsing in distilled water. Observe the feeding reaction, tentacle types, and nematocysts employed in the capture of prey.

▶   ii. ANTHOPOLYP.   Allow small, fairly transparent anemones such as *Metridium* or *Sagartia* to attach by their bases in finger bowls filled with seawater. Rub small fragments of clam or fresh liver in powdered carmine and present this food to the tentacles. What are the responses? Are nematocysts discharged? Remove a small piece of food and see whether nematocysts are present. Place the food on the oral disc. How is food moved toward the mouth? Do tentacles stuff food into the mouth? After food has been ingested, observe the activity of the acontia at the base of the actinopharynx. What is the role of the acontia?

▶   Place fragments of aluminium foil, filter paper, etc. on the oral disc and tentacles. Are these substances rejected? Can you offer a reason for rejection? Would you say that anemones are selective feeders?

b. Medusa

▶   If living medusae are available, it should be possible to observe the feeding process by the techniques described for polyps. Prey is subdued by the batteries of nematocysts on the tentacles and transferred to the mouth.

*This part of the exercise should be started at the beginning of the laboratory period to allow time for completion.

c. Nematocyst Discharge

Previously you observed the predaceous habits of cnidarians and the role of the nematocyst in the capture of prey. Now examine the mechanism of nematocyst discharge. First, familiarize yourself with the component parts of a nematocyst (Figure 3.13**B**). Nematocysts are formed in **cnidoblasts** and consist of an intracellular capsule and a trigger, the **cnidocil,** which projects from the nematocyst to the exterior. When discharged, the coiled thread within the capsule is everted (Figures 3.13**A** and **C**). Nematocysts may contain adhesive substances and/or toxins.

▶   Place a hydra or one of its tentacles* into a syracuse watch glass containing water and attempt to discharge the nematocysts by mechanically stimulating the tentacle with pieces of lens tissue, a glass rod, or cotton thread. What happens? What do you conclude about the role of mechanical factors and nematocyst discharge?

▶   Using the same experimental material (tentacles or whole animals) introduce small amounts of clam juice, ketchup, or reduced glutathione into the watch glass, but apply no mechanical stimulation. Do nematocysts discharge? What do you conclude about chemical control of nematocyst discharge?

▶   Combine two stimuli by taking a glass rod and crushing some *Artemia* with it, or dipping it in clam juice, and then touching it to the tentacle. What happens? What is the explanation?

▶   You have seen that it is possible to discharge nematocysts in portions of the body that have been isolated from the rest of the organism. There still remains the question of whether nematocyst discharge is under nervous control. To test for this, slightly smash and smear *Corynactis* tentacles (or *Metridium* acontia) between microscope slides, separate the slides, and allow to dry. Add 0.5% methylene blue or 5% acetic acid while observing under the microscope and watch discharge. Is nematocyst discharge under nervous control? What is meant by the term **independent effector?** Why do nematocysts react to acetic acid and not to clam juice?

3–3. DIGESTION AND CIRCULATION†

The coelenteron‡ is a simple, blind digestive sac that has a single opening and is typically lined by three types of cells: **mucous cells, gland cells,** and flagellated or ciliated **gastrodermal (digestive) cells.** Gland cells are presumed to secrete enzymes for digestion, and the gastrodermal cells phagocytize the products of digestion. Digested materials are circulated in the coelenteron by flagellated gastrodermal cells, and undigested matter is ejected through the mouth. The secretions of the mucous cells assist in digestion by lubricating the food materials.

a. Polyp

▶   i. HYDROPOLYP.   To study digestion in hydra, allow *Daphnia* or *Artemia* larvae to swim in a thick suspension of carmine and then feed them to starved hydras. Note the sequence in

---

*Good substitutes are the tentacles of the anemone *Corynactis californica,* which has large, easily observable nematocysts, or the acontia of *Metridium.*

†Portions of this exercise should be started at the beginning of the laboratory period to allow time for completion.

‡The terms used by different authors for the coelenteron vary. We have reserved *coelenteron* for the digestive cavity of cnidarians and *gastrovascular cavity* for the digestive cavity of Platyhelminthes (Section 4–3); the term *gut* has been used only for complete digestive tracts with both a mouth and an anus. This usage is arbitrary but at least consistent.

the feeding response: tentacle waving, capture of prey, movement of tentacles toward the hypostome, opening of the mouth, and engulfment of the prey. After some minutes place these fed hydras on a slide and compress under a cover slip. Where are the carmine particles located? Can you see "food droplets" inside gastrodermal cells? Where do you think digestion of this "stained" food is taking place?

▶ Take another hydra and feed in the manner previously described, but do not compress under a cover slip. Allow the animal to complete the digestive process.* What happens to undigested materials?

▶ ii. ANTHOPOLYP. Obtain a living sea anemone, e.g., a *Nematostella* or a small, fairly transparent *Sagartia,* and allow the animal to attach itself in a bowl of seawater. As the oral disc expands, you will see the centrally placed mouth. Trace the path of food into the coelenteron by rubbing some pieces of clam in carmine powder and placing them on the oral disc. Attempt to locate the food after it has been ingested. Does food get into the tentacles? What happens to undigested food? What role do the acontia play in digestion?

▶ The actual sites of digestion may be determined a few hours after feeding.* Use these specimens or, if available, you may use the animals fed previously when you studied the feeding reaction (Section 3–2). Relax the fed specimens in 7.2% $MgCl_2$ and open the coelenteron by a longitudinal incision from base to oral disc. Are the carmine particles in cells? Does digestion take place extracellularly? Intracellularly?

Because polyps are small, distribution of food materials and respiratory exchange can take place across the body surface by diffusion. This is assisted by the movement of fluid by the flagellated gastrodermal cells. Although the body of the sea anemone appears externally to be rather thick and bulky, this condition is more apparent than real. The sea anemone is a saclike organism with many radial partitions. (Recall the cross sections you made in Section 3–1.) Your studies with "carmine food" should have shown that seawater enters the mouth, is directed down the actinopharynx by the flagellated siphonoglyph, and inside the coelenteron is moved by cilia on the septa. The thin, platelike septa ensure that the living tissue is nowhere very thick, and again diffusion paths remain short. At the oral end of the primary and secondary septa there are **ostia** through which the chambers of the coelenteron communicate freely with one another. Below the actinopharynx, the coelenteron is not partioned completely.

▶ Verify these anatomical details by referring to the longitudinal section you prepared previously (Section 3–1) or to prepared slides, or repeat the tracing of circulatory pathways by means of a thick carmine suspension. Can you detect respiratory currents? Are there currents of rejection?

b. Medusa

Digested food is circulated in medusae by means of ciliated cells which line the radial and circular canals. Because most medusae are small, distribution of food materials and respiratory exchange takes place by means of diffusion with the assistance of ciliary currents and body movements.

▶ Examine the complex arrangement of the coelenteric canals in *Aurelia*. Opening out from each of the four gastric pouches are two straight canals (**adradial** canals), which run to the margin of the bell. The fluid moves by ciliary action, and the return flow from the margin of the bell is via the eight branched canals (**interradials** and **perradials**) (Figure 3.14). Examine a mount of *Aurelia* and trace the canals of this system. Is it a true circulatory system?

*These portions of the exercise should be started at the beginning of the laboratory period to allow time for completion.

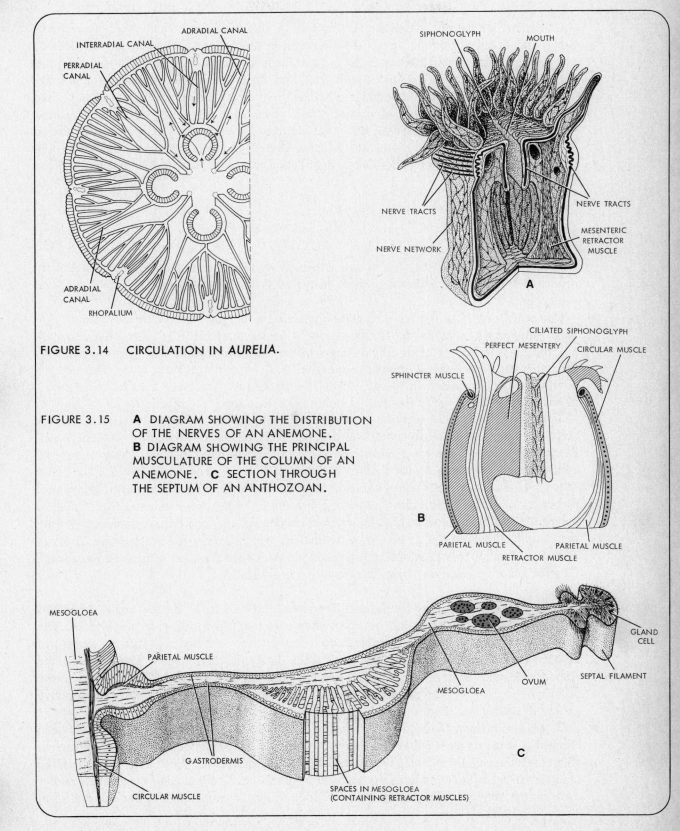

FIGURE 3.14   CIRCULATION IN *AURELIA*.

FIGURE 3.15   **A** DIAGRAM SHOWING THE DISTRIBUTION OF THE NERVES OF AN ANEMONE. **B** DIAGRAM SHOWING THE PRINCIPAL MUSCULATURE OF THE COLUMN OF AN ANEMONE. **C** SECTION THROUGH THE SEPTUM OF AN ANTHOZOAN.

3–4. NEUROMUSCULAR COORDINATION AND LOCOMOTION

Cnidarians are the first organisms you have so far encountered in which muscular contraction is largely under nervous control. The nerves and muscles are rather diffusely distributed, and because most muscular contractions are extremely slow, the sessile members of the group appear immobile most of the time. However, they do in fact exhibit continuous muscular activity, which forms part of coordinated behavior patterns taking place on a very slow time scale. They are also capable of making "fast" facilitated responses, as in feeding, and specialized "startle" responses, as when the entire polyp contracts suddenly in response to stimulation. The free-floating medusae, though wafted passively hither and thither by the currents of the ocean, exhibit coordinated contractions of the bell and may have fairly complex sense organs.

a. Polyp

i. HYDROPOLYP.   Although *Hydra* is usually sessile, it can glide on its base, float by means of a gas bubble which it secretes in the region of the basal disc (Figure 3.4C), and somersault.

▶ Obtain a living *Hydra* in a watch glass containing water and observe its movements. Are these movements spontaneous? Do they occur at regular intervals? The body movements are accomplished by means of contractile fibers in the epitheliomuscular and gastrodermal cells (Figure 3.4D). *Hydra* can shorten or lengthen its body by means of these fibers. How are the contractile fibers oriented in the epidermis? Gastrodermis? Are these muscles antagonists? Contraction of which set of muscles produces body extension? Contraction?

▶ The muscle tissues are fairly uniform in distribution, so that responses to external stimuli (i.e., not spontaneous movements) will depend solely on the distribution of the nerve net. Responses will be more rapid and more frequently encountered in areas where nerves are concentrated. Allow the *Hydra* to relax and then, with a glass rod drawn out to a fine point, touch the animal at various points along the body column and the tentacles. Record the type of response. What does this indicate about the distribution of nervous elements? Can you show **polarity?** (That is, is the wave of contraction propagated in only one or in both directions from the point of stimulation?) Can you show **facilitation?** (That is, are repeated shocks necessary to produce a reaction?) Is there spread of excitation? Which are the areas of greatest sensitivity? Of what adaptive value are these features to the organism?

ii. ANTHOPOLYP.   The neuromuscular system of the anemone is more highly developed and localized than that of *Hydra,* but can be demonstrated using similar simple methods. The anemone nervous system consists of a two-dimensional neuronal net without ganglionic centers. Conduction is outward in a circle from the center of stimulus. The extent of the response depends both on the intensity and the duration of the stimulus. In addition to the nerve net there are well-developed tracts containing elongate neurons; these serve for selective and rapid conduction (Figure 3.15A).

▶ Obtain specimens of individual sea anemones (*Metridium* or *Anthopleura*) that have been allowed to settle on glass plates some days previously. Using animals that are fully expanded, stimulate a tentacle by touching the tip with a pointed glass needle. What is the reaction of the tentacle? In what direction does the tentacle bend? Continue stimulating the tentacle and/or increase the stimulus strength. Note the reaction in neighboring tentacles and the oral disc.

Such reactions are normal components of the anemone **feeding response.** If stimulation is continued further, localized raising up of the edge of the oral disc extends around the disc until it is finally drawn down by contractions of the longitudinal mesenteric muscles (Figure 3.15**B**) and covered over by contraction of the sphincter muscle.

▶ A generalized body response, all-or-none in character, can be produced by applying a quick, firm stroke of a glass rod against the pedal disc.

Do these experiments confirm the existence of main through-conducting pathways?

▶ Examine the cross sections you prepared at the beginning of the exercise (Section 3–1) and/ or prepared slides of cross sections through the anemone body; locate the muscles in the body column and in the mesenteries (Figure 3.15). The well-developed muscular system is mainly derived from the gastrodermis. Identify the **parietal muscles,** which run in the mesenteries close to the body wall, and the longitudinal **mesenteric retractor muscles,** which run in the mesenteries between the oral and the pedal discs, and notice their arrangement (Figures 3.5**B** and **C**). In a whole intact living anemone identify the **sphincter muscle,** which closes the oral disc in purse-string fashion. Circular muscles are also present in the column, pedal disc, and tentacles, but cannot be identified in the cross sections except with difficulty. The contraction of the parietal muscles shortens the body column, whereas the mesenteric muscles pull down the oral disc and tentacles. If the body column is stimulated with a probe or electric shock at a frequency of one in 5 seconds, the parietal muscles contract; if the frequency is one in 2 seconds, the mesenteric muscles contract; and finally if one per second is the shock interval, the sphincter muscle contracts. Note that facilitation is required, and each muscle has its own **excitability threshold** (frequency of stimulation necessary to effect contraction). Coordination is effected by this graded response.

▶ As pointed out, although anemones appear immobile most of the time, they do make almost continuous but practically imperceptible movements. Place some anemones in a small bowl of seawater and allow them to attach by their bases. On the underside of the bowl, mark the position of each sea anemone by drawing the outline of the pedal disc with a marking pen. Note positional changes during the course of a week. How do anemones move from place to place?

▶ iii. COLONIAL HYDROZOANS.   Obtain some living specimens of the marine hydroid *Tubularia* and place them in a small bowl of seawater. Allow the animals to relax. Note the spontaneous movements. Does the entire animal contract? Do the tentacles show spontaneous movements? What do coordinated and uncoordinated tentacle movements imply about the nature of physiological coordinating mechanisms in the polyp? Attempt to determine the arrangement of the neuromuscular system by testing the reaction of the hydranth to tactile stimuli, as described for *Hydra*.

b. Medusa

The medusoid form is adapted to a free-swimming motile existence. Coordinated swimming movements and capture of food while afloat are reflected by the presence of more specialized muscular and nervous systems than those of hydropolyps. The gastrodermal cells are not contractile, and the main musculature, found in the subumbrellar surface of the bell, is epidermal in origin. The bell margin is well supplied with specialized sensory organs. In hydrozoan medusae there are usually **ocelli** (photoreceptors) and **statocysts** (organs of balance or equilibrium). Without the latter, orientation and coordination in swimming is lost. Why are these sensory devices lacking in polyps?

If live medusae are available, for example, *Gonionemus, Aurelia,* note the manner of medusoid motion by "jet propulsion," using the musculature of the bell. Capture of food by *Gonionemus* is quite interesting to observe. By pulsations of the bell the animal moves near the surface of the water with its convex side upward, then it turns over and floats downward, concave side upward and tentacles outstretched.

▶ In preserved specimens of *Gonionemus* or *Polyorchis* and *Aurelia,* observe that there are swellings on or between the bases of the marginal tentacles; these contain **statocysts** (Figure 3.16A) or more complex sensory receptors called **rhopalia** (Figure 3.16B). Rhopalia are found in eight indentations on the margin of the bell of *Aurelia* (Figure 3.3A). Each rhopalium consists of a statocyst and an ocellus. Remove a rhopalium from a preserved specimen, transfer it to a syracuse watch glass containing some seawater, and study its organization under the dissecting microscope.

▶ If living *Aurelia* are available, make cuts along the margin of the bell and remove the rhopalia one at a time. Note the change in swimming pattern after each removal. What is the indicated function of the rhopalium?

▶ The subumbrellar muscle of *Aurelia* provides the main locomotive force in swimming. Make a doughnut-shaped preparation of this muscle by cutting around the margin of the bell to remove all rhopalia and the center of the bell. Put the preparation in a bowl of seawater. Touch the circular muscle and observe the circulation of the generated impulse. Does the muscle become fatigued?

Noting the differences in locomotion between polyp and medusa, can you suggest a reason for the existence of polyp and medusoid forms in the same species?

3–5. SKELETON

Depending on their nature and composition, skeletons may give support and/or protection, serve for muscle attachment, and/or transmit muscle action. The hard skeletons of cnidarians give both support and protection, and the fluid-filled coelenteron, acting as a hydrostatic skeleton, transmits the action of muscles.

a. Chitinous Skeleton

Many hydrozoans secrete a chitinous exoskeleton, the **perisarc** or cuticle, which is protective in function. It may or may not extend around the reproductive polyps (as a **gonotheca**) and the feeding polyps (as a **hydrotheca**), and its extent and arrangement have taxonomic significance (p. 88).

▶ Review the form and extent of the perisarc in hydrozoans available in the laboratory (see p. 69). On stems, it forms groups of rings at points related to the branching. What is the functional significance of this?

▶ Observe living specimens of *Obelia* under the dissecting microscope and allow them to relax. With a needle, touch the tentacles of an extended hydranth and watch retraction into the hydrotheca.

b. Hydrostatic Skeleton

The body fluids of the sea anemone are contained in a muscular tube, the fibers of which run in several directions. There are usually longitudinal bands in the septa and a circular layer in

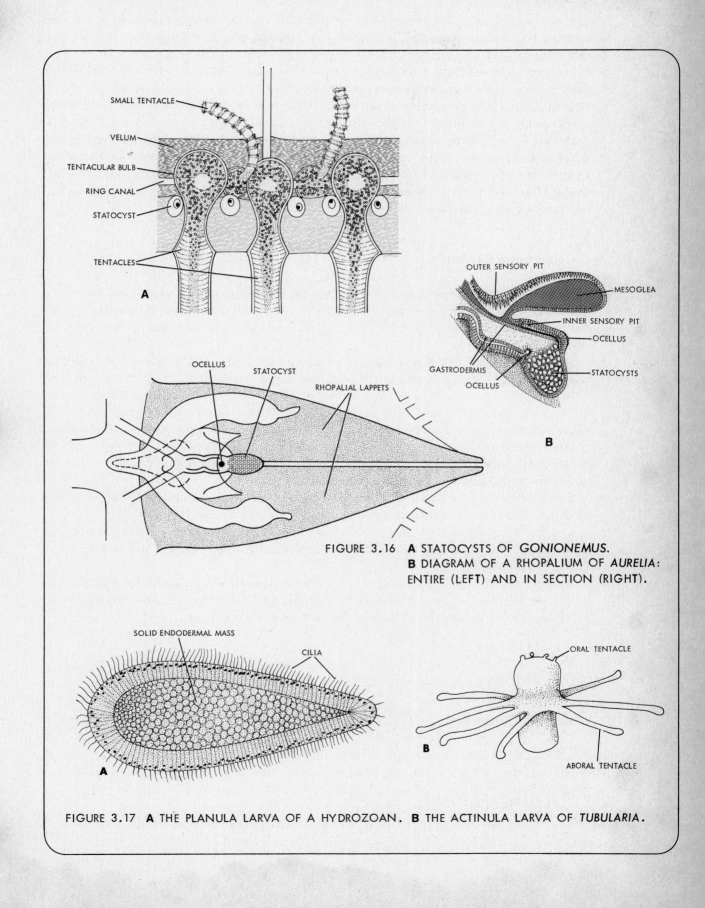

FIGURE 3.16 **A** STATOCYSTS OF *GONIONEMUS*.
B DIAGRAM OF A RHOPALIUM OF *AURELIA*:
ENTIRE (LEFT) AND IN SECTION (RIGHT).

FIGURE 3.17 **A** THE PLANULA LARVA OF A HYDROZOAN. **B** THE ACTINULA LARVA OF *TUBULARIA*.

the body wall. Contraction of one set of muscles acts on the incompressible fluid, and the resulting pressure is transmitted in all directions, enabling other muscles to alter their length, that is, to relax or stretch.

▶ Lift an anemone from its base or disturb it by violent prodding. Contraction of longitudinal muscles produces a shortening of the column, withdrawal of the disc, and bulging outward of the entire body. Allow the animal to relax and extend itself. Water is taken into the coelenteron via the siphonoglyph, the walls of the actinopharynx act as a valve, and the anemone maintains a constant volume. Elongation of the column is produced by contraction of the circular fibers in the wall, and compressed coelenteric fluid produces an extension of the longitudinal muscles. Observe the spontaneous contraction and relaxation in an anemone. Which muscles relax during shortening of the column? Place a small anemone in some 7.2% $MgCl_2$ for 20 minutes, and then stimulate the animal. What happens? Why?

c. Calcareous Skeleton

Massive calcareous skeletons are formed by some Hydrozoa, such as members of the orders Milleporina and Stylasterina, and some Anthozoa. In the Hydrocorallina (hydrozoan corals) the calcareous mass is secreted by the epidermis and corresponds to the perisarc of other hydrozoans.

▶ Study the skeleton of *Millepora*. Close inspection of the surface reveals pores that were once the points of emergence of the polyps and gonophores. Contrast this with the skeleton of *Obelia*.

The anthozoan coral is essentially an anemone that forms a skeleton composed of calcareous spicules combined with a horny substance. Anthozoans that form skeletons are the Alcyonaria (Octocorallina) and the Zoantharia (Hexacorallina). Alcyonarian coral skeletons are secreted by the mesogloeal cells of the polyp. These corals are distinguished from all other cnidarian polyps by the possession of eight pinnate tentacles.

▶ Study the form of the skeleton in *Tubipora* (organ pipe coral), *Corallium* (precious coral), *Gorgonia* (sea fan), *Pennatula* or *Stylatula* (sea pens), and *Renilla* (sea pansy).

The true stony corals belong to the Madreporaria and, in contrast to the alcyonarians, the hard exoskeleton is secreted by the epidermis and is a true exoskeleton. These forms are the reef and island builders.

▶ Examine the skeletons of the solitary coral *Fungia* and the small colonial form *Astrangia*. Examine specimens of the reef builders *Meandrina* (brain coral) and *Acropora* (antler coral). How does the coral reproduce? How is this related to reef building?

3–6. REPRODUCTION

a. Asexual Reproduction

i. BUDDING AND COLONY FORMATION. The most common method of asexual reproduction in cnidarians is by budding. A bud is an outgrowth of the body, consisting of both layers of body tissue (gastrodermis and epidermis), which ultimately differentiates into a polyp or medusa.

▶ Study preparations and slides of *Hydra* with buds (Figure 3.4A and C) as well as fresh material, if available. Is budding restricted to one particular area of the body column? How is budding regulated? Solitary polyps such as *Hydra* exist because the buds leave the parent body.

▶ Colony formation occurs when buds remain attached to the parent body. This pattern is common among hydrozoans and anthozoans. In studies of polymorphism (Section 3–1c) a variety of colonial hydroids were examined. Review the organization of the hydroid colony or examine specimens of *Obelia* (Figure 3.8A), *Pennaria* (Figure 3.8B), and *Tubularia* (Figure 3.7A). Concentrate on the branching patterns; construct a diagram to show where the youngest and oldest hydranths are located. How are feeding and reproductive polyps arranged in the colony?

Colony formation is rare among anemones, but common in the stony corals that may form large reefs. The coral polyp is an anemone that forms a hard calcareous exoskeleton, secreted by the epidermis and wholly outside the polyp body (Section 3–5c). The mode of asexual reproduction is not entirely understood. There is never longitudinal fission through the pharynx, but new mouths may appear in the connective tissue outside the original disc and acquire septa continuous with the parent body, or two or more mouths may arise on the oral disc of the parent polyp, within the parental circle of tentacles. Depending on the number and organization of buds, the colony may be branching or encrusting.

▶ Examine stony corals and branched anthozoans such as *Alcyonium, Gorgonia,* or *Pennatula*. What is the branching pattern?

ii. PRODUCTION OF MEDUSAE.   Depending upon the particular species, medusae can arise asexually by budding from a reproductive polyp, from almost any part of a feeding polyp, by development from a larval polyp (scyphistoma), or even from other medusae.

▶ In studies of polymorphism (Section 3–1) some of these methods of asexual reproduction were examined. Review the types of medusae and their method of formation in the species studied earlier.

b. Sexual Reproduction

i. HYDROZOA.   All hydrozoan medusae can reproduce sexually and are dioecious. They shed their eggs and sperm into the sea where fertilization occurs, or fertilization may take place *in situ* and the fertilized egg may develop to a later stage before it is released. Cleavage is complete, a blastula forms, and following gastrulation a solid-bodied **planula larva** is produced (Figure 3.17A). The ciliated larva eventually undergoes metamorphosis into a tiny polyp, sometimes called an **actinula larva** (Figure 3.17B).

▶ Examine prepared slides of the planula larva (Figure 3.17A). Review the location of the gonads in hydrozoan medusae (Section 3–1b).

Fertilization and development to the actinula stage take place in the gonophores in *Tubularia*. If available, remove some gonophores from a hydranth of a specimen of *Tubularia* (Figure 3.7A) and examine under the compound microscope. Some of the hydranths may have planula and/or actinula larvae (Figure 3.17). After some time, actinula larvae may be observed swimming in the dish.

Hydra is an unusual hydrozoan in that the polyp shows both asexual and sexual phenomena. The physiological mechanisms underlying the onset of sexuality are not completely understood, but seem to be linked to environmental conditions, such as high concentrations of CO_2, stagnation of water, and temperature changes.

▶ Obtain prepared slides showing the form of the ovary and testis in *Hydra,* and then examine the histology of the ovary and spermary in sections through these organs (Figure 3.4C). Are gonads located at a specific site along the body column? Are they in the same location as the budding zone? What cells differentiate to form eggs and sperm?

ii. SCYPHOZOA.   Adult scyphozoa are dioecious and reproduce sexually. Developmental and reproductive patterns are the same as in hydrozoans, except that the ciliated planula metamorphoses into a **scyphistoma.**
▶   Review the location of the gonads in Scyphozoa.

iii. ANTHOZOA.   Sea anemones may be dioecious or hermaphroditic. Fertilization may take place in the coelenteron or externally. Development is similar to that of other cnidarians; the larva is a planula, which settles and metamorphoses into a small anemone.
▶   Review the location of gonads in *Metridium* (Figure 3.5A) by referring to longitudinal sections. What is the origin of these gonads?
Details of cnidarian embryology and development are beyond the scope of this course. Students interested in these topics should consult the reference by Costello.

3–7. REGENERATION

The cnidarians show considerable capacity for regeneration, and many of them are capable of regenerating complete organisms from fragments of the parents. Hydrozoans and anthozoans regenerate better than scyphozoans. Regeneration in nature often parallels the process of **budding** (Section 3–6a).

a. Regeneration in Polyps

▶   *Hydra* is so named because of its resemblance to the many-headed creature of mythology, which grew two new heads whenever one of its existing ones was cut off. The regenerative powers of *Hydra* can be demonstrated by cutting a specimen into little pieces with a sharp scalpel, razor blade, or glass needle and then keeping the isolated fragments in watch glasses containing pond water. Note the pattern of regeneration. Which parts of the body regenerate most rapidly?
▶   Relax some small sea anemones in 7.2% MgCl$_2$ and cut into a few pieces with swift clean cuts, using a razor blade. Place in clean, running seawater and follow the pattern of regeneration. In some anemones, for example, *Metridium* and *Haliplanella*, pedal laceration and regeneration is a natural form of asexual reproduction.

b. Chimeras

▶   Cut up some brown hydras (*Pelmatohydra* sp.) and green hydras (*Chlorohydra* sp.) into four or five pieces, mix together, and centrifuge slowly for 15 minutes to obtain a fused pellet. Gently remove to a watch glass containing pond water and observe the growth of the organism. Does a **chimera** (regenerated organism composed of a mixture of cells of both species) result? What does this indicate about recognition of self? What does this indicate about the degree of organization and integration in hydras?

c. Polarity in Hydroids

Some examples of polarized regeneration (regeneration along a fixed oral-aboral axis) are demonstrated by the experiments described so far in this section. Another demonstration of this phenomenon can be performed using *Tubularia*.

► Take a piece of the stem of a healthy tubularian, cut it into sections of 2, 4, and 10 mm, and return them to running seawater. Record the number of bipolar and unipolar regenerates. Is there a difference in the time of regeneration of proximal and distal segments? The pieces can be marked and distinguished from each other by making the cuts in different directions, as in the following illustration.

References

Barnes, R. D. *Invertebrate Zoology*. 3rd ed. Philadelphia: W. B. Saunders Co., 1974.

Barrington, E. J. W. *Invertebrate Structure and Function*. Boston: Houghton Mifflin Co., 1967.

Borradaile, L. A., and F. A. Potts. *The Invertebrata*. London: Cambridge University Press, 1961.

Brown, F. A., Jr. (ed.). *Selected Invertebrate Types*. New York: John Wiley & Sons, 1950.

Buchsbaum, R. *Animals Without Backbones,* 2nd ed. Chicago: University of Chicago Press, 1972.

Bullough, W. S. *Practical Invertebrate Anatomy*. New York: St. Martin's Press, 1962.

Burnett, A. L. A model of growth and cell differentiation in *Hydra, Amer. Natur.* **100**:165–89, 1966.

Burnett, A. L. (ed.) *Biology of Hydra.* New York: Academic Press, 1973.

Carter, G. S. *A General Zoology of the Invertebrates*. London: Sidgwick and Jackson, 1951.

Costello, D. P., M. E. Davidson, A. Eggers, M. H. Fox, and C. Henley. *Methods for Obtaining and Handling Marine Eggs and Embryos*. Woods Hole, Mass.: The Marine Biological Laboratory, 1957.

Hadzi, J. *The Evolution of the Metazoa*. New York: Macmillan Publishing Co., 1963.

Hyman, L. H. *The Invertebrates,* Vol. 1. *Protozoa Through Ctenophora*. New York: McGraw-Hill Book Co., 1940.

Lane, C. The Portuguese man o' war, *Sci. Amer.* **202**(3):158–68, 1960.

Lenhoff, H. M., L. Muscatine, and L. V. Davis. *Experimental Coelenterate Biology*. Honolulu: University of Hawaii Press, 1971.

Lenhoff, H. M., and W. F. Loomis. *The Biology of Hydra and of Some Other Coelenterates*. Coral Gables: University of Miami Press, 1961.

Meglitsch, P. *Invertebrate Zoology*. New York: Oxford University Press, 1967.

Nicol, J. A. C. *The Biology of Marine Animals*. New York: Interscience Publishers, 1960.

Prosser, C. L. *Comparative Animal Physiology,* 3rd ed. Philadelphia: W. B. Saunders Co., 1973.

Ramsay, J. A. *Physiological Approach to the Lower Animals*. London: Cambridge University Press, 1952.

Rees, W. J. (ed.). *The Cnidaria and Their Evolution. Symp. Zool. Soc. London,* No. 16, New York: Academic Press, 1966.

Rees, W. J. Behavioral physiology of coelenterates, *Amer. Zool.* **5**:337–589, 1965.

Russell-Hunter, W. D. *A Biology of Lower Invertebrates*. New York: Macmillan Publishing Co., 1968.

Smith, R. I., and J. T. Carlton (eds.). *Light's Manual: Intertidal Invertebrates of the Central California Coast,* 3rd ed. Berkeley: University of California Press, 1975.

Welsh, J. H., R. I. Smith, and A. E. Kammer. *Laboratory Exercises in Invertebrate Physiology*. Minneapolis: Burgess Publishing Co., 1968.

An Abbreviated Classification of the Phylum Cnidaria (After Hyman)

Phylum **Cnidaria**

> *Class I* **Hydrozoa**
>
> Polymorphic with polyp and/or medusa present in the life cycle. Coelenteron without stomodaeum and not divided by septa. Mesogloea acellular. Sex cells ripen in the epidermis. Medusae with a velum (craspedote).

> *Order 1* **Hydroidea**
>
> Polyp generation well developed; medusae free or abortive.

> *Suborder 1* **Gymnoblastea**
>
> Hydrotheca and gonotheca absent. Ocelli, no statocysts, gonads on manubrium, e.g., *Hydra; Tubularia, Clava, Campanularia, Hydractinia.*

> *Suborder 2* **Calyptoblastea**
>
> Hydrotheca and gonotheca present; statocysts, gonads on radial canals, e.g., *Obelia, Pennaria, Bougainvillea.*

> *Order 2* **Milleporina**
>
> Hydroid colony with calcareous skeleton. Dactylozooids hollow, e.g., *Millepora.*

> *Order 3* **Stylasterina**
>
> Similar to Milleporina, but dactylozooids solid.

> *Order 4* **Trachylina**
>
> Craspedote medusae with or without a reduced hydroid phase, e.g., *Gonionemus, Polyorchis.*

> *Order 5* **Siphonophora**
>
> Highly polypmorphic, free-swimming or floating colonies, e.g., *Physalia.*

> *Order 6* **Chondrophora**
>
> Radially symmetrical, free-floating organisms represented by a single

large *Tubularia*-like polyp bearing unbranched tentacles. The chitinous, gas-filled float is the homolog of the perisarc, e.g., *Velella*.

Class II Scyphozoa

Acraspedote medusae, coelenteron without stomodaeum, mesogloea cellular, gonads endodermal, polyp generation lacking or reduced (scyphistoma).

Order 1 Stauromedusae

Attached by aboral stalk, e.g., *Haliclystus*.

Order 2 Semaeostomae

Corners of mouth prolonged into lobes. Margin scalloped, 8 or 16 rhopalia, e.g., *Aurelia*.

Order 3 Cubomedusae

Free-swimming forms, four perradial tentacles and four interradial rhopalia. Margin of bell turned inward as a velarium.

Order 4 Coronatae

Free-swimming, margin scalloped, 4 to 32 rhopalia.

Order 5 Rhizostomeae

Oral lobes fused, obliterating mouth; many small mouths and canals in oral lobes; no tentacles; margin scalloped; eight or more rhopalia.

Class III Anthozoa

Medusa absent from the life cycle altogether. Hexamerous or octamerous, radial or biradial symmetry. Stomodaeum. Coelenteron divided by septa; septa with nematocysts at the edge, mesogloea fibrous and cellular, gonads endodermal in septa.

Subclass 1 Alcyonaria

Polyps with eight pinnate tentacles, one siphonoglyph. Colonial cnidarians with endoskeleton.

Order 1 Stolonifera

Polyps not fused, skeleton of spicules fused into tubes, e.g., *Tubipora*.

Order 2 Alcyonacea

Lower part of polyp fused to fleshy mass from which only oral end protrudes. Not axial, e.g., *Alcyonaria*.

Order 3 Gorgonacea

Axial skeleton of calcareous and/or horny material, e.g., *Gorgonia, Corallium*.

Order 4 Pennatulacea

Colony of one long axial polyp with many lateral polyps, e.g., *Stylatula, Pennatula, Renilla*.

Subclass 2 Zoantharia

Tentacles simple, rarely branched, not eight.

Order 1 Actinaria

Septa paired, no skeleton, solitary, e.g., *Metridium, Sagartia, Corynactis.*

Order 2 Madreporaria

Compact calcareous exoskeleton, e.g., *Acropora, Fungia, Meandrina, Astrangia.*

Order 3 Ceriantharia

Equipment and Materials

Equipment

Microscope and lamp
Microscope slides and cover slips
Dissecting instruments
Silver foil or filter paper
Pointed glass rod
Blunt glass rod
Scalpel or razor blades
Centrifuge and centrifuge tubes
Finger bowls
Syracuse watch glasses

Solutions and Chemicals

Carmine powder
5% acetic acid
10^{-5} M reduced glutathione (freshly prepared)
0.5% methylene blue
7.2% $MgCl_2$
Ketchup

Living Material*

Pelmatohydra (4,6,11)
Tubularia (7,12)
Gonionemus (12)
Artemia larvae (4,11)
Daphnia (4,11)
Fresh mussel or clam or liver (9,12)
Metridium or *Sagartia* (7,12)
Corynactis californica (9)
Nematostella (12)
Aurelia (12)

Astrangia (7,9,12)
Renilla (7,9)
Chlorohydra (3,4,11)
Stylatula or *Muricea* (9)
Hydractinia (7,12)
Obelia (7,12)
Anthopleura (9)

Preserved Material

Metridium or *Sagartia*
Hydractinia
Aurelia
Gonionemus
Polyorchis
Haliclystus or *Lucernaria*
Alcyonium
Gorgonia
Pennatula
Millepora
Tubipora
Corallium
Meandrina
Fungia
Acropora

Prepared Slides

Aurelia scyphistoma
Aurelia strobilization
Aurelia ephyrae
Aurelia adult
Hydra whole mount
Hydra cross section

*Numbers in parentheses identify sources listed in Appendix B.

Hydra longitudinal section
Metridium cross section
Metridium longitudinal section
Tubularia
Clava
Campanularia
Obelia whole mount
Obelia medusae

Pennaria
Bougainvillea
Physalia
Hydra with ovary
Hydra testis
Hydra with buds
Planula larva

PLATYHELMINTHES

Contents

The Platyhelminthes (*platy,* flat; *helmin,* worm; Gk.) or **flatworms,** as their name suggests, comprise the parasitic **flukes** and **tapeworms,** and the free-living **planarians.** They are flattened, ribbon-like organisms adapted for crawling in the interstices of both living and nonliving material.

These are multicellular animals (Metazoa) that have their tissues organized into complex functional units called **organs.** Situated between the epidermis and the digestive tract, these organs may be arranged into organ systems, and the flatworms possess a complicated reproductive system. With increasing complexity, simple diffusion no longer suffices for elimination of nitrogenous wastes; thus, the platyhelminths exhibit a specialized excretory system. However, they resemble the cnidarians in that they lack true circulatory and respiratory structures; when present, the body cavity (**gastrovascular cavity**) functions in digestion and circulation; they lack a definitive anus; and they are unsegmented.

The formation of organs and organ systems is closely related to the presence of a true **mesoderm.** Although the middle layer of cnidarians may be fairly extensive, gelatinous, and in some cases filled with fibers, it never reaches the extensive development attained in the platyhelminths, and it has a different embryonic origin. The middle layer of flatworms consists of a **parenchyma** containing cells, fibers, and muscles together with various organs and organ systems. Because there are no spaces in this layer, the worms lack a mesodermally lined body cavity (**coelom**) and are spoken of as being **acoelomate.**

Flatworms are highly motile. Associated with their well-developed and coordinated locomotory system are their bilateral symmetry and anterior concentration of nervous tissue and sense organs (**cephalization**).

4–1. BODY ORGANIZATION

Most animals display some sort of symmetry, which is related to the pattern of locomotion or, in a broader sense, to the manner in which the animal orients itself in the environment.

Sessile or **sedentary** animals tend to show an orientation toward gravity and to exhibit **radial symmetry.** The body generally has the form of a cylinder with an attached end (down, basal, or aboral) and a free end (up, distal, or oral). The other body axes, which are arranged around this main longitudinal axis, form a series of similar **radii.** This geometrical design is seen in the cnidarian polyp and free-swimming medusae. The latter, although motile, spend a great deal of time passively floating or moving up and down as they orient themselves toward gravitational forces.

Motile animals usually enter the environment with the same end of the body forward at all times. Recall examples of highly motile Protozoa and their form. Did they always move with the same end forward? Such animals are usually elongate and display **bilateral symmetry;** the body can be divided along a longitudinal plane into right and left halves, which are mirror images of each other. These animals have an anterior end (head, front), a posterior end (rear, back), and dorsal (top) and ventral (bottom) surfaces.

a. External Anatomy

The phylum Platyhelminthes is divided into three classes: the Turbellaria, the Trematoda, and the Cestoidea. (See Classification of Platyhelminthes, p. 114.) Platyhelminths are bilaterally symmetrical, highly motile, and elongate; these wormlike animals are adapted for creeping, crawling, and swimming.

i. TURBELLARIA.   These are free-living platyhelminths that are flat and leaflike in shape.
▶ Examine a living specimen of the freshwater planarian *Dugesia* in a syracuse watch glass containing pond water, and notice the form of the body and the symmetry (Figure 4.1A). Can you recognize the anterior end? What structures are associated with the anterior end? Posterior end? Two body surfaces are clearly distinguishable, not only by differences in pigmentation but by the habitual behavior of the animal, which moves with the same surface always facing the

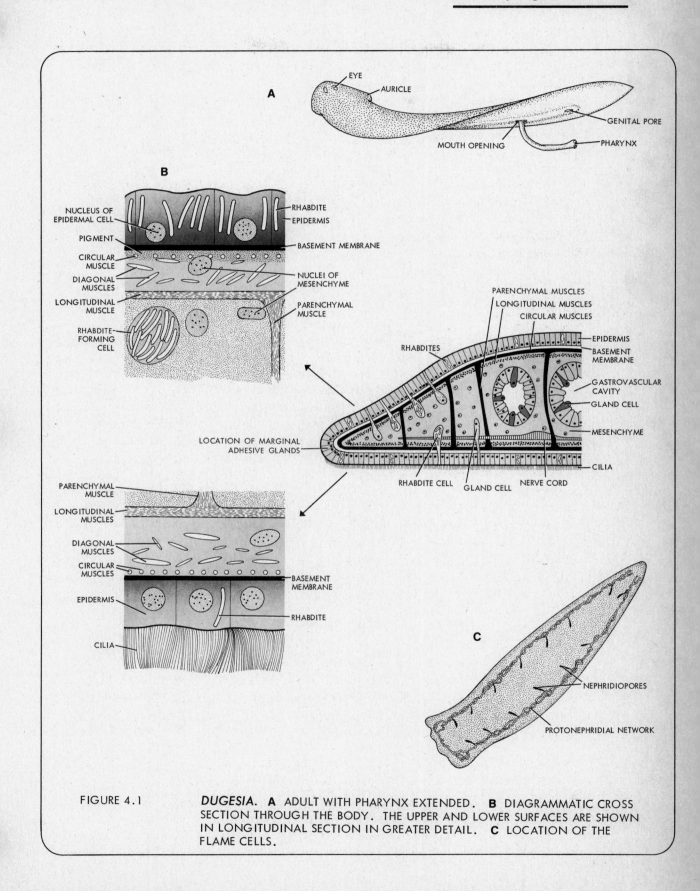

FIGURE 4.1   *DUGESIA.* **A** ADULT WITH PHARYNX EXTENDED.   **B** DIAGRAMMATIC CROSS SECTION THROUGH THE BODY. THE UPPER AND LOWER SURFACES ARE SHOWN IN LONGITUDINAL SECTION IN GREATER DETAIL. **C** LOCATION OF THE FLAME CELLS.

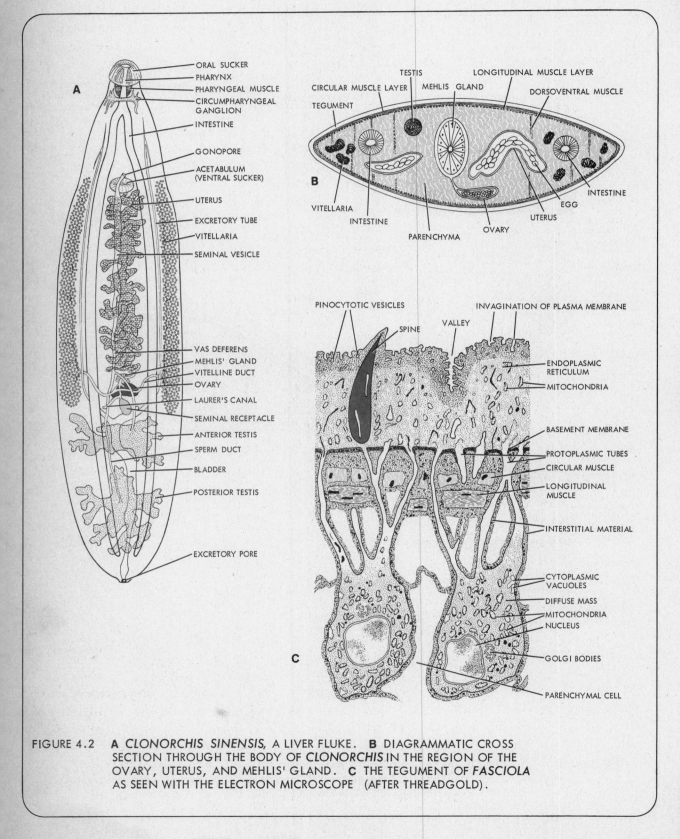

FIGURE 4.2 **A** *CLONORCHIS SINENSIS,* A LIVER FLUKE. **B** DIAGRAMMATIC CROSS
SECTION THROUGH THE BODY OF *CLONORCHIS* IN THE REGION OF THE
OVARY, UTERUS, AND MEHLIS' GLAND. **C** THE TEGUMENT OF *FASCIOLA*
AS SEEN WITH THE ELECTRON MICROSCOPE (AFTER THREADGOLD).

substratum. Can you tell the dorsal surface from the ventral surface? What associated structures characterize these surfaces? Turn the animal so that the ventral side faces you and observe the central position of the **mouth** and the muscular **pharynx,** which may protrude through the mouth opening. Small, relatively transparent specimens may show branches of the **gastrovascular cavity** which run anteriorly and posteriorly from the pharynx.

ii. TREMATODA.*   Trematodes, or **flukes,** are all parasitic as adults; the body form is leaflike.

▶   Examine a specimen of the adult liver fluke *Clonorchis* (Figure 4.2A). This flatworm lives in the bile ducts of humans and feeds on the living tissues of the host. Under the dissecting microscope observe the shape of the body. The anterior end of the worm tapers to a point and bears a sucker at its tip, whereas the posterior end is blunt. Does a single imaginary plane divide the animal into identical halves? What is the orientation of such a plane? The sucker at the anterior end is called the **oral sucker** and surrounds the **mouth.** Running posteriorly and laterally are the two branches of the **gastrovascular cavity.** Most of the tubular network in the center of the body comprises the **reproductive system** and will be examined later. In what direction is the liver fluke flattened? Can you see any difference between dorsal and ventral surfaces? How? Are **eyes** present on any part of the body? Is there any trace of body segmentation?

iii. CESTODA.   Cestodes are **tapeworms;** as their name indicates, they are ribbon-like in form, and as adults are all endoparasites in the guts of vertebrates.

▶   Examine a plastic whole mount of the pork tapeworm *Taenia solium* (Figure 4.3A), which shows the typical morphology of this class of flatworms. Notice that the body consists of a linear series of **proglottids** that become progressively smaller, so that the most anterior tip of the animal consists of a microscopic pointed **scolex** (Figure 4.3**B**). Study the scolex under the dissecting microscope and note the **suckers** and **hooks.** Is a mouth present? Just posterior to the scolex or holdfast region is the **neck,** which buds off proglottids containing male and female reproductive organs. Are there any obvious sense organs? Where are the youngest proglottids found? Oldest? Although tapeworms appear to be segmented, can you offer an argument against this?

▶   List criteria for distinguishing anterior from posterior for all three specimens. Which features are common? What differences in body form are most apparent between free living and parasitic flatworms?

b. Internal Anatomy

▶   i. TURBELLARIA.   Examine a prepared slide of a cross section of *Dugesia* (Figure 4.1**B**). Notice the one-layered, cellular **epidermis;** in other species the epidermis may be syncytial. Characteristic of the epidermis are the **rhabdites,** deeply staining curved rods that occur singly or in bundles. Little is known of their function; however, the fact that they discharge to the exterior when the animal is placed under adverse conditions suggests a protective role. In favorable preparations stained with eosin, a small region devoid of rhabdites and cilia may be seen on the margins of the body. These are the locations of the adhesive glands, whose secretions aid the animal in clamping to the substrate, especially when attempts are made to pick up the planarian. Observe the difference in coloration between the dorsal and ventral

*In this exercise the term *Trematoda* refers to members of the order Digenea.

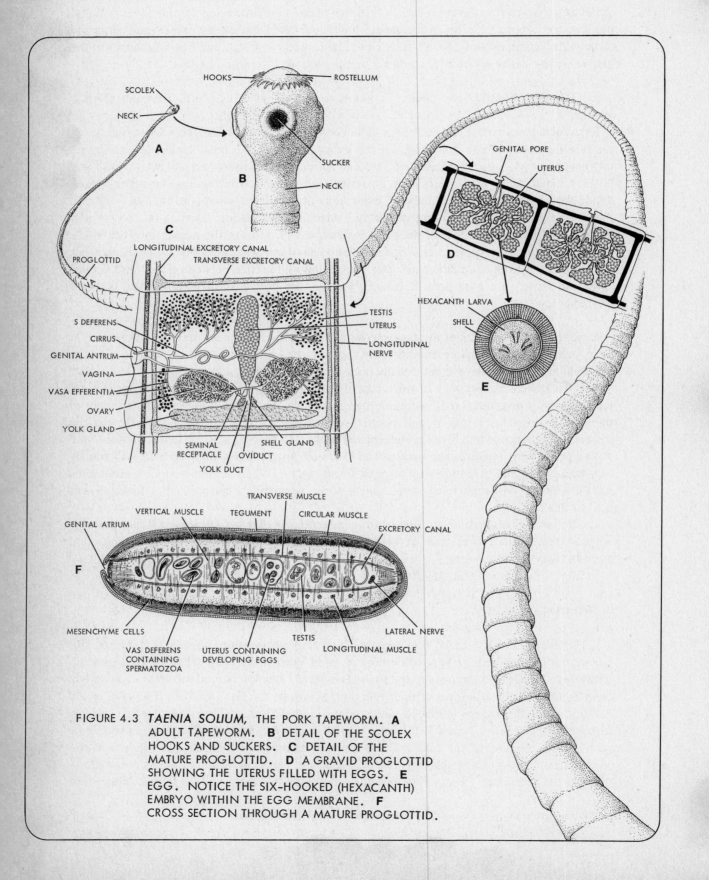

FIGURE 4.3 *TAENIA SOLIUM*, THE PORK TAPEWORM. **A** ADULT TAPEWORM. **B** DETAIL OF THE SCOLEX HOOKS AND SUCKERS. **C** DETAIL OF THE MATURE PROGLOTTID. **D** A GRAVID PROGLOTTID SHOWING THE UTERUS FILLED WITH EGGS. **E** EGG. NOTICE THE SIX-HOOKED (HEXACANTH) EMBRYO WITHIN THE EGG MEMBRANE. **F** CROSS SECTION THROUGH A MATURE PROGLOTTID.

surfaces. To what is this difference due? Which surface is ciliated? Is the ciliation uniformly distributed?

▶ If the section passes through the anterior portion of the body, you will see one of the main branches of the **gastrovascular cavity;** if the section passes through the posterior region, you may find a portion of the gastrovascular cavity and the **pharynx.** The walls of the pharynx are muscular. Contrast their structure with that of the walls of the gut diverticula. Is there a space between the gut and the body wall? While you are looking at this cross section, identify the **circular, longitudinal,** and **oblique muscles** that cross the parenchyma. Ventrally you may be able to identify two white circular areas, which are sections of the **ventral nerve cords.**

▶ ii. TREMATODA. The body surface of adult digenetic trematodes is somewhat different from that of turbellarians. Examine prepared sections of *Fasciola* and/or *Clonorchis* (Figure 4.2**B**). Notice the absence of cilia and that the surface is uniform and appears to be acellular. The external covering of flukes was formerly thought to be a nonliving cuticle, but recent cytological studies with the electron microscope have shown that the cuticular surface is in fact composed of the protoplasmic processes of living cells (Figure 4.2**C**). Therefore, it is suggested that the term **tegument** be used to designate the living surface of flukes, and the term **cuticle** be reserved for those nonliving acellular substances found on the external surface of other organisms (e.g., insects, nematodes).

▶ Cross sections of *Fasciola* and *Clonorchis* (Figure 4.2**B**) show a complicated arrangement of circular profiles, which are the **gastrovascular cavity** and **reproductive organs.** The branches of the gastrovascular cavity can be identified by their lateral position and the presence of finely dispersed materials in the lumen. Are the walls of the digestive cavity muscularized? Are the internal organs embedded in parenchymatous tissue? How are the **circular** and **longitudinal muscles** arranged?

▶ iii. CESTODA. Examine prepared slides of cross sections through a proglottid of *Taenia* (Figure 4.3**F**) under the low and high power of the compound microscope. Concentrate first on the surface and then on the internal organs.

▶ Note that the surface of the tapeworm is not ciliated and resembles that of an adult fluke. This external covering has also been called a cuticle, but recent observations with the electron microscope show that the surface of tapeworms is similar to that of flukes and is composed of protoplasmic extensions of living cells; thus it is better referred to as a tegument.

▶ Below the tegument identify the **tegumental cell bodies,** and notice the arrangement of **longitudinal** and **circular muscles.** A layer of circular muscles surrounds a large number of centrally located circular profiles. These are not portions of the gut, but consist of sections through the complicated winding tubular network of the **reproductive tract** (e.g., uterus, ovary, testes) and the **excretory canals** (empty). Are these structures free or embedded in parenchymatous tissue? Are dorsoventral muscles present? What similarities in arrangement of muscles are found in flukes and tapeworms? What are the differences?

▶ Make three fully labeled schematic sketches of cross sections through the representatives of the Turbellaria, Trematoda, and Cestoda that you have examined. Which structures are present in all three representatives?

4–2. NEUROMUSCULAR SYSTEM

The nervous system of the most primitive Turbellaria (acoels) is much like the epidermal nerve net of cnidarians. In other Turbellaria, Trematoda, and Cestoda the main nervous system has become subepidermal and is arranged in the form of **longitudinal cords** in the parenchyma with **transverse connections.** The cords are thickened anteriorly to form the bilobed **cerebral ganglia** ("brain"). The structure of the nervous system in most flatworms is best seen in prepared slides of stained sections.

▶ i. TURBELLARIA. You have already noted the ventral nerve cords in sections of *Dugesia* (Section 4–1b). If not, examine a prepared section (Figure 4.1**B**) and locate them now.

▶ The cerebral ganglia of planarians may be studied in living specimens by placing small, lightly colored organisms on a cover glass that has been ringed with petroleum jelly. Put another cover glass on top of this so that the planarian is quite compressed but not broken. This "planarian sandwich" can now be observed under dissecting and compound microscopes by placing the sandwich on a blank microscope slide and examining each surface in turn. The cerebral ganglia have the form of an inverted V with the arms running toward the "eyes." Portions of the longitudinal nerve cords may be seen running posteriorly from the brain. Save this compressed planarian for further studies (in Sections 4–3 and 4–4).

▶ The function of the nervous system in flatworms is most conveniently studied by observing behavior. With a paintbrush transfer a living specimen of the free-living freshwater planarian *Dugesia* to a watch glass with some pond water, and observe the polarity of the animal's movements. Which surface of the animal is down? Note the gliding motion and determine how it is accomplished. (Recall the structures you have seen in cross section, see Section 4–1b.) What muscles are involved in lengthening the body? Does the animal swim freely or does it attach? How is adhesion accomplished?

▶ The relative roles played by cilia and muscles in locomotion may be determined by treating worms with a solution of 1 to 2% lithium chloride, which inhibits ciliary action, or with a solution of 1 to 2% magnesium chloride, which paralyzes the muscles. Which chemical affects locomotion in planarians?

▶ Add some talcum powder to a dish of fresh pond water, mix well, and allow the suspension to settle to the bottom of the dish. Pour off the water. Transfer a fresh planarian to the talc suspension dish and note the tracks produced as the organism moves. What produces the tracks? Based on your observations, what are the roles of cilia and mucus in locomotion?

▶ Using a glass rod that has been drawn out to a fine point, note the reaction of a planarian to gentle strokes on the head and body. Now stroke the anterior end of the worm more strongly and observe the reaction. Allow the worm to relax between each set of touches. What happens if you repetitively touch the anterior end? Repeat these strokes at the posterior end and observe the reaction. Does the brain receive stimuli and effect a total body reaction? How are local responses mediated? Remove the planarian to a piece of cork, and with a sharp razor blade cut off the anterior end just posterior to the "eyes." Using a paintbrush, transfer the worm to a watch glass containing water and observe its behavior. Does the worm remain quiescent? Can you stimulate the decapitated worm by touching it with the glass rod? What do these experiments tell about the role of the brain in coordinated locomotion and initiation of motor activity?

Students with a bent toward neurophysiology may extend these observations in the section on behavior (Section 4–7). However, remember not to spend an excessive amount of time with these studies to the neglect of the rest of the exercise.

ii. TREMATODA. Adult flukes usually have a bilobed "brain," which surrounds the pharynx (**circumpharyngeal ganglion**). From the "brain" arise two longitudinal nerve cords that proceed posteriorly.

▶ Examine whole mounts of trematodes and attempt to identify the components of the nervous system (Figure 4.2**A**).

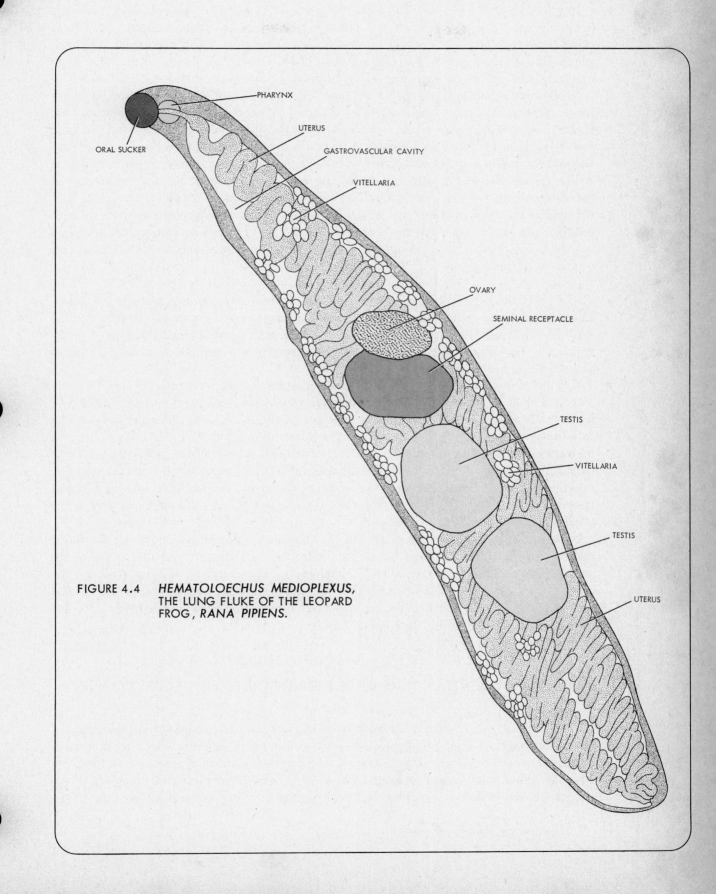

FIGURE 4.4   *HEMATOLOECHUS MEDIOPLEXUS,*
THE LUNG FLUKE OF THE LEOPARD
FROG, *RANA PIPIENS.*

► Neuromuscular activity in flukes can be demonstrated by using the frog lung fluke *Haemato-loechus* (Figure 4.4). Living adult flukes are easily obtained by pithing leopard frogs *(Rana pipiens)*, opening the body cavity by means of a ventral incision, and removing the lungs to a watch glass containing frog Ringer's solution. Usually the small brown, black, and white mottled flukes emerge on their own, although some teasing apart of the lungs may help.

► Observe the locomotory pattern of the lung flukes in a dish of Ringer's solution. How are these movements accomplished? Are cilia involved in locomotion? How many suckers are present? Do suckers play a role in movement? Note that flukes belong to the class Trematoda and *trema* means sucker (Gk.). What is the function of each of the fluke suckers in the natural habitat? Stroke the anterior and posterior ends of the worm with a glass probe in a fashion similar to that described for *Dugesia*. Does *Haematoloechus* respond to tactile stimulation in the same way as the planarian did? Are there any obvious sensory structures associated with the anterior end? With a sharp razor blade, cut off the anterior end just posterior to the anterior (oral) sucker. What happens to the locomotory habits? Why? Does a decapitated fluke respond to tactile stimuli? What effect do solutions of lithium chloride or magnesium chloride have on trematode locomotion?

► iii. CESTODA. Nervous tissue in the adult tapeworm is concentrated chiefly in the scolex. In favorable whole mounts the "brain" or **cephalic ganglia** may be seen just posterior to the suckers. Running posteriorly from these ganglia are a pair of lateral longitudinal nerve cords, which may be seen as clear lateral lines in whole mounts of proglottids (Figure 4.3C) and as circular profiles in cross sections (Figure 4.3F).

► Examination of living tapeworms will demonstrate the role of the neuromuscular system in locomotion. Rats infected with the tapeworm *Hymenolepis diminuta* will be provided. Transfer the rat to a screw cap jar containing a large wad of cotton soaked in ether and screw on the cap. Etherize the rat until it is dead and transfer to a dissecting tray. Open the body cavity by a midventral incision and remove the small intestine (that portion of the gut posterior to the stomach). Transfer the intestine to a bowl containing 0.85% NaCl. With scissors, open the intestine longitudinally and remove the white tapeworms. Transfer these to a finger bowl containing clean saline. Make sure you have the microscopic anterior scolex as well as the proglottids.

► Note the movements of the body and the individual proglottids. Under the dissecting microscope look at the scolex, noting its "searching" movements and the action of the suckers. Relate what you observe to the anatomy seen in the cross section. What are the functions of the suckers in flukes and tapeworms and of the hooks in tapeworms? Do isolated proglottids move? Does the scolex react to touching with the glass probe? What does this indicate about the role of the cestode "brain" in locomotion?

4–3. FEEDING AND THE STRUCTURE OF THE GASTROVASCULAR CAVITY*

Diffusion is an efficient means of distributing nutrients, wastes, and gases in small animals, but it is no longer effective with increasing animal size and the concomitant decrease in surface-to-volume ratio. The process of diffusion can still be utilized in the fulfillment of these functions when an animal grows larger, provided that there is a means of internal circulation and/or a change in body shape. Both these conditions are satisfied in flatworms by the form of the

*See note on terminology (Section 3–3).

digestive tract and the shape of the organism. Platyhelminths are flat and leaflike; thus the distance between the digestive cavity and the body tissues is short and nutrients are not far from the body cells. The multiple branching of the tract is an adaptation for internal circulation and there is close proximity between the site of digestion and absorption and the body tissues. Because of this the flatworm digestive tract is aptly called a **gastrovascular cavity.** In addition, respiratory exchange can take place across the entire body surface without special organs or tissues.

i. TURBELLARIA. The digestive tract of turbellarians (except acoels) consists of a midventral **mouth,** a muscular **pharynx,** and a blind-ending **intestine.**

▶ The "planarian sandwich" previously prepared for study of the nervous system (Section 4–2) can also be used to study the structure of the digestive tract. Turn the "sandwich" so that the ventral surface of the planarian faces you. Observe the centrally located mouth and the protrusible pharynx. Blind-ending branches of the intestine run anteriorly and posteriorly from the pharyngeal cavity. How many branches are there?

▶ The branching arrangement of the gastrovascular cavity provides the main taxonomic basis for separating the free-living flatworms into orders. Obtain whole mounts of *Bdelloura* and/or *Dugesia; Leptoplana* and/or *Planocera; Polychoerus, Stenostomum,* and/or *Syndesmis.* Diagram the main branches of the gastrovascular cavity in the different species, and using the classification provided on page 114, attempt to place each worm in its correct taxonomic group.

▶ Feeding in *Dugesia* may be studied in animals that have been starved for a few days. Place some organisms in a dish of water containing small bits of fresh liver. (The liver of the frog used to obtain *Haematoloechus* in Section 4–2 will be adequate.) Planarians should not be disturbed when feeding and the light should be subdued. Do not jar the dish—and be patient. Observe the behavior of the animal in the presence of food and note how the pharynx is protruded and applied to the liver. Try this same feeding experiment with decapitated worms.* Do they find food as quickly as uninjured animals? What does this tell you about **chemotaxis?** The protrusion of the pharynx can also be induced by the presence of meat juices or injurious chemicals, for example, dilute acetic acid. How is food moved into the digestive cavity?

▶ Refresh your memory of the microscopic anatomy of the pharynx by looking again at the cross section of *Dugesia* (Figure 4.1**B**).

ii. TREMATODA. In adult digenetic trematodes an anterior **oral sucker** surrounds the **mouth,** which leads via a muscular **pharynx** and a short **esophagus** into two blind-ending elongate **caeca.**

▶ Examine whole mounts of *Clonorchis* (Figure 4.2**A**) and identify the various parts of the digestive cavity. Note that the gastrovascular cavity of *Fasciola* is highly branched and atypical of digenetic trematodes.

▶ Flukes are endoparasitic and feed on the tissues of the host. What are the sources of nutrition for *Haematoloechus, Clonorchis, Fasciola,* and *Fasciolopsis?* What is the functional role of the oral sucker and pharynx in feeding? Compare these functional systems with those of *Dugesia.*

iii. CESTODA. The gastrovascular cavity is absent from all tapeworms. How then do tapeworms obtain their nutrients? Contrast the role played by suckers in tapeworms to that in flukes. Why do you think that adult tapeworms, living as they do in the gut of vertebrates, are not themselves digested?

*Directions for decapitating worms are given in Section 4–2.

4–4. EXCRETION AND OSMOREGULATION

Platyhelminths are the first Metazoa you have studied that have a discrete excretory system. It consists of a series of **tubules** that open to the outside of the body via pores (**nephridiopores**); internally, the tubules end blindly in ciliated **flame bulbs** (Figure 4.6). The mechanism of action is thought to be as follows: the ciliary flame in the bulb maintains a flow of liquid along the tubule, which is replaced by water and other substances that pass across the membrane into the cell. Although its excretory function has not been unequivocally demonstrated, the system is described as a **protonephridial** excretory system. Its prime function may be in **osmoregulation;** i.e., it regulates the internal fluid content of the organism. It is often absent in acoels and other marine Turbellaria, which tends to confirm this thesis.

i. TURBELLARIA.   The excretory system in most freshwater and land turbellarians consists of a branched network of tubules composed of flame cells and excretory canals. These run along the sides of the body (Figure 4.1**C**).

▶   Observation of ciliary activity in the flame cells is the most obvious way to locate the excretory system of flatworms. This is more easily said than done, however, and takes considerable skill and refinement in microscope technique. With luck, flame cells may be seen in small planarians that have been allowed to remain in the cover glass sandwich (see Section 4–2), or in planarians squashed between slide and coverslip. View these preparations under the high dry or oil immersion objectives.

ii. TREMATODA.   Trematodes also have a protonephridial excretory system. The excretory canals are situated along the sides of the body and end in a fine network of flame cells in the region of the ventral sucker. The main lateral excretory trunks run posteriorly and join to form a median duct or **bladder,** which opens ventrally via an **excretory pore** (Figure 4.2**A**).

▶   Most of the structures of the excretory system can be seen in prepared whole mounts; rarely, you may observe the flickering of the cilia in the flame cells of larger living specimens. Flame cells may often be more easily observed in cercariae (trematode larvae—see the following section). Place some cercariae obtained from the liver of *Cerithidea* (see Section 4–5b) in a drop of 0.1% neutral red on a slide and cover with a cover slip. Careful examination of the area lateral to the suckers (Figure 4.5) may show the flickering cilia of the flame bulb.

iii. CESTODA.   Flame bulbs are located in the scolex, and running posteriorly from these are the lateral excretory vessels. The excretory system is not easily studied, however. The most conspicuous elements are the longitudinal excretory canals.

▶   Examine a cross section of *Taenia* and whole mounts of proglottids (Figures 4.3**C** and **F**) and attempt to identify the components of the excretory system.

4–5. REPRODUCTION AND DEVELOPMENT

a. Anatomy of the Reproductive System

Almost all flatworms are **hermaphrodites,** both male and female organs existing in the same individual, although cross-fertilization is generally the rule. Reproductive organs are usually complex in anatomy and arrangement. The male system usually consists of a pair of **testes, vasa**

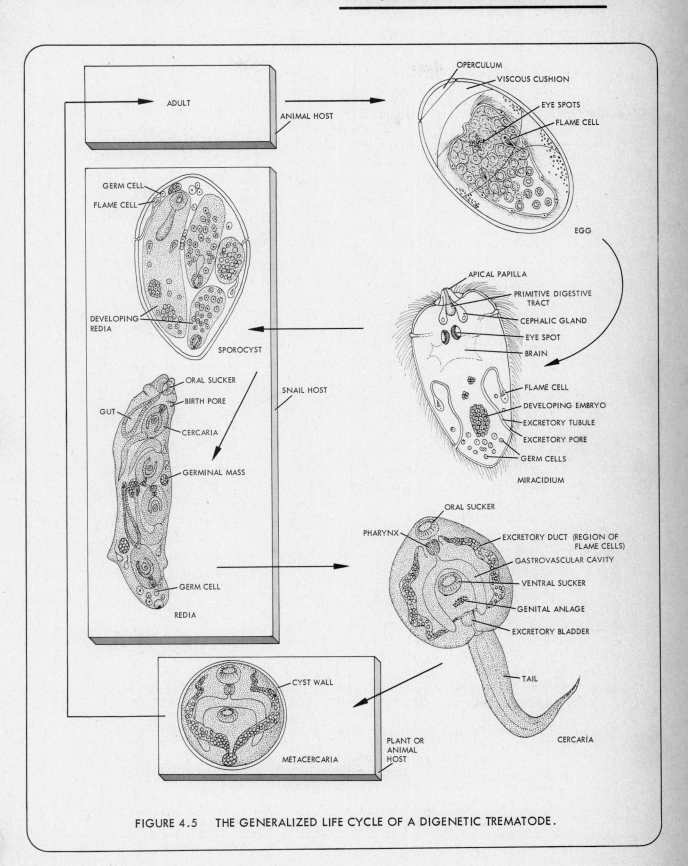

FIGURE 4.5   THE GENERALIZED LIFE CYCLE OF A DIGENETIC TREMATODE.

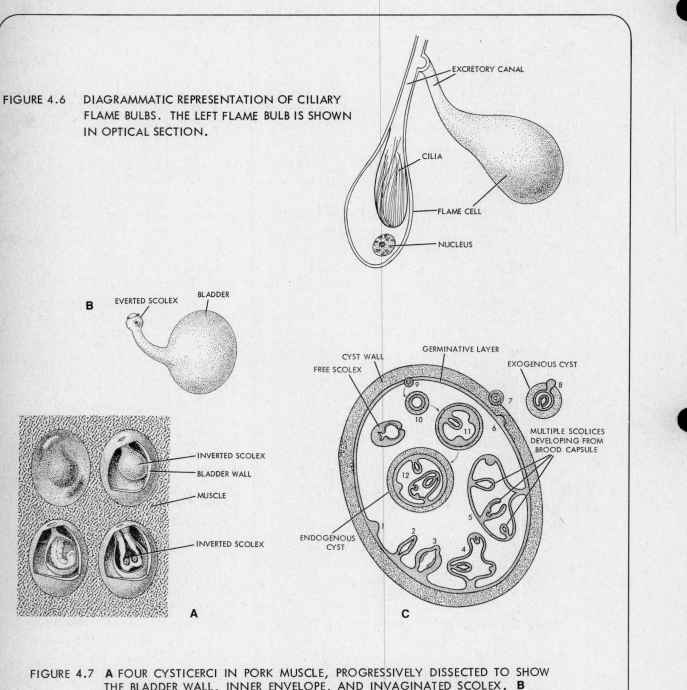

FIGURE 4.6   DIAGRAMMATIC REPRESENTATION OF CILIARY
FLAME BULBS. THE LEFT FLAME BULB IS SHOWN
IN OPTICAL SECTION.

EXCRETORY CANAL

CILIA

FLAME CELL

NUCLEUS

B

EVERTED SCOLEX   BLADDER

INVERTED SCOLEX

BLADDER WALL

MUSCLE

INVERTED SCOLEX

A

CYST WALL   GERMINATIVE LAYER

FREE SCOLEX   EXOGENOUS CYST

MULTIPLE SCOLICES
DEVELOPING FROM
BROOD CAPSULE

ENDOGENOUS
CYST

C

FIGURE 4.7 **A** FOUR CYSTICERCI IN PORK MUSCLE, PROGRESSIVELY DISSECTED TO SHOW
THE BLADDER WALL, INNER ENVELOPE, AND INVAGINATED SCOLEX. **B**
CYSTICERCUS WITH SCOLEX AND NECK EVAGINATED AND READY TO BECOME
ATTACHED TO THE INTESTINAL WALL. **C** DIAGRAMMATIC REPRESENTATION
OF THE HYATID CYST OF *ECHINOCOCCUS*. (1–5) STAGES IN THE DEVELOPMENT
OF SCOLICES FROM THE GERMINATIVE LAYER; (6–8) THE EXOGENOUS BUDDING
OF A DAUGHTER CYST FROM THE PARENT CYST; (9–12) THE ENDOGENOUS
BUDDING OF A DAUGHTER CYST FROM THE PARENT CYST.

efferentia, **vas deferens,** a **seminal vesicle,** an **ejaculatory duct,** and a copulatory organ (**penis** or eversible **cirrus**). The female system typically consists of a single **ovary,** an **oviduct,** a **seminal receptacle,** and paired **vitelline** (yolk) **glands.**

i. TURBELLARIA. Turbellarians have considerable powers of asexual reproduction. Some species of the freshwater triclad *Dugesia* rarely reproduce sexually, more commonly reproducing asexually by transverse fission. *Dugesia* is unusual in that when it does reproduce sexually, the reproductive system appears only during the spring and summer months and is resorbed following the reproductive phase. The reproductive systems of most Turbellaria are complex and difficult to locate. Some of the reproductive organs may be identified in stained whole mounts of the commensal marine triclad *Bdelloura,* however.

▶ Examine a stained whole mount of *Bdelloura* (Figure 4.8) and attempt to identify the numerous rounded **testes** and **yolk glands,** located laterally and extending posteriorly from the level of the **ovaries.** These are small, round, paired bodies lying between the second and fourth side branches of the anterior intestine. From the ovaries, paired **ovovitelline ducts,** which will not be visible, pass back to the paired **seminal bursae,** lateral and posterior to the large dark mass of the pharynx. Each seminal bursa opens to the exterior via a ventral **bursal pore.** From the bursae, the ovovitelline ducts pass back and join as a short duct that enters the posterior end of the **genital antrum,** located behind the pharynx. The **penis** is a conical body behind the pharynx from which two **sperm ducts** run forward, either side of the pharynx. The penis lies in the cavity of the genital antrum, which opens to the exterior via a **gonopore.**

ii. TREMATODA. The reproductive system of the trematodes differs from that of the Turbellaria chiefly in the enlargement of the main canal of the female system into a coiled uterus, in which a great many fertilized eggs are stored. Many exhibit a copulation canal with its own external pore. The male copulatory organ is usually an eversible cirrus.

Examine a whole mount of *Clonorchis* (Figure 4.2A); identify the pair of branched **testes** posteriorly. From each testis a **sperm duct** passes forward; these join as the **vas deferens,** which enlarges to form a **seminal vesicle** opening by a **gonopore** just in front of the ventral sucker. A sperm-transferring organ is absent in *Clonorchis,* but an eversible **cirrus** can be identified in whole mounts of *Fasciola* near the ventral sucker. The single median **ovary** is rounded and gives rise to a short **oviduct** joined by the **vitelline duct** and the copulatory **Laurer's canal.** Laurer's canal gives rise to a seminal receptacle and opens to the exterior via a dorsal **genital pore.** As the oviduct passes forward to the large coiled **uterus,** it widens slightly to form the **ootype,** surrounded by a **shell gland (Mehlis' gland).** The uterus, full of egg capsules, opens to the exterior via the **genital antrum** in the region of the ventral sucker. The large laterally situated yolk glands, the **vitellaria,** give rise to small ducts, which join a main duct on each side. The two ducts meet to form a common duct that passes to the oviduct at the level of the ovary.

Self-fertilization can probably occur in trematodes, but cross-fertilization is the general rule. Copulation involves the insertion of the penis or cirrus into the terminal portion of the Laurer's canal or the uterus of another fluke. Sperm are stored in the seminal receptacle and released when required.

▶ Although the preceding description applies to most trematodes, flukes in the family Schistosomatidae live in the hepatic portal system of birds and mammals and are noteworthy because of their sexual dimorphism. Examine demonstration slides of the human blood fluke *Schistosoma* that show the male and female worms *in copula.* Observe the **gynecophoric canal** in the male worm. What is the adaptive significance of the gynecophoric canal in schistosomes? Why is it absent from other flukes?

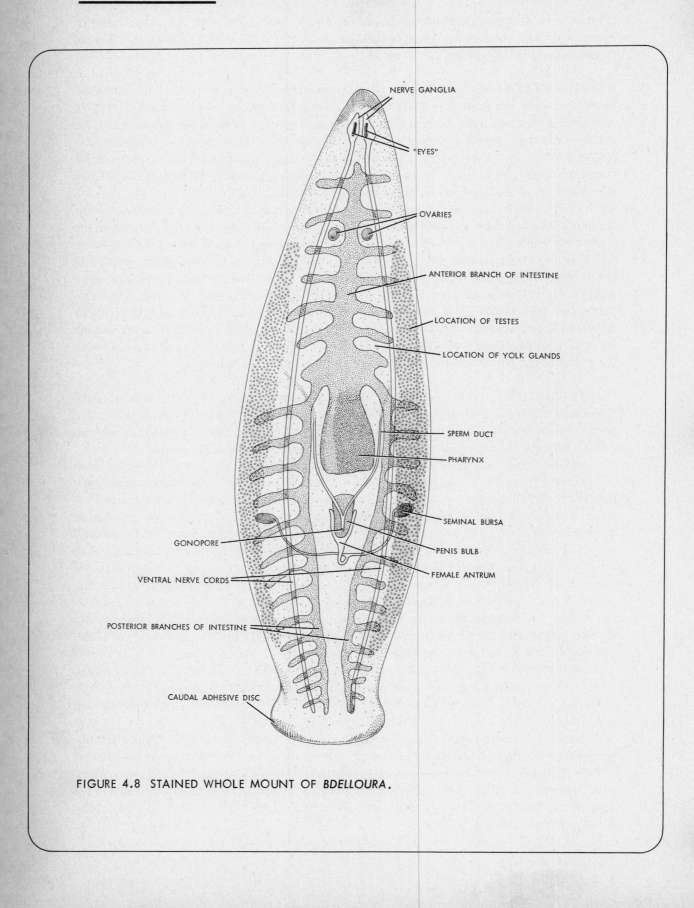

FIGURE 4.8 STAINED WHOLE MOUNT OF *BDELLOURA*.

iii. CESTODA. Tapeworms are hermaphrodites. The adult tapeworm consists of a microscopic anterior holdfast (the **scolex**) and a long, ribbon-like **strobila.** The strobila is composed of serially repeated **proglottids,** which increase in size from their place of origin at the neck to their place of detachment at the posterior end. At least one complete set of reproductive organs is found in each proglottid. Immature proglottids are found nearest the scolex and show little structural organization; further along the strobila the proglottids begin to mature, and the reproductive organs can easily be seen.

The male system consists of a large number of **testes** scattered throughout the proglottid; from each testis runs a **vas efferens.** The **vasa efferentia** run together and finally connect to a common **vas deferens.** The vas deferens leads to the common **genital antrum** where it enlarges to form a **cirrus.** The female system consists of a bilobed **ovary,** an **oviduct, yolk gland** with duct, **shell gland, uterus** and **vagina** with a **seminal receptacle.** The vagina opens into the common genital antrum, which in turn opens to the exterior via a genital pore. Self-fertilization is common among tapeworms. Gravid proglottids, found at the posterior end, show little in the way of reproductive organs except for the uterus, which is completely distended with eggs. Gravid proglottids separate from the posterior end of the strobila and are shed with the feces; on drying outside the body, the embryonated **eggs** are liberated.

▶ Study a prepared slide of a mature proglottid of *Taenia* (Figure 4.3**C**) and identify the structures of the male and female reproductive system. A quick glance at a prepared slide of gravid proglottids of *Taenia* (Figure 4.3**D**) should show the uterus packed with eggs. If living *Hymenolepis* are available (see Section 4–2), tear apart a gravid proglottid and transfer some eggs to a watch glass containing saline. Alternatively, examine prepared slides of fluke and tapeworm eggs (Figures 4.5 and 4.3**E**). Study the structure of the egg. How do these eggs hatch? Is the egg of a flatworm an ovum?

Can you explain the hypertrophied nature of the reproductive systems found in trematodes and cestodes?

b. Larval Stages

Flukes and tapeworms, as with many parasitic organisms, have life cycles that are complicated by the presence of a variety of larval stages. Why do you think this is so? During these larval stages asexual reproduction may take place (**budding** or **polyembryony**), further multiplying the number of infective stages.

i. TREMATODA. Trematodes are divided into two major groups, depending on the number of hosts in their life cycles: the Monogenea, whose members have a single host, and the Digenea, whose members have more than one host. Monogenetic trematodes are ectoparasites and lack distinct larval stages (they exhibit direct development). The endoparasitic digenetic trematodes have complex life cycles (Figure 4.5). Typically, the **egg** hatches to a ciliated, free-swimming **miracidium,** which penetrates the body of an invertebrate (usually a snail) and becomes a saclike **sporocyst.** The sporocyst gives rise to generations of **rediae** and/or **cercariae.** The latter have tails, and, on their release from the snail, swim, find, and penetrate the vertebrate host, or they encyst in or on another host or on vegetation as **metacercariae.** In the latter case, they finally enter the body of a vertebrate by ingestion.

▶ Living digenetic trematode larvae may be found in the liver of the common mud snail *Cerithidea californica.* Crack the pointed tip of the snail shell in a vise and carefully remove the terminal portion of the hepatopancreas (liver). Orange, tan, and mottled livers are usually infected; deep brown or green ones are not. Examine the infected livers in a watch glass under the dissecting microscope. The elongate, transparent fibrous structures are usually sporocysts. Cercariae are most easily recognized by their erratic swimming movements. Transfer a drop of liquid containing larvae to a microscope slide, cover with a cover glass, and examine under high magnification.

▶ Miracidia can often be obtained from frog bladder flukes *(Gorgodera)* in fresh water. Take the leopard frog from which you removed the lungs to find *Haematoloechus* (Section 4–2). Extend the ventral incision into the body cavity to the posterior limits of the abdomen and locate the bladder. Remove the entire bladder to a syracuse watch glass of fresh water and tease it apart. Examine under the dissecting microscope. If bladder flukes are present, miracidia may emerge from embryonated eggs. Make wet mounts of the miracidia or examine prepared slides. Why is this stage not found in the snail?

ii. CESTODA. Tapeworms such as *Taenia* have a simple life cycle, usually involving one intermediate host; however, there are tapeworms that have more complicated life cycles (e.g., pseudophyllideans).

In the life cycle of *Taenia* the embryonated eggs inside the uterus of a ripe proglottid pass out with the feces of the definitive host. Upon ingestion by an intermediate host, a six-hooked embryo (**oncosphere**) emerges, penetrates the gut of the host, and encysts in the tissues as a **bladder worm** or **cysticercus.** The cysticercus remains in this stage of development until the intermediate host is eaten by a definitive host; in the gut of the latter the scolex emerges and a young tapeworm develops (Figure 4.7**B**).

▶ Study measly pork or beef containing cysticerci of the pork *(Taenia solium)* and beef *(Taenia saginata)* tapeworms, respectively (Figure 4.7**A**).

Asexual budding occurs in some cysticercus stages but not in *Taenia*. The adult tapeworm *Echinococcus granulosus* lives in the gut of the dog. Chance ingestion of the tapeworm eggs by a suitable mammal results in the hatching of an oncosphere, which penetrates the intestinal wall and is carried to the liver or other organs via the blood; here it forms a cysticercus with multiple bladders called a **hydatid cyst.** Brood capsules are budded off from the inner surface of the cyst wall, and as many as 30 invaginated scolices may be formed from localized thickenings of the inner surface of each brood capsule. Brood capsules may break loose and lie free within the interior of the fluid-filled cyst. Daughter cysts can also be produced by internal or external budding. Hydatid cysts produce new brood capsules over a period of years and increase enormously in size, causing pressure and damaging vital organs. When swallowed by a carnivorous host the brood capsule is digested away, this releases the scolices which evaginate, attach to the intestinal wall, and form adult tapeworms. Through too great a familiarity with infected dogs, humans may ingest *Echinococcus* eggs and develop hydatid cysts; in this way humans become an intermediate host for the dog tapeworm.

▶ Examine a slide of a hydatid cyst (Figure 4.7**C**).

4–6. REGENERATION

Powers of regeneration among flatworms are quite variable. They are high in some of the Turbellaria and absent from adult trematodes and cestodes (except for the scolex and neck region of tapeworms). The turbellarian *Dugesia* exhibits remarkable regenerative powers and is probably one of the best studied of all organisms in regard to morphogenesis and polarity. We can only make some preliminary studies at this stage, but the interested student is referred to the extensive references in Brønsted (1955) and Child (1941). Absorbing and fascinating as this topic is, students are advised not to spend the major part of their efforts on these studies to the neglect of other aspects of flatworm biology.

a. General Directions

▶ Break a double-edged razor blade in two and break off small pieces of the sharp edge. Clean thoroughly in acetone. Cut pieces of 7 mm glass tubing 10 to 12 cm long. Heat one end of a piece of tubing until it is soft and press it against a metal plate at an angle of 45°. Clean this oblique surface and glue a chip of razor blade onto it with Du Pont waterproof cement or epoxy cement. (Wooden applicator sticks are also suitable for making knife handles.) Allow the cutting edge of the razor blade to project a few millimeters beyond the handle of this "planaria knife."

▶ Prepare labeled petri dishes for operated specimens and a syracuse watch glass for discarded pieces. Operate while observing the specimens under a binocular dissecting microscope.

▶ With a brush, place a specimen of *Dugesia* that has been starved for a week in a drop of water on a clean slide. Allow the animal to expand maximally and place the slide on top of an ice cube. This will slow the movements of the worm. Make cuts with your "planaria knife." The knife cuts must be perpendicular, rather than oblique. The cut surface should be sharp and clean. After the operation transfer the organisms to the petri dish. Do not feed the regenerating animals. Keep them in a cool dark place. Regeneration should be complete in 8 to 14 days. Why should starved animals be used, and why should you not feed the organisms during the regeneration period?

b. Suggested Experiments

Note that these operations require considerable skill; the results may not always be everything that you desire.

▶ i. TWO HEADS. Cut off the head below the eyes, discard the head, and slit the body longitudinally. The slit may have to be reopened in 12 to 24 hours. Each half should reconstitute a head.

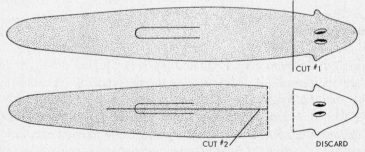

▶ ii. DOUBLE HEAD. Cut a piece from the central region of the body, including the posterior half of the pharynx; discard the head and tail and preserve the central portion. Each side should reconstitute a head.

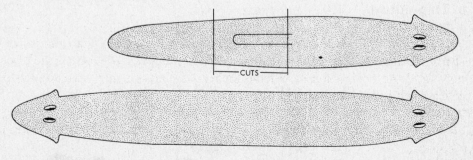

▶ iii. TWO TAILS.   Slit an organism posteroanteriorly down the center of the body as far as the eyes. Twenty-four hours later, reopen the slit and cut off the head. Regeneration should take place as in the diagram.

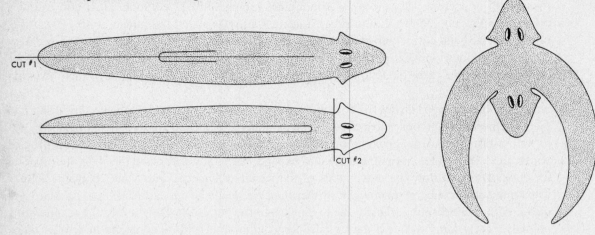

4–7. BEHAVIOR

Because of their uncomplicated nervous system the flatworms have been favorite subjects for experimental analysis of sense organs, orientation, and the mechanics of learning. *Dugesia* has been most often used because of its accessibility, its hardiness, its ability to regenerate, and its well-developed sense organs.

Some observations have already been made on the movements of planarians and their responses to touch (thigmotaxis) and to food (Section 4–2). The present section will amplify these studies. (For directions on how to perform surgery on these organisms, including preparation of a "planaria knife," see Section 4–6a.)

a. Chemoreception

The auricles of planarians are ciliated grooves situated on either side of the head (Figure 4.1A). They are capable of discriminating the presence of chemicals in the water.

▶ Test the ability of planarians to detect the juice of liver extract or bouillon (1) with auricles intact, (2) with one auricle removed, and (3) with both auricles removed. Those surgically inclined should attempt to extirpate the brain while leaving the auricles intact. What is the subsequent behavior toward food? Design adequate controls for all your experiments and, of course, base your conclusions on more than one case.

b. Thigmotaxis

▶ Cut a planarian in half across the middle and test its reaction to touch. Using a glass probe, stroke the anterior section of both pieces. What is the response? Allow the pieces to relax and now stroke the posterior ends. What is the resultant movement? What do such experiments tell of the role of the brain in thigmotaxis?

▶ Planarians have a negatively thigmotactic dorsal surface and are positively thigmotactic ventrally. This is true regardless of gravity. Thus, planarians will not come to rest if both surfaces are contacted. Try to demonstrate this by placing a piece of tin foil on the dorsal surface of a planarian—observe its reaction. Does placing the tip of a paintbrush on the dorsal surface, elicit similar response?

▶ Planarians will perform righting reactions. Turn a worm ventral surface upward and observe the response. Perform the same experiment on worms that have been cut in two. Time the righting reaction of pieces of planarians and of intact organisms. Drop planarians into water 8 to 10 cm deep and observe which end contacts the surface of the container first. What is the role of the brain in the righting reaction?

c. Rheotaxis

▶ Planarians usually react to water currents. By means of a Pasteur pipet, determine the reactions of planarians to a stream of water from a pipet directed (1) at the anterior end, (2) at the side, and (3) at the posterior end of the body. Do isolated halves react similarly? Do reactions to strong and weak currents differ?

d. Phototaxis

The eye spots of *Dugesia* have a cuplike arrangement of pigmented cells and sensory receptor cells that are partially enclosed by the cup. The light stimulus enters the cup at an angle.

▶ Planarians are generally negatively phototactic. Allow a planarian to glide along in subdued light and then illuminate the animal frontally. In which direction does the animal turn? Again allow the animal to crawl in subdued light and illuminate laterally. What is the reaction? Repeat these experiments with animals that have one eye removed and then with animals that have both eyes removed. What can you conclude about the role of the eye in phototaxis? How does the structure of the eye allow planarians to discriminate the direction of the light?

▶ The ability of planarians to discriminate between light and dark areas may be demonstrated by placing them in a petri dish over a piece of paper that has a checkerboard or bull's-eye pattern of dark and light areas. Observe the movements of animals against different backgrounds, both with and without eyes.

e. Learning

Time does not permit studies of conditioning, although such work is extremely interesting. Interested students will find a review of the subject in Best (1963).

References

Best, J. Protopsychology, *Sci. Amer.* **208**(2):54–62, 1963.

Borradaile, L. A., and F. A. Potts. *The Invertebrata*. London: Cambridge University Press, 1961.

Brønsted, H. V. Planaria regeneration, *Biol. Rev.* **30**:65–126, 1955.

Brown, F. A., Jr. (ed.). *Selected Invertebrate Types*. New York: John Wiley & Sons, 1950.

Buchsbaum, R. *Animals Without Backbones*, 2nd ed. Chicago: University of Chicago Press, 1972.

Bullough, W. S. *Practical Invertebrate Anatomy*. New York: St. Martin's Press, 1962.

Chandler, A. C., and C. P. Read. *Introduction to Parasitology*. New York: John Wiley & Sons, 1961.

Child, C. M. *Patterns and Problems of Development*. Chicago: University of Chicago Press, 1941.

Dawes, B. *The Trematoda*. London: Cambridge University Press, 1946.

Erasmus, D. A. *The Biology of Trematodes*. London: Arnold, 1974.

Hamburger, V. A. *A Manual of Experimental Embryology*. Chicago: University of Chicago Press, 1943.

Hyman, L. H. *The Invertebrates*, Vol. II. *Platyhelminthes and Rhynchocoela*. New York: McGraw-Hill Book Co., 1951.

Lender, T. Factors in morphogenesis of regenerating freshwater planarians, *Advan. Morphol.* **2**:305–31, 1962.

Riser, N. W., and M. P. Morse (eds.). *Biology of the Turbellaria*. New York: McGraw-Hill Book Co., 1974.

Smyth, J. D. *The Physiology of Trematodes*. San Francisco: W. H. Freeman and Co., 1966.

Taliaferro, W. H. Reactions to light in *Planaria maculata* with special reference to the function and structure of the eyes, *J. Exp. Zool.* **31**:59–116, 1920.

Wardle, R. A., and J. A. McLeod. *The Zoology of Tapeworms*. Minneapolis: University of Minnesota Press, 1952.

Wolff, E. *Recent Researches of the Regeneration of Planaria. 20th Symposium for Developmental Biology and Growth,* D. Rudnick (ed.). New York: Ronald Press, 1961.

An Abbreviated Classification of the Phylum Platyhelminthes (After Hyman, Yamaguti, and Others)

Phylum **Platyhelminthes**

→ *Class I* **Turbellaria**

Free-living flatworms with ciliated epidermis, a ventrally directed mouth and with the gastrovascular cavity, when present, never bifid (two-branched).

Order 1 **Acoela**

Mouth, no gastrovascular cavity, no flame cells, marine, e.g., *Polychoerus*.

Order 2 **Rhabdocoela**

Gastrovascular cavity without branches, e.g., *Microstomum*.

Order 3 **Alloeocoela**

Gastrovascular cavity may have short diverticula.

Order 4 **Tricladida**

Gastrovascular cavity with three main branches, e.g., *Dugesia, Bdelloura, Bipalium.*

Order 5 **Polycladida**

Gastrovascular cavity with many diverticula, e.g., *Stylochus, Hoploplana*.

Class II **Trematoda**

Flukes, exclusively parasitic, definitive stage covered with a nonciliated tegument; ciliated epithelium confined to larva hatched from egg; suckers usually present; gastrovascular cavity usually present.

Order 1 **Monogenea**

Direct development with no asexual generations; large posteroventral adhesive apparatus with hooks or spurs and/or suckers. Ectoparasitic on cold-blooded aquatic vertebrates, single host.

Order 2 **Digenea**

Development complicated with an alternation of three or more generations and an alternation of hosts. Almost all species endoparasitic. Asexual generations interposed in life cycle, nearly always in molluscs. Larva hatched from egg is ciliated. One or two cuplike suckers for attachment, one of these circumoral, e.g., *Clonorchis, Fasciola, Haematoloechus.*

Class III **Cestoidea** Cestoda

Tapeworms, exclusively parasitic organisms, adults hermaphroditic, covered with a nonciliated tegument; ciliated epithelium, when present, confined to embryos; scolex provided with suckers and/or hooks; no gastrovascular cavity; body in most cases divided into separate, sexually complete units called proglottids.

Subclass 1 **Cestodaria**

Body not divided into proglottids; only a single set of reproductive organs; oncosphere (embryo) hatching from egg has ten hooklets.

Subclass 2 **Cestoda**

Body typically with scolex and proglottids, oncosphere typically possesses six hooklets.

Order 1 **Tetraphyllidea**

Scolex with four earlike outgrowths (bothridia); vitellaria scattered in two lateral rows, genital pores lateral. Found in spiral valves of elasmobranchs.

Order 2 **Trypanorhyncha**

Scolex with two or four bothridia and four long evertible proboscides armed with spines; vitellaria in continuous sleevelike distribution. Found in spiral valves of elasmobranchs.

Order 3 **Pseudophyllidea**

Scolex with two sucking grooves (bothria); eggs shed from uterine pore; vitellaria scattered in dorsal and ventral sheets. Found in teleosts and land vertebrates, e.g., *Dibothriocephalus.*

Order 4 **Cyclophyllidea**

Scolex with four cup- or saucer-shaped suckers; hooks and/or rostellum may be present, no uterine pore, genital pore usually lateral; vitelline gland a single or bilobed mass posterior to ovary. Found in birds and mammals, e.g., *Taenia, Hymenolepis, Echinococcus.*

Order 5 **Proteocephala**

Equipment and Materials

Equipment

Dissecting microscope
Compound microscope and lamp
Microscope slides and cover slips
Petroleum jelly
Vise
Glass rods 0.3 to 0.5 cm
7-mm glass tubing or wooden
  applicator sticks
Bunsen burner
Small paintbrush
Syracuse watch glasses
Talcum powder
Razor blades
Epoxy cement
Cork
Pipets
Tin foil
Finger bowl
Screw cap jar
Dissecting instruments
Cotton

Solutions and Chemicals

Carmine powder
Ether
Frog Ringer's solution
Fresh liver tissue
0.1% neutral red
Lithium chloride (2%)
Magnesium chloride (2%)
0.85% NaCl

Living Material*

Dugesia (6,11)
Cerithidea californica for digenetic trematode larvae † (9)

Frogs *(Rana pipiens)* for adult
  trematodes † (10)
Hymenolepis diminuta in rats† (3)

Preserved Material

Measly pork or beef (11)

Prepared Slides

Dugesia, whole or plastic mount
Dugesia, cross section
Bdelloura, whole mount
Clonorchis, whole mount
Clonorchis, cross section
Clonorchis, eggs
Fasciola hepatica, whole mount
Fasciola hepatica, cross section
Taenia solium, plastic mount
Taenia, mature proglottid, whole
  mount
Taenia, gravid proglottid, whole
  mount
Taenia, mature proglottid, cross
  section
Taenia, cysticercus
Leptoplana or *Planocera,* whole
  mount
Stenostomum, whole mount
Polychoerus or *Syndesmis,* whole
  mount
Schistosoma, male and female *in
  copula,* whole mount
Miracidia
Redia
Sporocysts
Cercaria
Hydatid cyst

*Numbers in parentheses identify sources listed in Appendix B.
†One frog, rat, or snail should provide enough material for a class of 20 to 30 students. See exercises on pages 102–104 and 109–110.

ASCHELMINTHES (PSEUDOCOELOMATES)

Contents

The Aschelminthes (*asco,* sac; *helmin,* worm; Gk.) comprise a heterogeneous group of organisms that includes one of the most successful groups in the animal kingdom, the ubiquitous **nematodes** or roundworms, as well as the **rotifers, gastrotrichs, kinorhynchs,** and **nematomorphs** or horsehair worms. These all have in common the presence of a complete digestive tract or **gut** with both mouth and anus, a body covering that consists of a nonliving, protective layer called the **cuticle,** a constant number of cells within any one species, and a body cavity (**pseudocoel**) unlined by mesoderm and embryonically derived from the blastocoel. The fluid-filled space of the pseudocoel separates the gut from the body wall. In this space are situated the organs of reproduction and excretion.

*Parts of this portion of the exercise should be started at the beginning of the laboratory period to allow time for completion.

In spite of these and other similarities, the Aschelminthes are an extremely diverse group. Many writers have emphasized the relative absence of unifying characteristics, questioning how closely related the aschelminths are to each other and whether the various divisions should be treated as separate phyla. However, the Aschelminthes probably arose from the acoelomates or a single group of coelomates, and it is very unlikely that the pseudocoelom evolved more than once. Therefore the aschelminths can for the time being be retained in one phylum. Those who prefer the alternate view may treat each of the aschelminth classes considered in this chapter as a phylum, and the whole assemblage of phyla grouped together under the general term *pseudocoelomates*.

5–1. BODY ORGANIZATION

a. External Anatomy

Most aschelminths are cylindrical and tapered at one or both ends. All are bilaterally symmetrical and have defined anterior and posterior ends, and the digestive tract is usually complete.

i. NEMATODA.   Nematodes are usually cylindrical and tapered at both ends, and nearly all species move rather actively. *Ascaris,* a nematode parasite of the intestines of humans, pigs, and horses, is considerably larger than most roundworms and therefore atypical; however, this feature makes it useful for illustration of nematode body structure, which is amazingly uniform from species to species.

▶   Obtain some preserved *Ascaris* (or living ones collected from the local slaughterhouse) and examine the external appearance (Figure 5.1 **A**). Note that the body is covered by a thin, transparent acellular layer. This is the **cuticle,** which is secreted by the syncytial layer of cells that underlies it (the **hypodermis**). The body is unsegmented, although sometimes the cuticle may be ridged. The sexes are separate in ascarids, as in all nematodes, and the female is larger than the male. The male is most easily recognized by its smaller size and curved posterior end with **copulatory spicules.** Can you distinguish anterior from posterior? Can you distinguish between the dorsal and ventral surfaces? Note that the dorsal and ventral midlines are marked by narrow white lines (dorsal and ventral **longitudinal lines**), and broader longitudinal lines are found laterally (**lateral lines**). Set this specimen aside under moist toweling for future work.

ii. ROTIFERA.   Rotifers are microscopic, and the majority are found in fresh water. The body of the rotifer is cylindrical in cross section (Figure 5.2**B**), and, like that of the nematodes, it is covered by a **cuticle;** in contrast to the nematodes, however, it may be thrown into sections so that telescoping of the body is possible. This cuticular segmentation is not reflected by the internal organs and is therefore referred to as **pseudosegmentation.** The anterior end of the body is expanded into a crown of cilia (**corona**), which serves as the principal food-gathering device and functions in locomotion (swimming). Posterior to this is the **trunk,** which may be tapered to form a **foot.**

▶   Examine wet mounts of living specimens of freshwater rotifers, e.g., *Philodina* (Figure 5.2**A**), *Asplanchna* (Figure 5.2**C**), or *Epiphanes,* or specimens mounted on glass slides, and compare their external anatomy to that of the nematode. Note that *Asplanchna* is an example of a rotifer with a saclike form showing no cuticular segmentation.

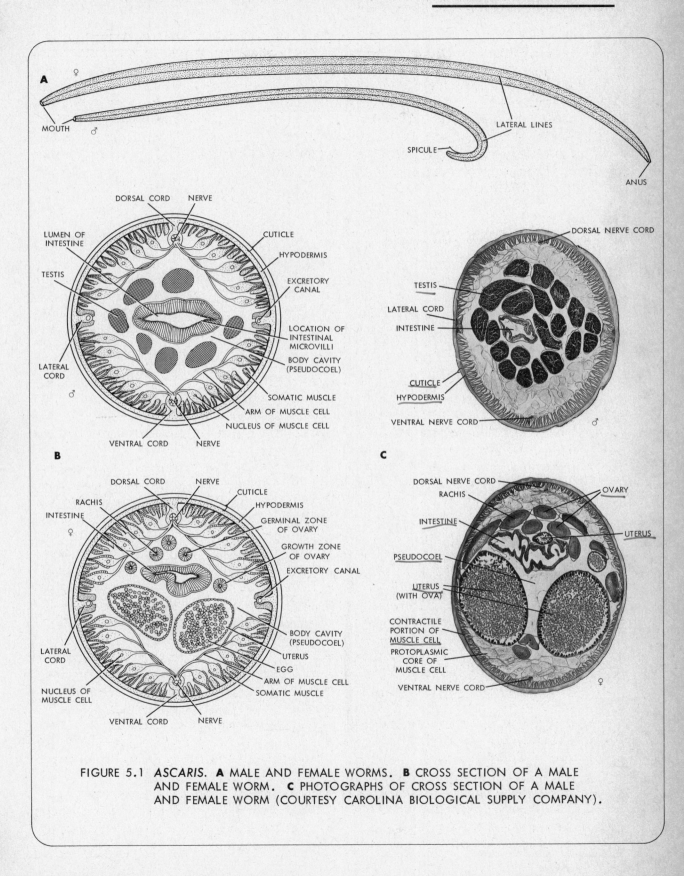

FIGURE 5.1 *ASCARIS.* **A** MALE AND FEMALE WORMS. **B** CROSS SECTION OF A MALE
AND FEMALE WORM. **C** PHOTOGRAPHS OF CROSS SECTION OF A MALE
AND FEMALE WORM (COURTESY CAROLINA BIOLOGICAL SUPPLY COMPANY).

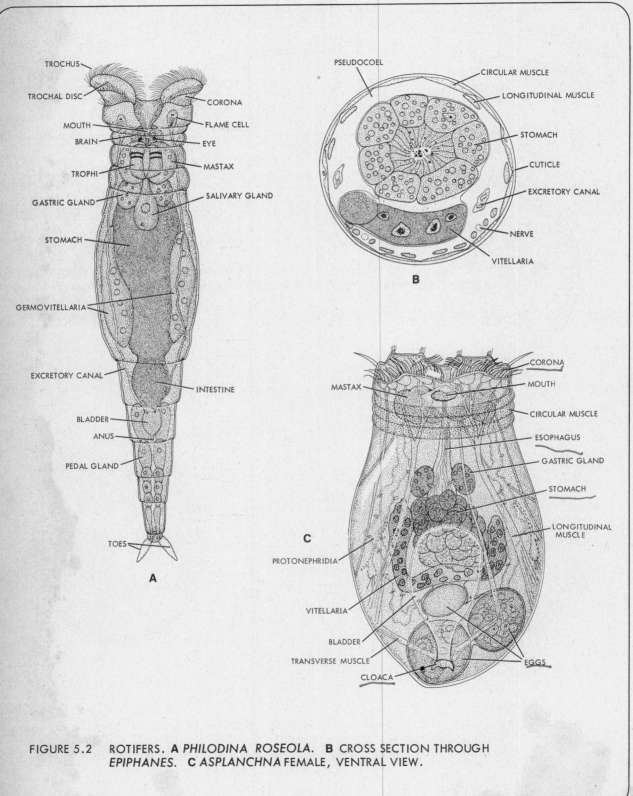

FIGURE 5.2   ROTIFERS. **A** *PHILODINA ROSEOLA.* **B** CROSS SECTION THROUGH
EPIPHANES.   **C** *ASPLANCHNA* FEMALE, VENTRAL VIEW.

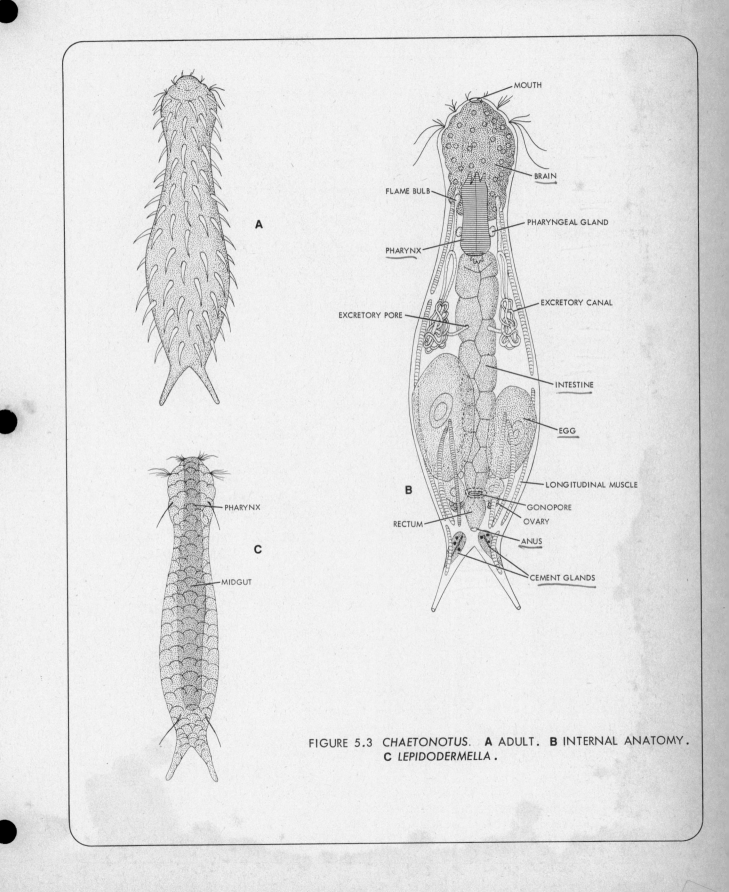

FIGURE 5.3 *CHAETONOTUS*. **A** ADULT. **B** INTERNAL ANATOMY.
C *LEPIDODERMELLA*.

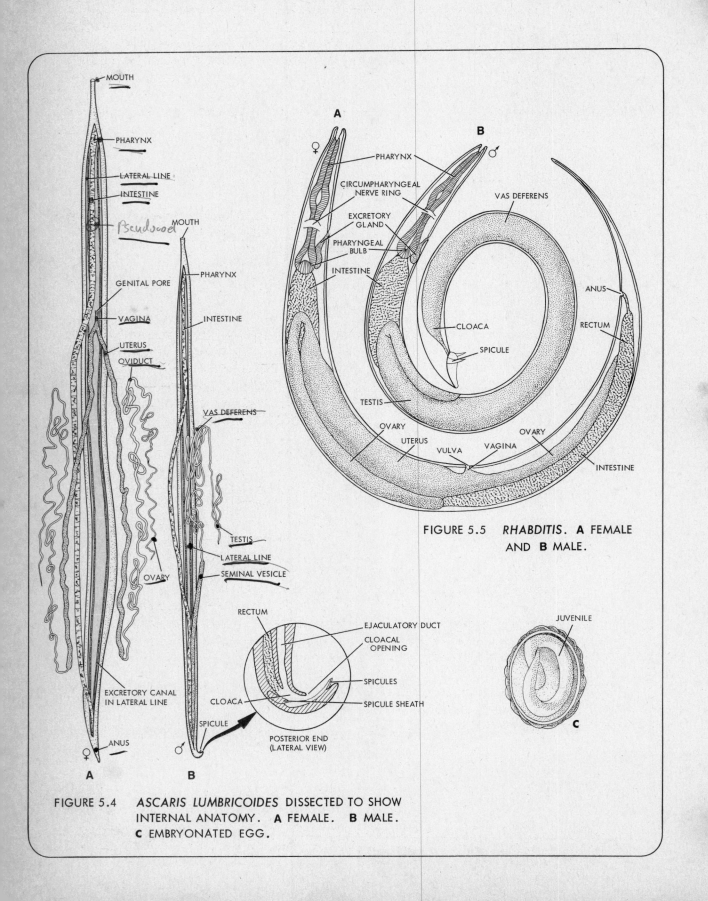

FIGURE 5.5 *RHABDITIS.* **A** FEMALE AND **B** MALE.

FIGURE 5.4 *ASCARIS LUMBRICOIDES* DISSECTED TO SHOW INTERNAL ANATOMY. **A** FEMALE. **B** MALE. **C** EMBRYONATED EGG.

iii. GASTROTRICHA.   The gastrotrichs resemble the rotifers to a certain degree. They are microscopic and elongate, but are flattened dorsoventrally and the cilia are restricted to certain areas of the body. The cuticle is highly modified to form spines, bristles, and scales.

▶   Examine wet mounts of common freshwater gastrotrich *Chaetonotus* or *Lepidodermella* (Figure 5.3), which has an enlarged head separated from the trunk by a neckline constriction. The posterior end is usually forked.

▶   If available, compare the form of a kinorhynch and a nematomorph with those organisms already studied.

b. Internal Anatomy

▶   i. NEMATODA.   Take a preserved specimen of *Ascaris* and make a l-mm cross section through the middle of the body, using a sharp razor blade. Examine the arrangement of the internal organs in this section under the dissecting microscope. Do the organs lie free in the body cavity? More details of the internal anatomy may be seen in prepared cross sections; study such a section, and with the help of the accompanying diagram (Figure 5.1**B**) identify the various parts, including all the circular profiles.

▶   Take another preserved specimen of *Ascaris* and, by means of a dorsal longitudinal incision in the body, expose the internal organs. Pin the body wall apart in a dissecting pan and cover with water, which will support the internal organs and facilitate examination (Figure 5.4). The space between the body wall and the alimentary tract is the **pseudocoel.** In life this space is filled with fluid, and it plays an important role in the alimentation and locomotion of nematodes. The complicated tubular network found internally comprises the digestive and reproductive systems, which will be discussed separately (Sections 5–4 and 5–7). Keep your dissected specimens under a layer of water for future work.

▶   Many of the internal structures seen in the dissected *Ascaris* can easily be identified in other species of living intact nematodes. Take a petri dish that was "seeded" a week previously with decaying pieces of the body wall of an earthworm. Among this decaying flesh you may observe white fiber-like objects. Look at these under a dissecting microscope. Pick up a few of these nematodes with a dissecting needle, and transfer to a drop of water on a microscope slide. Cover with a cover slip, examine, and note the tubular arrangement of the internal organs (Figure 5.5). Observe the transparent **cuticle,** the syncytial **hypodermis** that underlies it, the tubular **gut** with the pulsating **pharyngeal bulb** at the anterior end, and the convoluted tubes that comprise the **reproductive system.** Careful observation may reveal the **nerve ring,** which surrounds the region just anterior to the pharyngeal bulb. The worm you have been looking at is *Rhabditis;* it lives encysted in the body of the earthworm and young juveniles may be found in the excretory organs (nephridia). These nematodes are **ovoviviparous,** and you may find that some of the female worms contain eggs or developing juveniles.

ii. ROTIFERA AND GASTROTRICHA.   The internal anatomy of the rotifera and gastrotricha will be treated under the appropriate functional system.

Internal anatomy and physiology of kinorhynchs and nematomorphs will not be studied.

5–2. THE CUTICLE

i. NEMATODA.   It has been observed (Section 5–1) that aschelminths are clothed in a nonliving, acellular protective covering called the **cuticle.** The cuticle of nematodes is a complex structure consisting of four main layers (Figure 5.6). The main component is inelastic **collagen,** which imparts properties of impermeability and rigid strength to the cuticle. For movement to occur the cuticle must be elastic, and the collagen is laid down in the form of a spiral latticework of parallelograms with protein between. Movement of the worm results in a change in the angle between the fibers of the lattice, so that extension and contraction can take place. Because of its rigid strength combined with elasticity and relative impermeability, the cuticle plays important roles in movement and protection.

▶ Peel pieces of cuticle from a preserved specimen of *Ascaris*. Make a wet mount and examine under the microscope.

▶ To demonstrate the ability of the nematode cuticle to withstand environmental hazards, prepare a series of buffers ranging from pH 1–2 to 10–11. Take a drop from a culture of the vinegar eel *Turbatrix,* and place it in a small volume of each of these buffers contained in depression slides. Observe the motility of the worms at the various pH values. Is the cuticle impervious to all substances? What is the pH of vinegar, the usual habitat of the worm?

Although nematode growth is usually preceeded or accompanied by shedding of the cuticle **(molting),** this process is not always an indispensable prerequisite for growth. However, molting is probably necessary for allowing rapid changes in the thickness of the cuticle as the diameter of the worm increases during growth. Note that growth in nematodes normally involves enlargement of cells rather than increase in cell numbers.

ii. ROTIFERA AND GASTROTRICHA. The cuticular form in rotifers and gastrotrichs is discussed in Section 5–1; further details will not be given here.

5–3. LOCOMOTION

▶ i. NEMATODA. Recall the arrangement of the **pseudocoel** of the nematode *Ascaris,* or re-examine a cross section of this worm and locate the pseudocoel (Figure 5.1**B**).

▶ In life the pseudocoel is filled with fluid. A small puncture in the body of living *Ascaris* will produce a sharp jet of pseudocoelomic fluid, demonstrating the high internal turgor pressure (e.g., 16 to 225 mm Hg) under which the fluid is held. To withstand such pressures the body wall and the cuticle are considerably thickened. The fluid-filled muscular tube thus forms a highly developed **hydrostatic skeleton,** which maintains the body shape and is also involved in movement.

The muscle cells of the body wall are attached to the basal layer of the cuticle **(hypodermis),** and the remaining portion of each cell projects inward. Each muscle cell consists of three parts: an elongate fibrous portion with the fiber parallel to the long axis of the worm, an expanded baglike portion containing the single nucleus, and an extension from the nuclear bag to the dorsal or ventral nerve cord. On reaching the nerve cord, each arm divides and interlaces with divisions from the other arms to form a syncytial strip of muscle tissue that runs parallel to and synapses with the nerve cords. The electrical properties of these somatic muscle cells have been well worked out. Because of their relatively large size they lend themselves to investigative experiments. Students with an interest in this direction should consult the reference by Castillo and Morales.

▶ The muscle cell layer is one cell thick, and the fibers run longitudinally. Four rigid projections of the hypodermis called **cords** divide the longitudinal musculature into four functional fields. The lateral cords are larger than the dorsal and ventral cords and are seen externally as the **lateral lines.** Recall the arrangement of the muscle cells, hypodermis, and cords in sections of *Ascaris,* or re-examine a prepared slide of a cross section.

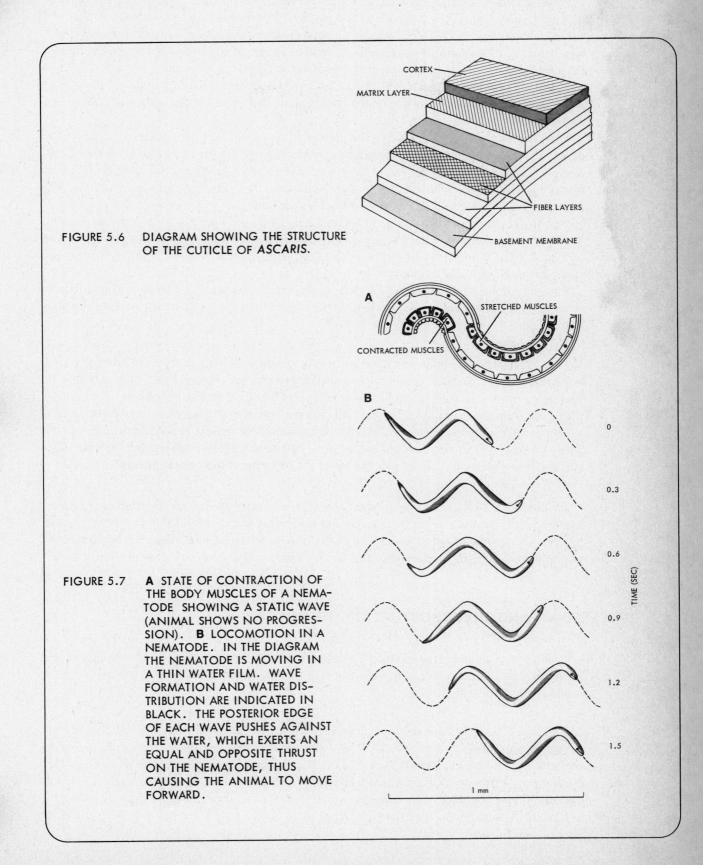

FIGURE 5.6   DIAGRAM SHOWING THE STRUCTURE OF THE CUTICLE OF *ASCARIS*.

FIGURE 5.7   **A** STATE OF CONTRACTION OF THE BODY MUSCLES OF A NEMATODE SHOWING A STATIC WAVE (ANIMAL SHOWS NO PROGRESSION). **B** LOCOMOTION IN A NEMATODE. IN THE DIAGRAM THE NEMATODE IS MOVING IN A THIN WATER FILM. WAVE FORMATION AND WATER DISTRIBUTION ARE INDICATED IN BLACK. THE POSTERIOR EDGE OF EACH WAVE PUSHES AGAINST THE WATER, WHICH EXERTS AN EQUAL AND OPPOSITE THRUST ON THE NEMATODE, THUS CAUSING THE ANIMAL TO MOVE FORWARD.

Unlike ordinary striated muscles, which act between two points of attachment, the muscles of *Ascaris* may contract in groups, called **fields,** which produce a local shortening; because the volume of internal fluid is constant and incompressible, the internal pressure is increased, which causes extension of muscle cells in another region of the body. There are no circular muscles, and the high and changeable hydrostatic pressure antagonizes the longitudinal muscles and provides the restoring force. Through this system the dorsal and ventral muscles act as antagonists, producing sinusoidal waves along the length of the body and, ultimately, movement of the whole worm (Figure 5.7).

▶ Most nematodes lie on their sides, the dorsoventral undulations thus being in the horizontal plane. For progression to occur, the medium in which the worm lies must be relatively viscous. Because relative viscosity increases with decrease in size, smaller forms can swim in less absolutely viscous media than larger forms. Confirm this by placing some living *Ascaris* in 30% seawater at 37°C. A traveling wave passes across the body, but does the worm move forward? Now transfer the worm to 30% seawater containing methyl cellulose at 37°C. What happens to the locomotory pattern? What happens to locomotion if the worm is placed in a glass tube with 30% seawater? Why? How does this relate to the movement of *Ascaris* in its natural habitat, the intestine of pigs and humans?

▶ Attempt similar studies on *Rhabditis* and *Turbatrix*. Try slowing down the body undulations by use of 5% methyl cellulose. Do these worms lie on their sides? In what plane are the body undulations?

ii. ROTIFERA. Definite muscle layers are absent from the body wall of rotifers. A bewildering array of fiber bundles traverse the pseudocoel (Figure 5.2**C**) and can best be studied by observing the variety of movements performed by living organisms.

▶ Make a wet mount of a living rotifer and observe how it can twist, bend, flex, shorten, and lengthen; attempt to correlate movements with positions of muscle attachments. Observe the ciliary corona and the mechanics of swimming. Feet are usually for walking, but for what other purpose does the rotifer use its foot? What is the function of the **cement glands?**

iii. GASTROTRICHA. Gastrotrichs possess six sets of paired longitudinal muscles, together with variously arranged circular and transverse tracts that closely resemble those of rotifers. Gastrotrichs depend on ventral tracts of cilia for movement (Figure 5.3). Notice the gliding motion of living gastrotrichs and the similarity of movement to that of planarians.

5–4. FEEDING AND THE STRUCTURE OF THE GUT

In most aschelminths the gut is a complete tube that opens anteriorly as the mouth and posteriorly as the anus. What is the advantage of this system compared to the arrangement found in the flatworms?

i. NEMATODA. Cross sections of *Ascaris* show that the gut is a straight tube, essentially devoid of musculature (Figure 5.1**B**). The walls are only one cell thick, and in life under the pressure of the pseudocoelomic fluid, the gut is collapsed. Counteracting the effects of a high internal turgor pressure, which would tend to expel the contents of the gut when the mouth was opened, a muscular **pharynx** is present, which pumps food into the mouth during feeding.

▶ Examine a section through the pharynx (Figure 5.8**A**), which reveals it to be a triradiate

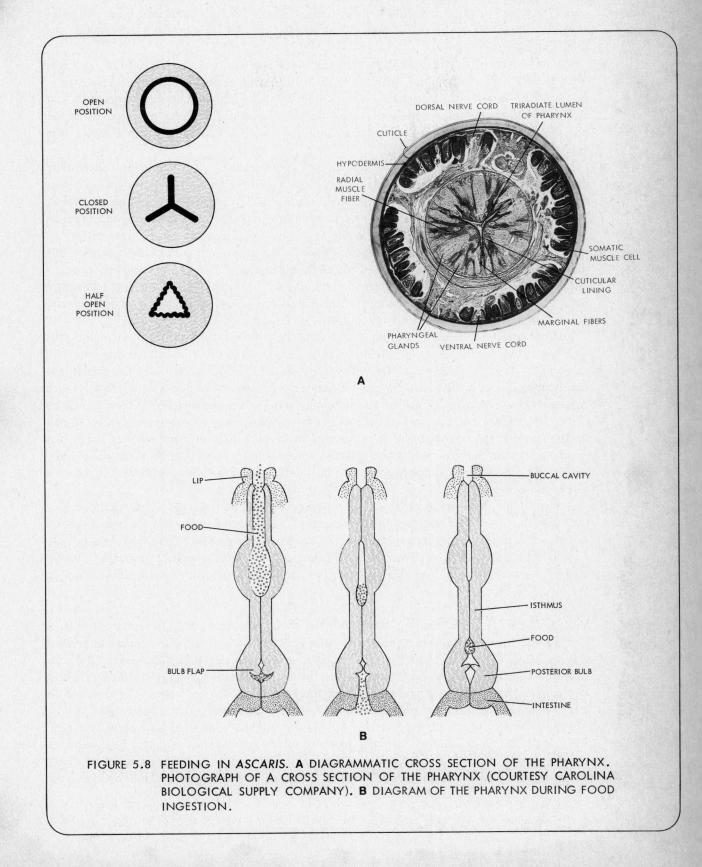

FIGURE 5.8 FEEDING IN *ASCARIS*. **A** DIAGRAMMATIC CROSS SECTION OF THE PHARYNX. PHOTOGRAPH OF A CROSS SECTION OF THE PHARYNX (COURTESY CAROLINA BIOLOGICAL SUPPLY COMPANY). **B** DIAGRAM OF THE PHARYNX DURING FOOD INGESTION.

structure. Large bundles of radially arranged **muscle fibrils** extend between the inner and outer surfaces. Three highly branched **pharyngeal glands,** one dorsal and two ventrolateral, are present in the pharynx wall.

▶   Recall the muscular pumping of the pharyngeal bulb in wet mounts of *Rhabditis*, or make fresh mounts and observe this. The sequence of positions of the muscles during feeding is indicated in Figure 5.8**B**. Pharyngeal opening is accomplished through muscle contraction, and closure through the turgor pressure acting on the elastic walls of the pharynx. Thus the pharynx can be described as a functional sphincter, although anatomically it is not a true sphincter.

▶   Most of the modifications in gut structure among nematodes involve the region between the mouth and the pharynx (**buccal capsule**). Some show well-developed **lips,** e.g. *Ascaris*. The lips of *Ascaris* can be seen clearly by taking a razor blade and cutting off the anterior tip of a worm. Place this on a slide in a drop of water so that you look directly into the mouth, and examine under the microscope. The largest lip is dorsal and the other two lips are ventrolateral.

▶   The pharynx and buccal capsule are both lined with cuticle, and in the buccal capsule this may be thickened to form **ridges, plates,** and **teeth.** In the case of hookworms, the buccal plates cut the flesh of the host. In some groups, there may be protrusible **stylets** present to puncture plant and animal prey. Demonstrations of the buccal apparatus of *Necator, Dorylaimus*, and *Haemonchus* will be available for comparative purposes.

The remainder of the gut is the **intestine;** its wall consists of a single layer of epithelial cells with numerous microvilli on the luminal surface; little is known of its role in digestion and absorption. There is a valve between the pharynx and the intestine, which regulates the unidirectional flow of materials from the former to the latter. Note that cilia would not work efficiently in a tube with collapsible walls; they are absent from the intestine as from the entire nematode group.* The most posterior part of the intestine is the **rectum,** which has a slitlike opening to the exterior. Defecation occurs by means of turgor pressure, that is, when the pumping pressure in the intestine is unable to overcome the turgor pressure of the pseudocoelomic fluid. The actual opening is controlled by a muscle from the body wall acting as a dilator (**depressor-ani** muscle).

▶   Return to the dissected specimen of *Ascaris* (Section 5–1) and identify the straight digestive tract that runs the entire length of the body (Figure 5.4). Carefully lay aside the coiled ovaries or testis and observe that the anterior end of the gut is differentiated into a pharynx, but the remainder appears as an undifferentiated, flattened tube. Trace this tube posteriorly to the slightly muscular rectum. The anus is located on the ventral side of the worm, subterminal to the posterior end.

ii. ROTIFERA.   The feeding mechanism of rotifers is distinctly different from that of nematodes. The anterior or head end bears a flat, ciliated surface called the **corona** or **trochus** (Figure 5.2A). The coronal (trochal) cilia create a current of water, which draws food particles toward the ventroanteriorly positioned **mouth.** Because the beating cilia give the impression of two wheels turning, these organisms are sometimes called wheel animalcules. The mouth leads into the highly muscularized **pharynx (mastax),** which bears cuticularized jaws called **trophi.** Muscular activity of the pharynx causes the jaws to produce a grinding and chewing action. This part of the gut is usually in constant motion and was mistakenly thought by Leeuwenhoek to be the heart. **Salivary glands** open into the pharyngeal cavity. Posterior to the pharynx are the narrow **esophagus** and the saclike **stomach.** Paired anterolateral **gastric glands** open into the

---

*Recent studies with the electron microscope have demonstrated the presence of structures of ciliary origin in certain organs (reproductive, sensory papillae, intestine) of some nematodes.

stomach. Cilia line the digestive tract and move food into the **intestine** and waste material into the **cloaca** and out through the **anus.** The anus is situated dorsally, near the base of the foot.

▶ Observe some living rotifers as they feed on a drop of Congo-red-stained yeast suspension or a dense culture of *Chlamydomonas*. A detailed study of the trophi can be made by placing rotifers in a drop of 5% sodium hypochlorite. The soft parts of the body are dissolved, and only the more resistant trophi remain.

5–5. EXCRETION

Two principal types of excretory and/or osmoregulatory organs are found in the Aschelminthes: (1) the nonciliated type which occurs in nematodes, and (2) the ciliated, protonephridial flame bulb type seen in the rotifers and gastrotrichs.

i. NEMATODA. The excretory systems of nematodes may be glandular and/or tubular. The more primitive glandular type is found principally in the free-living marine nematodes. It consists of one or two large **renette cells** lying in the pseudocoel ventral to the pharynx; a cellular extension of variable length called the **neck** runs forward from the renette cell and opens to the exterior via a midventral **excretory pore.** Freshwater, terrestrial, and parasitic nematodes usually have renette cells associated with tubular **longitudinal ducts,** which run in the lateral lines. The longitudinal canals are linked anteriorly by a **transverse canal,** giving the system an H-shaped configuration. The transverse canal opens by a median **excretory pore** just posterior to the pharynx. The ducts are intracellular, and probably derived from an altered renette cell. The structure of this system is quite variable; for example, some groups lack the renette cell altogether and some have it only in the parasitic juvenile stages. *Rhabditis* has a two-celled renette and H-shaped excretory canal system, whereas in *Ascaris* the renette is unrecognizable and the anterior canals of the H-shaped system are reduced. The consensus of opinion today is that the excretory system functions primarily in osmoregulation.

▶ The excretory system is difficult to see in both living and preserved material, but can be traced in serial sections. Examine a cross section of *Ascaris* (Figure 5.1**B**) and locate the excretory canals.

▶ ii. ROTIFERA. Take some living rotifers and arrest their movement with a drop of 5% methyl cellulose. If the illumination is carefully controlled, it may be possible to observe the flickering of the **flame bulbs** on either side of the mastax (Figure 5.2**A**). Alternatively, crush the rotifer by exercising controlled pressure on the cover slip, and examine the broken rotifer body for flame cells. **Excretory canals** run along the sides of the body and empty into a **bladder** in the posterior part of the trunk. The bladder periodically empties its contents into the **cloaca,** from which they are expelled to the exterior. The system is primarily osmoregulatory in function and some freshwater species may eliminate fluid in amounts equal to their own volume every 10 minutes.

iii. GASTROTRICHA. A flame bulb arrangement similar to that of rotifers exists in some gastrotrichs, but they lack a bladder. On each side of the body a ciliated **flame bulb** leads via an extensively coiled **excretory canal** to a midventral **excretory pore** (Figure 5.3**C**). Flame bulbs are absent from marine gastrotrichs, a fact that emphasizes their osmoregulatory function.

▶ Make wet mounts of living freshwater gastrotrichs and attempt to identify the components of the excretory system.

5–6. NERVOUS SYSTEM

Much of the work on the nervous system of the Aschelminthes has been restricted to studies of morphology. The details of function have been sadly neglected, largely because of the small size of the organisms involved.

i. NEMATODA.   The anterior end of most nematodes, especially the parasitic forms, is devoid of elaborate sense organs. **Papillae,** or free nerve endings, serve for chemoreception, tactoreception, and photoreception. The cuticle is usually thinner over the papillae than over the rest of the body. The main concentration of nervous tissue is in the anterior region surrounding the pharyngeal bulb (**circumpharyngeal nerve ring),** and from here the main **longitudinal trunks** and the minor **lateral trunks** run dorsally and ventrally in the hypodermal cords.

▶   Examine wet mounts of living *Turbatrix* or *Rhabditis*. The only part of the nervous system that you will be able to recognize will be the circumpharyngeal nerve ring.

ii. ROTIFERA AND GASTROTRICHA.   The **"brain"** of the rotifer consists of a bilobed mass anterior to the mastax, from which run two main **ventral cords** (Figure 5.2A). Some species have **"eyes"** and other well-developed sensory devices. Similarly, the gastrotrich "brain" surrounds the pharynx and sends nerves posteriorly and anteriorly to innervate the **cephalic bristles.**

▶   Examine rotifers and gastrotrichs whose movements have been arrested with methyl cellulose (Section 5–5) and attempt to identify these structures.

5–7. REPRODUCTION

The Aschelminthes vary considerably from group to group both in the arrangement of their reproductive organs and in their life histories.

i. NEMATODA.   Most nematodes are dioecious, and males are usually smaller than females. The reproductive system consists of an elongate tube, which loops about many times and ends blindly. Some parts of the tube are collapsible. The sperm are not flagellated but most often amoeboid. Certain portions of the reproductive tract, for example, the uterus and vagina, are capable of peristaltic action. The amoeboid sperm and peristaltic extrusion of eggs are both functional adaptations, which overcome the effects of the high turgor pressure in the pseudocoel and the collapsed walls of the reproductive system.

The female reproductive system (Figure 5.4A) lies in the posterior two thirds of the body cavity. From the anterior **genital pore,** the single **vulva** and **vagina** run back to join paired **uteri,** which are full of eggs. The uteri run back and turn forward into a pair of coiled **oviducts** leading in turn to paired filamentous and blind-ending **ovaries,** which lie tangled among the oviducts.

The male system (Figure 5.4B) leads forward from the curved tail. Into the hindgut open two small sacs that house two **copulatory spicules,** mechanical devices for opening the vulva during copulation when the **ameboid sperm** are ejected under pressure from the **ejaculatory duct,** which also opens into the hindgut. The single male opening thus serves both the function of elimination of undigested materials and that of ejaculation, and the hindgut is therefore a **cloaca.** From the cloaca the ejaculatory duct leads forward and swells into a long **seminal vesicle** where spermatozoa are temporarily stored. The seminal vesicle narrows into the **vas deferens,** which coils back and forth and ensheaths the midgut, merging into a coiled blind-ending **testis.**

▶ Study the reproductive structures in dissected specimens and cross sections of *Ascaris* (Figures 5.4 and 5.1**B**) and identify all parts. To facilitate examination of the male system, remove one lateral wall of the curved tail. If you have a female worm, cut out a section of a uterus, open it and extract the **eggs.** Shed from the ovaries, these are fertilized in the **spermatheca,** located near the junction of the oviduct and uterus, and then surrounded by chitinous shells. Mount some eggs in a drop of water and examine their structure with the high power lens of your microscope. Wash your hands thoroughly following this exercise. *Ascaris* eggs can be infective to humans and juveniles that find their way to the lungs can cause symptoms of pneumonia.

ii. ROTIFERA.   The female reproductive system of rotifers consists of a single **ovary** and a yolk-producing **vitellarium** (Figure 5.2). The **eggs** pass from the **oviduct** into the **cloaca** and then to the outside. Males are extremely rare; they are smaller than females, they lack a gut, and the single **testis** discharges into the **cloaca.** During copulation the **penis** is inserted into the female oviduct.

There are two kinds of females. One kind always reproduces parthenogenetically, producing eggs that all develop into females, whereas the other kind may reproduce bisexually. The bisexual females produce males intermittently, as well as resting eggs. The cycle commences with parthenogenetic reproduction and is terminated by a short period of sexual reproduction, the timing of which depends to a large extent on environmental conditions. Some species are exclusively parthenogenetic. Further details of reproduction in rotifers may be found in the reference by Hyman.

▶ Return to the preparation of the rotifer in which movement was arrested by the use of 5% methyl cellulose (Section 5–5), or prepare a fresh wet mount, and try to identify the organs of reproduction (Figure 5.2).

iii. GASTROTRICHA.   All freshwater gastrotrichs reproduce parthenogenetically (e.g., *Lepidodermella, Chaetonotus*). In the middle of the body lie the paired **ovaries,** which discharge via the **gonopore** in the region of the anus (Figure 5.3). Marine gastrotrichs such as those in the order Macrodasyoidea are hermaphroditic. Further details of reproduction in gastrotrichs will be omitted from consideration here and may be found in the reference by Hyman.

▶ Attempt to identify the reproductive structures in wet mounts of living gastrotrichs or in prepared slides (Figure 5.3). Cultures of freshwater gastrotrichs may show females distended with eggs.

5–8. LIFE CYCLES AND DEVELOPMENT

i. NEMATODA.   The anatomy of the male and female reproductive systems is treated in Section 5–7; now consider some aspects of nematode development.

After sperm are transferred from the male to the female, they travel up the female reproductive tract and either fertilize the mature ova immediately or are stored in the seminal receptacle of the female for subsequent fertilization. As soon as a sperm has entered an egg cell, a **fertilization membrane** appears around the egg and a **shell** begins to form. A lipoidal **vitelline membrane** is produced on the inner side of the shell, and as the egg passes down the uterus,

protein is added on the outside of the shell. Thus, eggs of most nematodes are covered by three distinct membranes: (1) an external **protein coat** from the uterine wall, (2) a true **egg shell** formed by the egg itself, and (3) the **vitelline membrane.** While the membranes are being formed, meiotic divisions occur, the pronuclei fuse, and cleavage ensues.

▶   *Rhabditis* is ovoviviparous and the eggs lack a proteinaceous coat. *Ascaris* deposits eggs that have not undergone cleavage and the eggs have a proteinaceous outer coat. All variations are found between these extremes among the various nematode species. Observe demonstrations of these two kinds of eggs.

▶   Because the cells of *Ascaris* normally have only four chromosomes, the genus has been used extensively for studies of gametogenesis (meiosis). Using prepared sections of the uterus of *Ascaris*, attempt to identify the different stages (Figure 5.9).

▶   Make a wet mount* of some living adult *Rhabditis* and observe the arrangement of the reproductive organs (Figure 5.5). Living eggs may be obtained by teasing apart the body of a female. The fertilized egg is 45 μm long and can be used for studying early cleavage stages. Transfer eggs to a depression slide, but do not exclude oxygen (e.g., do not ring with petroleum jelly). Pro-nuclear fusion and cleavage should occur in 2 hours at room temperature. Cleavage in nematode eggs is of a highly modified spiral type and each stage is rigidly determined. Make wet mounts of the cleavage stages and observe under the microscope.

In the embryonic nematode all cell division soon ceases except for the reproductive cells. In the somatic cells of postembryonic nematodes the chromosomes are continually fragmenting, whereas the germ cells are the only ones that divide at all and retain their full chromosomal complement. It was this observation that helped Weissman to formulate his theory of the continuity of the germ plasm. Growth consists of vacuolation and extension of existing cells; there are a fixed number of very large cells which carry out all body functions. In *Rhabditis*, for example, there are 68 muscle cells, 200 nerve cells, 120 epidermal cells, and 172 cells composing the gut.

Having completed their development within the egg, the juvenile worms hatch. These juveniles are generally fully developed except for the gonads and body size. A significant feature of postembryonic development in nematodes is the occurence of molts. Successive juvenile stages (called larval stages by some authors) are separated by molts. There are always four juvenile molts, and the adult emerges from the fourth molt. In some cases one or two molts take place within the egg. The third juvenile stage is characteristically the period of wandering in nematodes, and in the parasitic species it is generally the infective form.

Juvenile worms may eventually hatch from your preparation of *Rhabditis* eggs, although this occurrence is not easily obtained in the laboratory. These juveniles invade a living earthworm through the various body openings and eventually encyst in the body. Development is arrested until the earthworm dies and decay begins. The decaying meat apparently contains a substance that provides the physiological trigger necessary for further development.

It has been estimated that there are more than 500,000 species of nematodes, of which about 9000 have been described; thus, it is impossible to examine more than a few life cycles here. Most of the species are found in the marine environment, with less than 20% found in the soil and in fresh water. Approximately 50% of the described forms are parasitic, and it is these that have the most complex life histories.

▶   Study the life cycles of *Turbatrix, Rhabditis, Ascaris lumbricoides, Necator, Enterobius, Dracunculus, Wuchereria,* and *Trichinella.* Pay particular attention to location of the adult, location of the juveniles, route of migration, infective larval stages, physiological triggers for establishing infections, and intermediate hosts.

*This portion of the exercise should be started at the beginning of the laboratory period to allow time for completion.

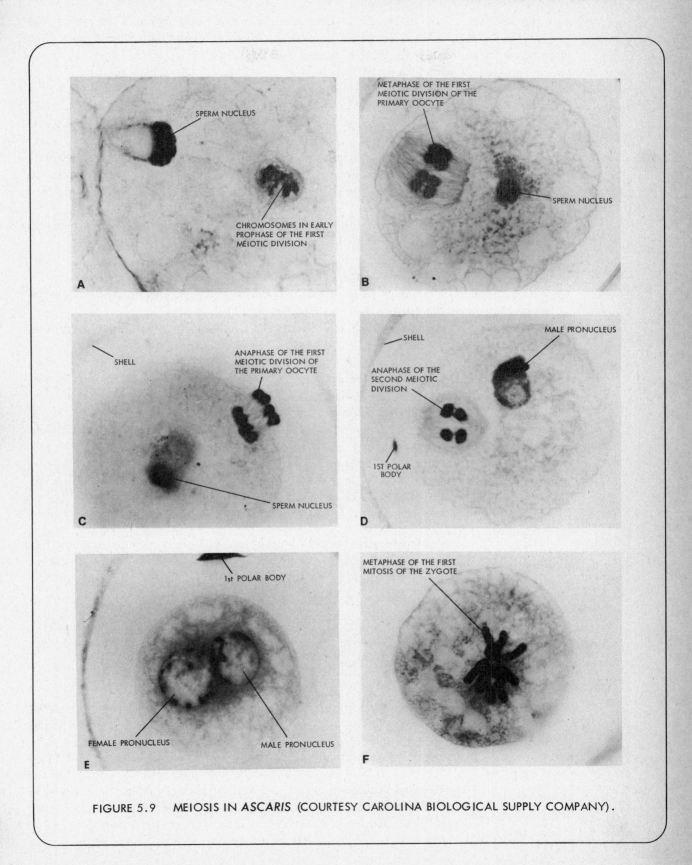

FIGURE 5.9   MEIOSIS IN *ASCARIS* (COURTESY CAROLINA BIOLOGICAL SUPPLY COMPANY).

ii. ROTIFERA AND GASTROTRICHA.   Life cycles and development of the rotifers and gastrotrichs are mentioned in Section 5–7; further details are beyond the scope of this course and may be found in the reference by Hyman.

References

Birky, C. W. Studies on the physiology and genetics of the rotifer *Asplanchna, J. Exp. Zool.* **155**:273–88, 1964.

Brown, F. A., Jr. (ed.). *Selected Invertebrate Types.* New York: John Wiley & Sons, 1950.

Castillo, J., and T. Morales. "Electrophysiological Experiments in *Ascaris lumbricoides*," in *Experiments in Physiology and Biochemistry,* Vol 2, G. A. Kerkut (ed.). London and New York: Academic Press, 1969.

Chandler, A. C., and C. P. Read. *Introduction to Parasitology.* New York: John Wiley & Sons, 1961.

Cheng, T. *General Parasitology.* New York: Academic Press, 1973.

Crofton, H. D. *Nematodes.* London: Hutchinson and Co., 1966.

Gilbert, J. J., and G. A. Thompson, Jr. Alpha-tocopherol control of sexuality and polymorphism in the rotifer *Asplanchna, Science* **159**:734–36, 1968.

Hyman, L. H. *The Invertebrates,* Vol. III. *Acanthocephala, Aschelminthes, and Entroprocta.* New York: McGraw-Hill Book Co., 1951.

Lee, C. *The Physiology of Nematodes.* Edinburgh: Oliver and Boyd, 1965.

Noble, E. R., and G. A. Noble. *Parasitology. The Biology of Animal Parasites,* 3rd ed. Philadelphia: Lea and Febiger Publishers, 1971.

Rogers, W. P. *The Nature of Parasitism.* New York: Academic Press, 1962.

Sasser, J. N., and W. R. Jenkins. *Nematology.* Chapel Hill: University of North Carolina Press, 1960.

An Abbreviated Classification of the Phylum Aschelminthes (After Hyman)

Phylum Aschelminthes

Pseudocoelomate, mostly vermiform, bilaterally symmetrical; unsegmented or pseudosegmented body covered with a cuticle; digestive tract lacks a definite muscular wall, anus posterior to the mouth.

Class 1 **Rotifera**

Aquatic, microscopic animals with anterior end modified into a ciliary crown; pharynx with internal jaws; protonephridia, e.g., *Philodina, Epiphanes, Asplanchna.*

Class 2 **Gastrotricha**

Aquatic, microscopic animals lacking a ciliary crown. Cuticle with plates or spines. Pharynx without jaws; cilia present on limited areas of the body, e.g., *Chaetonotus, Lepidodermella.*

Class 3 **Nematoda**

Cylindrical vermiform animals, devoid of cilia; musculature arranged in four longitudinal fields; pharynx long and triradiate; excretory system not protonephridial, e.g., *Ascaris, Rhabditis, Turbatrix, Haemonchus.*

Class 4 Nematomorpha (Gordiacea)

Long, slender vermiform animals without differentiated pharynx; parasitic juveniles and free-living adults.

Class 5 Kinorhyncha

Microscopic marine animals devoid of cilia; cuticle with spines; protonephridia present.

Equipment and Materials

Equipment

Microscope slides and cover slips
Razor blades
Dissecting microscope
Compound microscope
pH indicator paper
Depression slides
Pipets
Glass tubing
Finger bowls
Dissecting pans
Pins
Dissecting instruments

Solutions and Chemicals

pH solutions 1–11
30% seawater
5% methyl cellulose
5% sodium hypochlorite
Congo-red-stained yeast suspension

Living Material*

Ascaris (3)

Rotifers, e.g., *Philodina,
Asplanchna, Epiphanes* (4,6,11)
Gastrotrichs, e.g., *Chaetonotus,
Lepidodermella* (4,6,11)
Rhabditis in decaying earthworm
flesh† (4,6,11)
Turbatrix (4,6,11)
Chlamydomonas (4,6,11)

Preserved Materials

Ascaris

Prepared Slides

Philodina or *Asplanchna*, whole
mounts
Chaetonotus, whole mount
Necator, whole mount
Haemonchus, whole mount
Dorylaimus, whole mount
Ascaris, meiosis
Ascaris, eggs
(Kinorhynchs and nematomorphs)

*Numbers in parentheses identify sources listed in Appendix B.
†One week before needed, cut up earthworms in short sections and remove the gut. Pieces of body wall should be laid out on petri plates with 2% agar and left at 10°C. In one week a thriving culture of *Rhabditis* will be obtained.

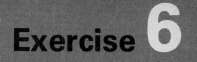

ANNELIDA

Contents

The phylum Annelida (*anellus,* ring; L.) includes the earthworms, leeches, and a variety of other wormlike organisms. Typically, they are bilaterally symmetrical, elongate in an antero-posterior direction, cylindrical in cross section, and divided externally by a linear series of rings (**annuli**) into body segments (**metameres**). The external segmentation is reflected by the arrangement of the internal organs, which are serially repeated, a condition known as **metamerism.** Metameric segmentation distinguishes the annelids from other wormlike forms. It provides a degree of plasticity in that certain segments can be modified and specialized to perform specific functions, and segments can respond either individually or collectively.

The gut is a straight tube supplied with its own musculature, so that it functions independently of muscular activity in the body wall. Like the nematodes, the annelids have a fluid-filled internal cavity, which separates the gut from the body wall and is similarly involved in locomotion. However, this space is not embryonically derived from the blastocoel, but is formed much later in development as a split in the mesoderm by which it is entirely lined. This fluid-filled internal space is a **true coelom.**

The excretory and circulatory systems are well developed, and some members of the phylum have respiratory organs. The nervous system is concentrated anteriorly **(cephalization)** into cerebral ganglia from which arises a ventral nerve cord with segmental ganglia.

Members of this phylum exhibit all the adaptations necessary for life in marine, freshwater, or terrestrial environments and may be free-living or parasitic.

6–1. BODY ORGANIZATION

a. External Anatomy

The Annelida are divided into three classes (see p. 164 for classification), the characteristics of which can best be demonstrated by an examination of three representatives: the terrestrial earthworm *Lumbricus,* an oligochaete; the marine worm *Nereis,* a polychaete; and the freshwater leech *Hirudo,* a hirudinean (Figures 6.1, 6.2, and 6.3). *Lumbricus* lives in moist soil; it remains underground in its burrow during the daylight hours but in the evening, especially after a rain, it may emerge from its burrow. *Nereis* lives in sand or mud tubes formed by the protrusible pharynx and mucus secretions; generally, only the anterior end of the worm emerges from the tube in search of food. *Hirudo* lives in lakes and ponds; it feeds by periodically attaching itself to a vertebrate host and sucking blood.

▶ Obtain preserved specimens of these three worms and lay them out side by side in a dissecting pan. All three are elongate, and the metameric segmentation of the body is conspicuous externally as a series of annuli.

▶ i. OLIGOCHAETA.   Examine the external features of the common earthworm, *Lumbricus* (Figure 6.1A). Note the well-marked external segmentation. Locate the nonannulated band **(clitellum);** the portion of the body with the fewest number of segments on one side of this band is the anterior end. Observe the subterminal position of the mouth. The preoral ring, which is considered not to be a true segment, is called the **prostomium** (*pro,* before; *stoma,* mouth; Gk.). The first segment of the body surrounds the mouth and is called the **peristomium** (*peri,* around; *stoma,* mouth; Gk.). Locate the anus at the posterior end of the body. Note the presence of bristles (**setae** or **chaetae**) on the body of *Lumbricus*. How many bristles are there per segment and on what surface? In which direction do the bristles point? Verify by stroking the animal lengthwise with your finger. Is there a difference in coloration between the dorsal and ventral surfaces? Is the pattern the same as that in the free-living planarian *Dugesia?* (See Exercise 4.)

▶ The entire surface of the body of *Lumbricus* is covered with a thin **cuticle.** Peel off a small piece of the cuticle by rubbing the worm gently with dry paper toweling; examine the cuticle under the dissecting microscope. Is it composed of cells? How is it similar to the cuticle of aschelminths?

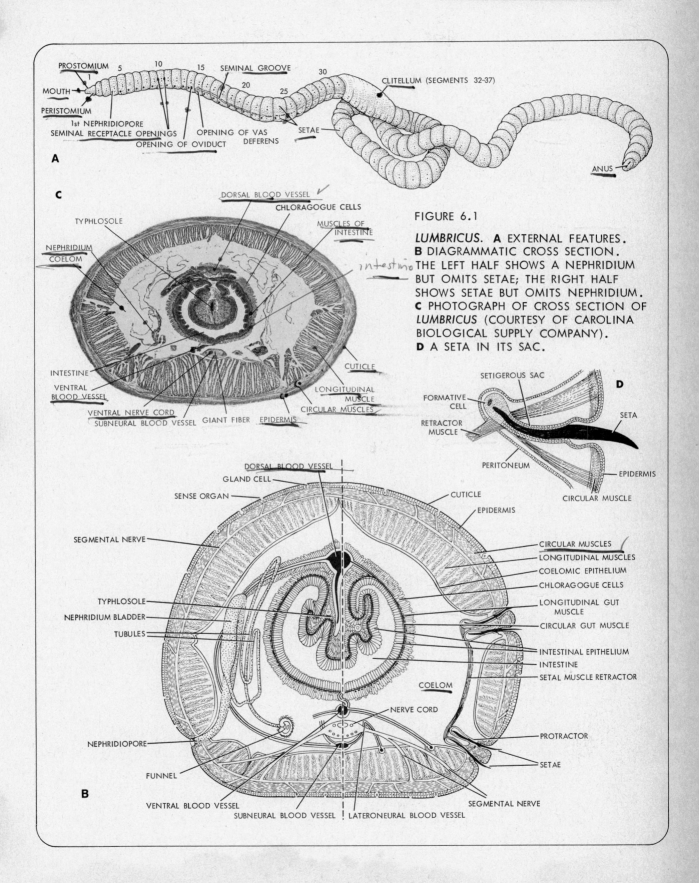

FIGURE 6.1

***LUMBRICUS.* A** EXTERNAL FEATURES.
B DIAGRAMMATIC CROSS SECTION.
THE LEFT HALF SHOWS A NEPHRIDIUM
BUT OMITS SETAE; THE RIGHT HALF
SHOWS SETAE BUT OMITS NEPHRIDIUM.
C PHOTOGRAPH OF CROSS SECTION OF
LUMBRICUS (COURTESY OF CAROLINA
BIOLOGICAL SUPPLY COMPANY).
D A SETA IN ITS SAC.

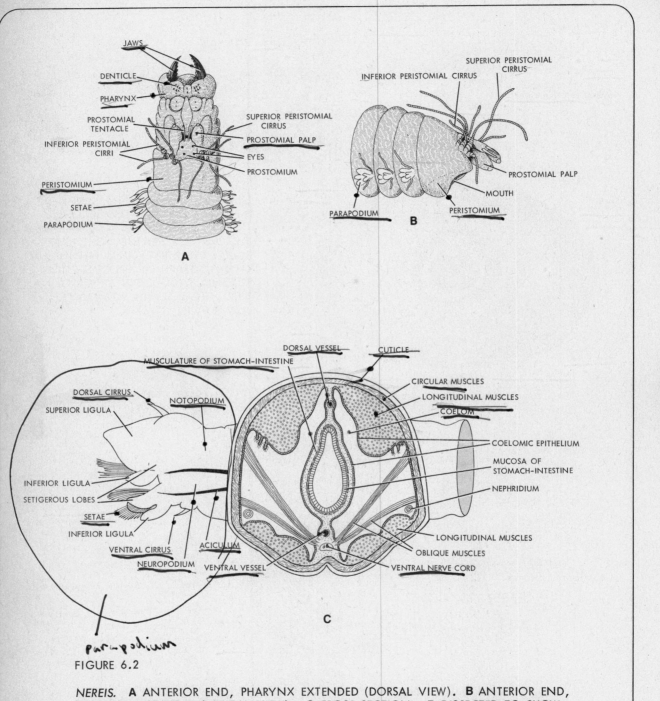

parapodium

FIGURE 6.2

NEREIS. **A** ANTERIOR END, PHARYNX EXTENDED (DORSAL VIEW). **B** ANTERIOR END, PHARYNX RETRACTED (LATERAL VIEW). **C** CROSS SECTION. **D** DISSECTED TO SHOW THE INTERNAL ORGANS. **E** DIAGRAM SHOWING THE ARRANGEMENT OF THE BLOOD VESSELS. **F** HETERONEREIS. **G** A PARAPODIUM FROM THE EPITOKE REGION OF A HETERONEREIS.

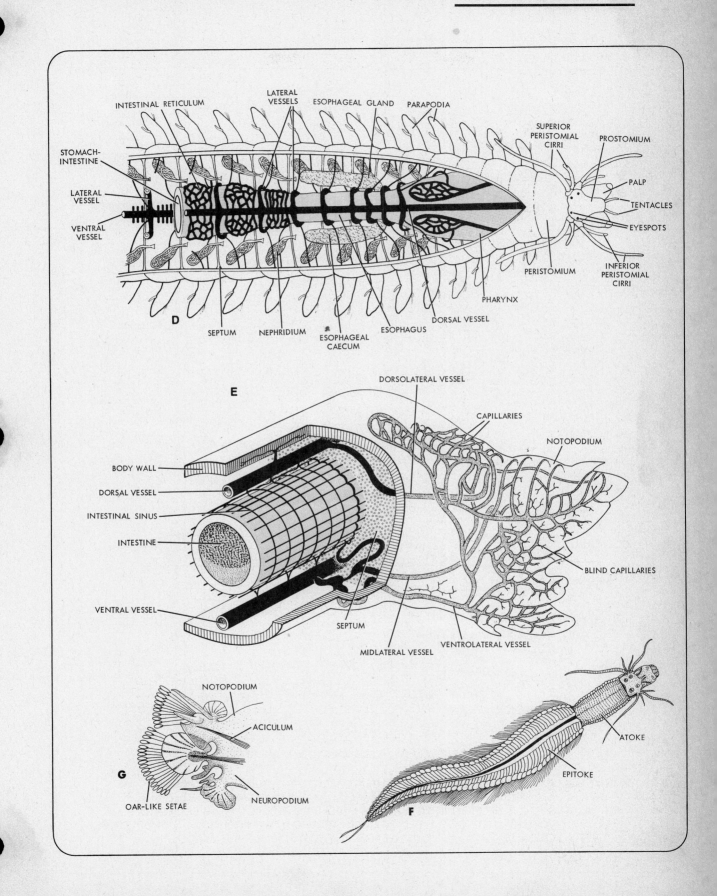

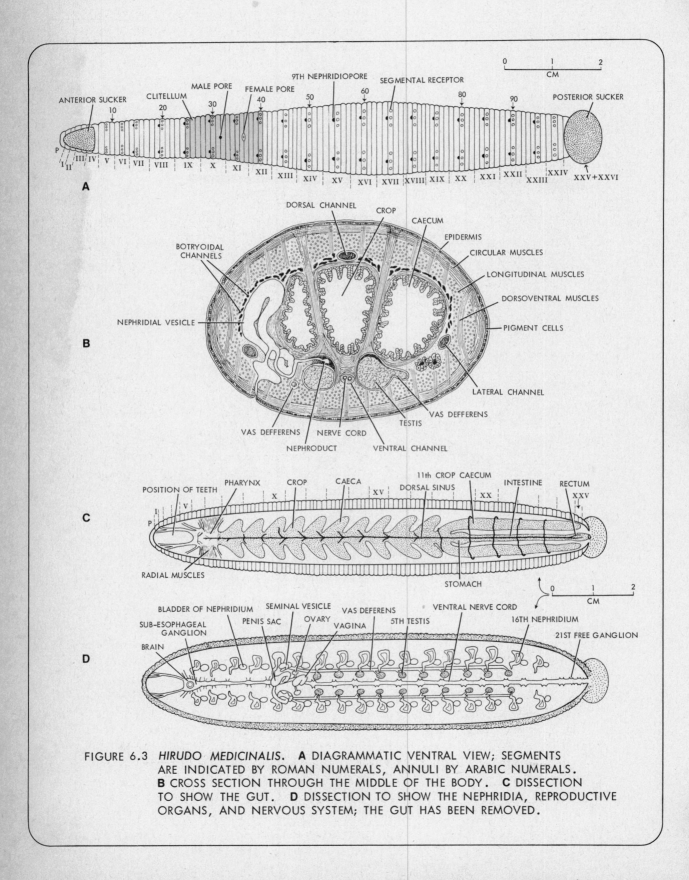

FIGURE 6.3 *HIRUDO MEDICINALIS*. **A** DIAGRAMMATIC VENTRAL VIEW; SEGMENTS
ARE INDICATED BY ROMAN NUMERALS, ANNULI BY ARABIC NUMERALS.
B CROSS SECTION THROUGH THE MIDDLE OF THE BODY. **C** DISSECTION
TO SHOW THE GUT. **D** DISSECTION TO SHOW THE NEPHRIDIA, REPRODUCTIVE
ORGANS, AND NERVOUS SYSTEM; THE GUT HAS BEEN REMOVED.

▶   Strip the cuticle from segments 9–15 and, using a lens or binocular microscope, attempt to locate the external genital openings. Earthworms are hermaphrodites. The paired **vasa deferentia** open lateroventrally on the fifteenth segment, and **seminal grooves** lead back from them to the anterior edge of the clitellum. During copulation, sperm pass along the grooves to the spermathecae of the partner worm. The **spermathecal openings** are located between segments 9 and 10 and 10 and 11 at the same level as the dorsal setae. The **oviducts** open lateroventrally on segment 14. Attempt to locate the **nephridial openings** on one or two of the stripped segments. These openings are dorsal to and in front of the ventral row of setae in all segments except the first three and the last. Middorsal **coelomic pores** open from the coelom in the segmental grooves of all segments following 10, but are difficult to see. In life they exude coelomic fluid and help to keep the worm moist. In sexually mature worms, segments 32–37 are swollen with glandular tissue forming a **clitellum.** The clitellum secretes the cocoon.

▶   ii. POLYCHAETA.   Examine the body of *Nereis* (Figure 6.2). If living *Nereis* are available, notice the iridescent **cuticle.** To what is this due? The most obvious difference between *Nereis* and *Lumbricus* is the presence of lateral segmental projections in the former. These fleshy appendages or **parapodia** are used in swimming and crawling movements. Remove a large parapodium and examine it in a bowl of water using a lens or a dissecting microscope. Each parapodium is biramous, consisting of a bilobed dorsal **notopodium** and ventral **neuropodium,** which function in respiration and locomotion. From each there arises a tactile **cirrus** and a bundle of jointed chitinous **setae.** They are fine bristles composed of a long blade jointed to a sturdy shaft. The setae arise from invaginated sacs in the parapodia, and each one is secreted by a single cell at the base of each sac. Remove one or two of the setae and examine under the microscope. The parapodia are supported internally by two modified setae called **acicula** (Figure 6.2C). Observe the number and arrangement of the setae. How does the number of setae per segment compare with the number per segment in *Lumbricus?* How is this reflected in the taxonomy of these representatives? (See Classification of Annelida, p. 164.)

▶   The anterior end of *Nereis* shows conspicuous cephalization and differs distinctly from that of *Lumbricus* in that it bears a pair of prostomial **tentacles** and paired **palps,** four prostomial **eyes,** and four pairs of tentacular **cirri** on the peristomium. The peristomium is probably composed of two fused segments and the peristomial cirri are modified dorsal and ventral cirri. Notice the eversible **pharynx** with its black serrated **teeth** or **denticles.** One pair of denticles is enlarged to form **jaws** (Figures 6.2A and B). The pharynx is normally everted during feeding and protrudes following death: do not confuse it with the head structures. Locate the ventrally directed **mouth** and the terminal **anus.** The elongated anal segment lacks parapodia, but has long ventral cirri.

▶   iii. HIRUDINEA.   Examine the external anatomy of the leech *Hirudo* (Figure 6.3A). Can you locate any setae? In what plane is the body flattened? Can you distinguish the ventral surface from the dorsal surface by differences in coloration? How many segments are there in a leech? Note the prominent **annulation** of each segment. How many annuli are there per segment? One annulus of each segment is marked by a row of **sensory papillae.** Five pairs of dorsal papillae in the first five segments are modified into **eyes.** The more pointed anterior end of the leech bears a small **ventral sucker,** probably formed from the prostomium and part of the peristomium. The posterior end bears a large disclike ventral sucker. Locate the **mouth** and **anus.** The single opening of the **vas deferens** is located midventrally at the second annulus of segment 10. The **penis** may protrude from the pore. The **vagina** opens midventrally at the second annulus of segment 11. The **nephridiopores** open lateroventrally at the last annulus of each segment from 6 to 22, but are difficult to see. There are abundant secretory cells in the hypodermis of segments 9–11, which form a functional **clitellum.**

From these brief observations, can you relate habitat to external form in each specimen? What does the absence of eyes in *Lumbricus* suggest to you about its habitat compared with that of *Nereis* and the leech? What is the function of setae in *Lumbricus?* What would be the disadvantage of parapodia to *Lumbricus?* Which segment of the body is most modified in *Nereis?* In *Lumbricus?*

b. Internal Anatomy

▶   With a sharp scalpel or razor blade, make a cross section of one of the posterior segments of *Lumbricus, Nereis,* and the leech four or five rings from the anus, and place each in a small labeled watch glass containing water. Using these cross sections and prepared slides, study the organization of the internal organs of the representative annelids.

   i. POLYCHAETA.   *Nereis* has the most "generalized" arrangement of internal organs (Figure 6.2**C**).

▶   Notice that the parapodium is a flattened extension of the body and is not solid. In a stained section, locate the supportive **aciculum** and the muscles attached at the proximal end; these effect protrusion or retraction of the parapodium. The bundles of jointed setae of the neuro- and notopodia can be moved by **setal muscles.**

▶   The entire surface of the parapodium and the body proper is covered by a thin, iridescent **cuticle** secreted by a single layer of cells that underlies it, the **epidermis** or **hypodermis.** Below the hypodermis of the body wall are a layer of **circular muscles** and four bands of **longitudinal muscles** (a dorsolateral pair and a ventrolateral pair). **Oblique muscles** originate on the median ventral surface and pass dorsolaterally to the body wall. This arrangement of musculature, acting on the incompressible coelomic fluid, permits the body to be shortened by contraction of the longitudinal muscles and lengthened by contraction of the circular muscles.

▶   Just beneath the muscles of the body wall is a thin continuous sheet of tissue called the **coelomic epithelium,** which surrounds the **dorsal** and **ventral blood vessels** and folds around the **gut.** The space enclosed by the coelomic epithelium is the **coelom;** in life it is filled with fluid. The wall of the gut is muscularized. How are the longitudinal and circular muscles of the gut oriented relative to each other? Identify the ventral nerve cord. Does it lie within the coelom? What holds the gut in place?

▶   ii. OLIGOCHAETA.   Examine the cross sections of *Lumbricus* (Figure 6.1**B**) and compare the internal anatomy with that of *Nereis.*

▶   The body surface of *Lumbricus* is covered by a **cuticle** secreted by the **hypodermis** that underlies it. Beneath the hypodermis is a layer of **circular muscles** and then a layer of **longitudinal muscles.** The longitudinal muscles are divided into seven bands: a dorsolateral pair, two pairs of ventrolaterals, and a single midventral muscle. How are setae positioned in the body? Are muscles associated with the setae? Are oblique muscles present in *Lumbricus?* What does this tell you about the function of the oblique muscles in *Nereis?*

▶   Beneath the longitudinal muscles is the **coelomic epithelium,** which surrounds the dorsal and ventral **blood vessels,** the ventral **nerve cord,** and the **excretory** and **reproductive organs** and envelops the **gut.** Identify as many of these organs as are present in your section. The gut (**stomach-intestine**) has a poorly developed musculature, shows a middorsal fold, the **typhlosole,** and is covered by a layer of yellowish **chloragogue** cells.

▶   iii. HIRUDINEA.   Examine cross sections of the leech (Figure 6.3**B**).

▶   The surface arrangement of the musculature in the leech is essentially the same as in other annelids. The coelomic space is reduced by mesenchymal tissue (also called **botryoidal tissue** because of its resemblance to a bunch of grapes). This tissue consists of fine capillaries of the coelomic blood sinus system surrounded by globular, pigmented cells, forming a connected system of sinuses. The gut (crop portion) is provided with branches called **caeca,** which increase the surface area and are used for storage of blood, the main diet of this leech. The reduction in the coelom is reflected in the method of locomotion and the habitually intermittent feeding of these worms. Identify the nerve cord and blood vessels.

6–2. LOCOMOTION

Flatworms move by means of cilia and muscles. Such a combination of methods is effective only in organisms of relatively small size. Increase in girth reduces the effectiveness of cilia, and muscular action becomes a more important locomotory device. Nematodes possess a fluid-filled body cavity surrounded by muscles (hydrostatic skeleton), and movement involves hydraulic action. In these roundworms, however, the pseudocoelomic space does not isolate the gut from the activity of the body wall musculature, and movements of the gut and body wall are intimately related (see Exercise 5).

Annelids also possess a hydrostatic skeleton, but the pressures involved are of a relatively low magnitude (1.5 to 20 mm Hg; cf. nematodes), and movements of the body wall are independent of the muscular activities of the gut. Segmentation of the body increases flexibility and allows localized and controlled locomotion. Part of this movement is regulated by the restricted movement of coelomic fluid between segments. You have already studied the anatomical arrangement of annelid musculature (see Section 6–1); if not, do so now before proceeding to see how it works.

►   i. OLIGOCHAETA.   Take a live earthworm and place it on a sheet of moist paper toweling that has been ruled with lines about 1 cm apart; allow the worm to crawl across the sheet. Notice how the worm moves forward as a wave of peristalsis passes along the body. Does the wave pass from anterior to posterior or vice versa? Do the "fat" or "thin" regions of the body move forward during a peristaltic wave? Now place the animal on a smooth surface such as a wet glass plate. Does the animal move forward? What role do frictional forces play in locomotion of the earthworm?

In order for the earthworm to move there must be some means of obtaining a grip on the substrate, and this is the function of the setae. The peristaltic wave is produced by the action of the longitudinal and circular muscles upon the incompressible fluid of the hydraulic skeleton. The body of the earthworm is internally partitioned into segments by septa. Each segment is thus isolated from the adjoining ones except for a small ventral sphincter which may be opened and closed, and the continuity of the gut, nephridia, and ventral nerve cord. During locomotion, the sphincter is kept closed and the septa act as bulkheads so that the volume of fluid within each segment remains constant. When the longitudinal muscles of a segment contract, the circular muscles relax, and because of the incompressibility of the coelomic fluid the segment becomes shorter and fatter. At the same time the setae are protruded and anchor the worm. When the circular muscles contract, the longitudinal muscles relax and the segments involved become long and thin; the setae are withdrawn, the body extends, and forward progression is obtained. Because the segmental septa limit the movement of coelomic fluid, body contractions and extensions may be localized, that is, each segment may contract or expand independently of every other segment. Coordination between segments is maintained by the ventral nerve cord, and segments contract or expand in peristaltic waves.

▶   The passage of peristaltic waves in the earthworm can be more easily observed by making a short longitudinal slit in the dorsal body wall, and noting the opening and closing of the slit as the animal crawls across moist paper toweling. Repeat this experiment but make a transverse slit. Diagram the opening and closing of the slit and explain its form in relation to the peristaltic wave and types of muscles that contract to produce the opening and closing of the slit. Diagram the movements of a crawling earthworm. Consult the papers by Gray (1938, 1939) for further details.

▶   The role of the coelomic fluid during burrowing can also be demonstrated by using this earthworm. Place the worm on some loose, moist soil and note the time required for burrowing. Remove about 25% of the worm from the posterior end by making a quick cut with a sharp razor blade, and again note the time needed to burrow. Is the time increased? Why? Place the worm in 0.2% chloretone for 3 to 5 minutes. Does the animal become limp? Why? Can the worm burrow? Why? What would be the effect on the burrowing time if coelomic fluid were removed? Why? What would be the disadvantage to the earthworm of support by a skeletal rod?

ii. POLYCHAETA.   The movement of *Nereis* in principle involves the same mechanisms as the movement of *Lumbricus,* but shows modifications because the internal septa are incomplete.

▶   Place a small, living nereid on a piece of wet paper toweling ruled with lines 1 cm apart and observe the progression of the worm. Does the body undulate in serpentine fashion? Record the flexures of the body and successive positions of the parapodia. Do parapodia on opposite sides of the body move synchronously in the same direction? Do all the parapodia on one side move simultaneously? If not, how do they move? How are the movements of the parapodia related to twisting of the body? Do all the longitudinal muscles of each segment contract at the same time or are the muscles of only one side contracted at any one time? What enables the parapodium to remain stiff during the effective stroke? How is gripping to the substrate achieved? Does the undulatory wave pass along *Nereis* in the same direction as the wave of contraction during movement in *Lumbricus?* How are the longitudinal muscles arranged and how do they operate during locomotion in *Nereis?* Contrast this with the situation in *Lumbricus.*

▶   *Nereis* can also swim. Place the worm in some seawater and observe the manner of movement. What systems are involved in swimming? What are the similarities between the movements of nematodes and nereids? What are the differences? Place a nereid in a glass tube of slightly larger bore than the diameter of the worm, and place the worm and the tube in a bowl of seawater. How is the anterior end of the worm protruded? By what means does the animal anchor itself to the tube? Note the undulatory movements of the body that produce ventilative (respiratory) currents. In what plane are these? What muscles are involved in the movements?

iii. HIRUDINEA.   You have observed that oligochaete and polychaete locomotion is brought about by the antagonistic action of two sets of muscles in the body wall (longitudinal and circular) acting on the incompressible coelomic fluid. In earthworms the segmental septa limit the movement of coelomic fluid so that local body contractions and extensions occur, whereas in many polychaetes the reduction or absence of septa permits several body segments to act as a unit. In leeches septa are absent (Figure 6.3C), and the entire body functions as a unit. The amount of coelomic fluid is reduced and to a large extent replaced by botryoidal tissue, which is also deformable and can function in a similar fashion to a hydraulic skeleton.

Leeches have an outer layer of circular muscles and an inner layer of longitudinal muscles in the body wall, but in addition they possess a double layer of oblique muscles between these two layers. The oblique muscles run as a helical spiral along the length of the body. When the body is long and thin, the spiral fibers reinforce the action of the longitudinal muscles as they contract and the body is shortened; when the leech is short and thick, they reinforce contractions of the circular muscles that produce elongation. In an intermediate position, that is, when the spiral muscles are neither lengthening nor shortening, they serve to increase the internal hydrostatic pressure and enable the leech to perch upright on its posterior sucker.

▶ Leeches swim and creep. Take a living leech, drop it into a container of fresh water, and observe the swimming motion. In what plane are the undulations? In what direction does the undulatory wave pass? Contrast the swimming movement of the leech with that of the nereid and relate each to the arrangement of muscles in the body.

▶ To study creeping patterns place a living leech on a glass plate and watch its movements; record the sequence of movements and the position of the suckers over a period of time. Place a leech in a small dish of water. Allow the posterior sucker of the worm to attach to a cover slip held above the bottom of the dish by means of forceps. What is the reaction? Release the cover slip. What happens? Why? Now present a second cover slip to the anterior sucker and allow the animal to attach as the cover slip is held above the bottom of the dish. Does the posterior sucker remain attached? Release the cover slip. What happens? Explain your observations. What movement occurs if the anterior end of the worm is supported by a piece of string and the sucker makes no contact? What body movements occur if both sucker areas are supported by loops of string? What muscles are used in extension of the body? What structures in the leech are analogues of the setae in *Lumbricus?*

▶ Relate locomotor patterns in all three organisms to their normal habit.

6–3. NEUROMUSCULAR SYSTEM

The neuromuscular system of annelids consists of a ventral nerve cord with segmental ganglia and a concentration of nervous tissue at the anterior end (Figure 6.4**B**). The dissection of the nervous system will be left until the gut and reproductive system have been studied; at this point, however, its function in coordinating movement will be examined.

a. Ventral Nerve Cord

The ventral nerve cord contains fast through-conducting nervous pathways called the **giant axon system,** although sometimes the term "giant" is an exaggeration of the size of the nerves involved. The system functions mainly in the rapid conduction of nerve impulses over relatively great distances, which facilitates rapid withdrawal reactions. (Note that the speed with which a nervous impulse is conducted depends on two factors: the insulation and the diameter of the nerve.) The anatomy and function of this system vary among the annelids.

▶ In *Lumbricus* three giant axons conduct impulses in both anterior and posterior directions, although for the most part stimulation of the anterior end activates the central fibers and stimulation of the posterior end activates the two lateral giant axons. Refer to prepared cross sections of the worm (Figure 6.1**B**) and attempt to locate the specialized giant fiber tracts in the ventral nerve cord. Allow a worm to crawl, strongly stimulate the anterior end by poking it with the tip of a pencil, and observe the reverse peristalsis. What is the "startle" reflex and how is it mediated?

Coordination of the peristaltic locomotor waves in the earthworm is a complex affair

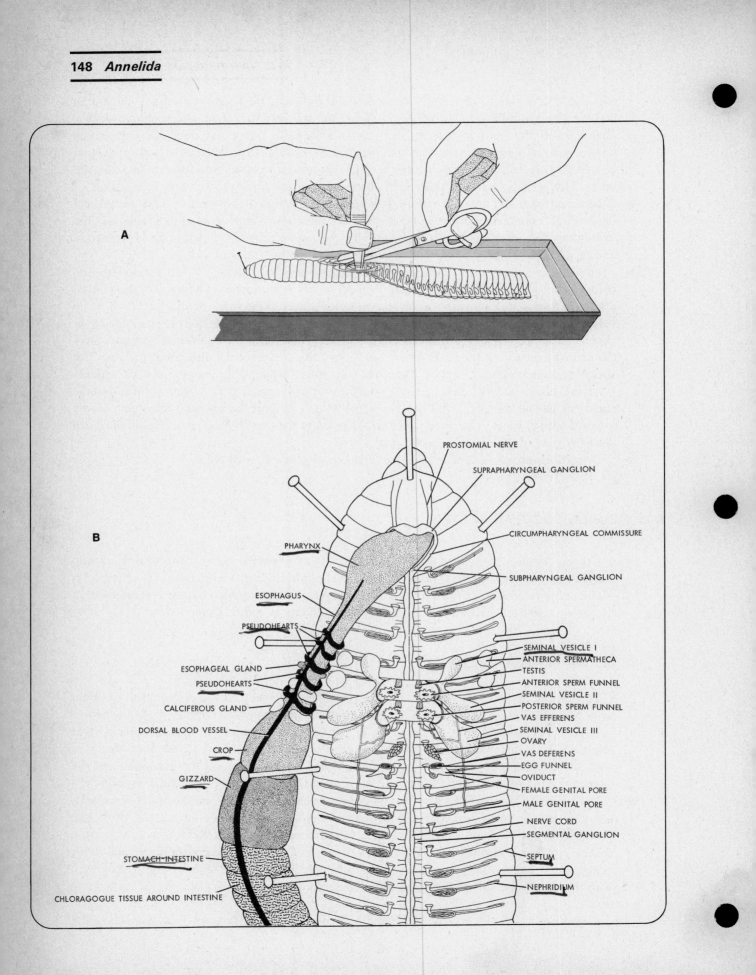

A

B

PROSTOMIAL NERVE

SUPRAPHARYNGEAL GANGLION

CIRCUMPHARYNGEAL COMMISSURE

PHARYNX

SUBPHARYNGEAL GANGLION

ESOPHAGUS

PSEUDOHEARTS

SEMINAL VESICLE I
ANTERIOR SPERMATHECA
TESTIS
ANTERIOR SPERM FUNNEL
SEMINAL VESICLE II
POSTERIOR SPERM FUNNEL
VAS EFFERENS
SEMINAL VESICLE III
OVARY
VAS DEFERENS
EGG FUNNEL
OVIDUCT
FEMALE GENITAL PORE
MALE GENITAL PORE
NERVE CORD
SEGMENTAL GANGLION
SEPTUM
NEPHRIDIUM

ESOPHAGEAL GLAND

PSEUDOHEARTS

CALCIFEROUS GLAND

DORSAL BLOOD VESSEL

CROP

GIZZARD

STOMACH-INTESTINE

CHLORAGOGUE TISSUE AROUND INTESTINE

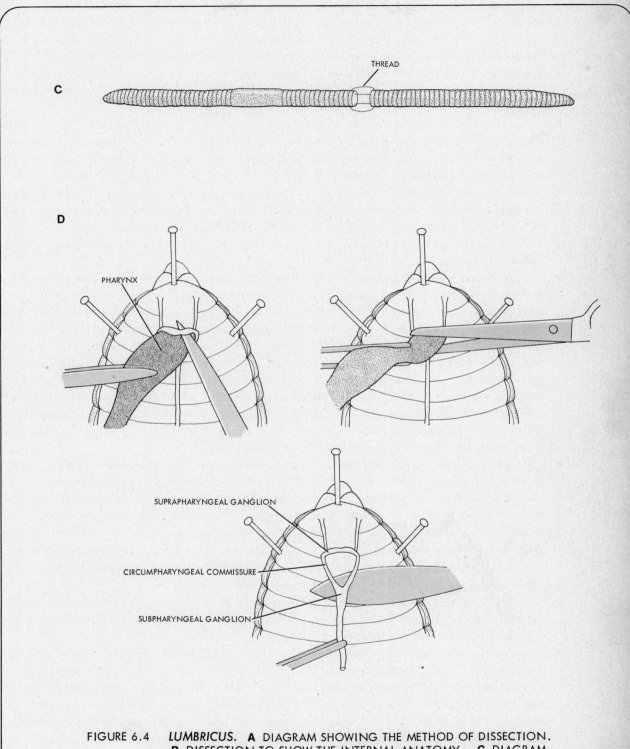

FIGURE 6.4   *LUMBRICUS.* **A** DIAGRAM SHOWING THE METHOD OF DISSECTION.
B DISSECTION TO SHOW THE INTERNAL ANATOMY. **C** DIAGRAM
SHOWING THE METHOD OF CUTTING AND TYING AN EARTHWORM.
D DISSECTION OF THE NERVOUS SYSTEM.

involving both the ventral nerve cord and local reflex arcs. When a wave of nervous activity is propagated along the length of the nerve cord, muscle contractions are set up in successive segments; when there are contractions of longitudinal muscles in one portion of the body, tension is exerted on adjacent regions, local reflex arcs are set up, and muscle contractions in these adjacent regions occur. To study the respective roles of the nerve cord and/or reflex arcs in locomotor coordination, elimination of the activity of one of these is necessary.

The role of the nerve cord in normal locomotion can be illustrated by the following experiments.

► Anesthetize a few intact worms by immersing them in 0.2% chloretone; wash in tap water after anesthetization and make a lateral incision so the nerve cord is exposed. Using fine scissors and needles, carefully free the nerve cord and sever the lateral nerves that run from the cord to the musculature of the body wall. This leaves a paralyzed section, with the anterior and posterior regions effectively coordinated only through the nerve cord. After the worms have recovered from the anesthesia, allow them to crawl and observe whether the peristaltic waves in both halves are coordinated. What do you conclude from your results?

To study the role of local reflex arcs in coordination of the crawling pattern, perform the following experiments.

► Anesthetize a few earthworms as described previously. After anesthetization is complete, make two lateral incisions of about 4 to 6 segments in length on either side of the ventral nerve cord. Remove the ventral nerve cord from these segments, carefully cutting all the nerve branches that run to the body wall. The worm is now denervated and paralyzed in one region of the body. Allow the animal to recover from anesthetization, after which you should observe the movements of the worm on moist toweling. Are the peristaltic locomotor waves coordinated with each other on either side of the paralyzed segments? Alternatively, cut an earthworm in two pieces and tie the cut ends loosely together again with sewing thread (Figure 6.4C). Such an experimental scheme eliminates nervous conduction but permits mechanical forces to be transmitted through the thread. When the operated worms move, is the peristaltic wave coordinated between the two halves of each worm?

► Suspend an intact worm vertically in a graduated cylinder filled with water. (The buoyancy of the water almost balances the weight of the worm and the longitudinal muscles will not be placed under tension). What is the locomotor pattern? Repeat your observations after the worm has had a weight attached while suspended in the water. Now place the animal on a rough horizontal surface. Are the locomotor waves that pass backward coordinated on both sides of the body as the worm crawls forward? Repeat with the animal crawling on a horizontal wet glass plate. (Muscular contractions in one part of the body now exert little tension on the adjacent segments because frictional forces have been reduced).

► Relate your findings to the statement: Rhythmic activity in the central nervous system initiates and coordinates locomotion but local reflex arcs acting in succession also produce coordinated locomotor movements.

► In *Nereis* the arrangement and function of the nervous system is more complex than in *Lumbricus*. Median dorsal fibers conduct impulses anteriorly and a pair of paramedial fibers conduct posteriorly. The large pair of ventrolaterals provides a through-conducting system; it fires only when the stimulation is strong. Try stimulating first the anterior and then the posterior end of a specimen of *Nereis* and observe the reactions. Repeat these experiments on the leech and compare the results with those obtained for both *Nereis* and *Lumbricus*.

Withdrawal reactions may be observed in tube-dwelling polychaetes such as sabellids and serpulids. (See Classification of Annelida, p. 164.) These worms usually receive stimulation only at the anterior end, and the principal direction of nervous conduction is from the anterior to the posterior. The giant axons of the nerve cord run the entire length of the body and directly

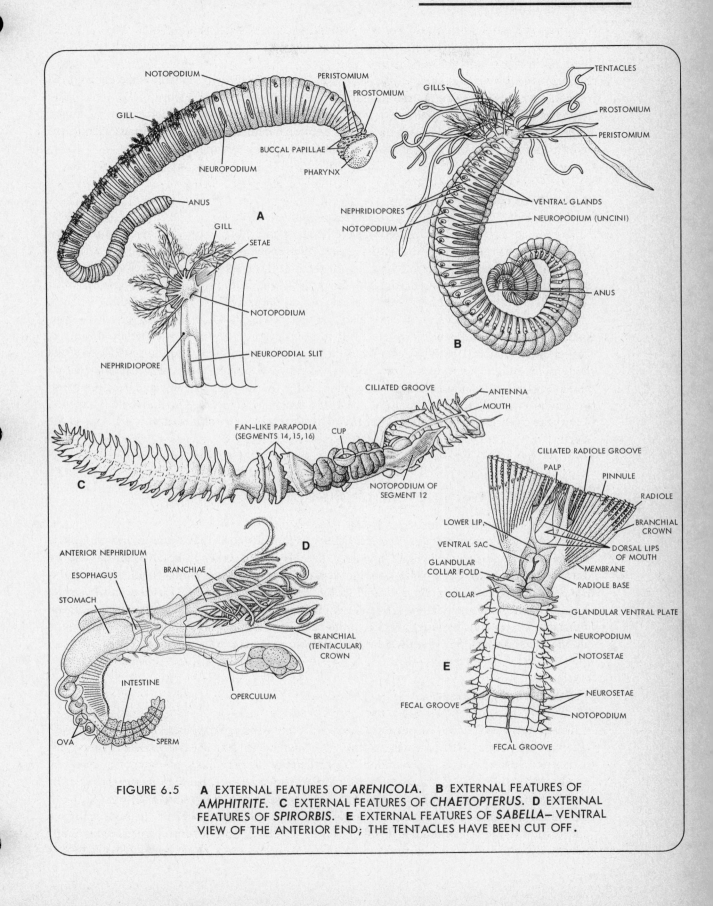

FIGURE 6.5   **A** EXTERNAL FEATURES OF *ARENICOLA*.   **B** EXTERNAL FEATURES OF *AMPHITRITE*.   **C** EXTERNAL FEATURES OF *CHAETOPTERUS*.   **D** EXTERNAL FEATURES OF *SPIRORBIS*.   **E** EXTERNAL FEATURES OF *SABELLA*— VENTRAL VIEW OF THE ANTERIOR END; THE TENTACLES HAVE BEEN CUT OFF.

innervate the longitudinal muscles, thus facilitating rapid withdrawal reactions. These feather duster worms expand a broad crown of tentacles (modified from the prostomium) which function in respiration and food gathering.

▶ Use available living material (e.g., *Eudistylia, Sabella, Hydroides*) and allow the animal(s) to remain covered by seawater and to expand the tentacular crown (Figure 6.5**D**). Touch the tentacles and note the response. How long does it take to withdraw? How long does the animal remain retracted in its tube? Does the animal become habituated? How many times will it react to stimulation? How are the nervous and muscular systems organized?

b. Cerebral Ganglia

▶ The role of the "brain" in locomotion can be determined as follows. The cerebral ganglia of the earthworm are located in the third segment (the second segment posterior to the peristomium). Decapitate an earthworm by cutting off the first three segments and observe its burrowing behavior. Now remove the next segment, which contains the subesophageal ganglia. What is the effect on burrowing behavior? What role(s) do the ganglia play in locomotion? Perform similar experiments on *Nereis* and compare the results with those obtained for the earthworm.

▶ If the body of the leech is severed but the nerve cord left intact, the animal swims in coordinated fashion. If only the ventral nerve cord is cut, however, and the remainder of the body is left intact, movements are uncoordinated between the two halves of the worm on either side of the cut. Decapitated leeches can swim, but if the subesophageal ganglia are removed, the animals can no longer creep. Removal of the anal ganglia prevents swimming movements. Try these experiments and verify that there are at least three centers of coordination in the leech.

6–4. FEEDING AND THE STRUCTURE OF THE GUT

The annelid gut is a straight tube that opens to the exterior via the mouth and anus. It is muscularized and independent of muscular activity in the body wall, although it may be anchored to the body wall by the segmental septa. The arrangement of musculature in the gut is the reverse of that in the body wall, that is, longitudinal on the outside, circular on the inside. The gut itself is not segmented, and generally the anterior portion of the body and the gut are considerably modified in direct relation to the diet. The most elaborate modifications are found among the sedentary polychaetes and the simplest among the burrowers.

a. Burrowing Herbivores

▶ The earthworm *Lumbricus* is a good example of a burrowing herbivore. To demonstrate the ingestion of "food," put some earthworms in a container with moist facial tissue. After a few days place the worms in 0.2% chloretone, and when they have relaxed, put them in a dissecting pan, ventral side down. Anchor a specimen to the pan by means of a pin through the prostomium, and make a *dorsal* longitudinal incision with a pair of scissors (Figure 6.4**A**). Keep the blade of the scissors flat to avoid injuring the tissues and organs beneath the body wall, and cut forward to the prostomium and back as far as the clitellum. Using forceps and a dissecting needle, open the cut wide and free the body wall from the septa. Note how the gut is held in place by the septa. Make sure that the worm is fully extended, and pin the animal to the pan through the body wall so that the gut and associated structures are clearly exposed. Cover with water.

▶ By means of the accompanying diagram (Figure 6.4**B**), identify the muscular **pharynx,** which is the principal organ of ingestion. This narrows to form the **esophagus,** which is covered by the five pseudohearts of the circulatory system. In segment 10 is a pair of small diverticula, the **esophageal glands.** In segments 11 and 12, similar pairs of esophageal diverticula contain calcium carbonate particles, giving them a whitish appearance. These **calciferous glands** are excretory organs and probably function in controlling the level of calcium ions in the blood. Posterior to the esophagus is the thin-walled **crop** (segments 14–16) and the thick-walled **gizzard** (segments 17–19). The remainder of the gut is the **stomach-intestine,** which is covered by yellowish **chloragogue** cells (site of glycogen synthesis, deamination of proteins, formation of ammonia and urea). Carefully slit open the gut without damaging any blood vessels and observe the condition of the facial tissue. Is there any evidence of localized compaction and digestion of the "food"? Note the modifications of the digestive tract for herbivorous habits, the presence of a pharyngeal pump, a storage stomach, a grinding gizzard, and calciferous glands for calcium ion regulation. Set this worm aside for future use.

▶ An entirely parallel situation is found in the burrowing marine polychaete *Arenicola* (Figure 6.5**A**). If specimens of this worm are available, note that the external form of its body is quite similar to *Lumbricus*. The number and size of the parapodia are reduced. The pharynx is protrusible as a proboscis and is used for ingesting mud. Posterior to this is the esophagus with its esophageal glands, and the remainder of the gut is the intestine surrounded by chloragogue tissue.

b. Tentacle Feeders

Among sedentary tubicolous polychaetes, the prostomium may be modified to form clusters of elongate threadlike tentacles. These tentacles are hollow extensions of the coelom, equipped with muscles, mucus-secreting cells, and cilia. The organisms creep over the sea bottom collecting bits and pieces of organic material.

▶ *Amphitrite* (Figure 6.5**B**) and *Terebella* are examples of species that employ this feeding mechanism. If living animals are available, observe how the movements of the tentacles appear to be uncoordinated. However, if you concentrate on an individual tentacle, you can see how it creeps along the substrate by ciliary action; observe how particles are caught in mucus and moved toward the mouth by ciliary tracts in a gutter of the tentacle or by muscular movements. The particles are then sorted by the peristomial lips and either discarded or ingested. For further details refer to the book by Dales (1963).

c. Filter Feeders

i. TENTACULAR METHOD.   The marine feather duster worms such as *Sabella, Eudistylia, Hydroides,* and *Spirorbis,* are permanent residents of tubes, which they secrete around themselves. The anterior end of the body is modified into a crown of ciliated **tentacles;** some species (the serpulids) have an **opercular cap,** which protects the animal when it is retracted in its tube (Figures 6.5**D** and **E**).

▶ Obtain a living specimen of a sabellid (*Sabella, Eudistylia*) or a serpulid (*Hydroides, Spirorbis)* and attempt to remove the worm from its tube without damage to the animal. How does the animal retain its position within the tube? Transfer the worm to a piece of Tygon or glass tubing of slightly larger bore than the diameter of the worm. If you are unable to remove the worm from its tube without damage, merely transfer the animal in its tube to a small finger bowl filled with seawater and allow it to relax. Note the manner in which the tentacular crown fans out. Without disturbing the animal, add a few drops of Congo-red-stained yeast suspension to the water in the vicinity of the tentacular crown and observe the movement of particles. Are the particles drawn in from below the mouth or from above? Is particle selection accomplished by ciliary mechanisms? How is the feeding current produced? For further details of feeding in these worms, refer to the paper by Nicol (1931).

ii. MUCUS BAG.   The tube-dwelling marine polychaete *Chaetopterus* is a most bizarre-looking annelid. The animal lives in a U-shaped parchment-like tube, buried in the sand, with only the chimney-like ends of the tube projecting from the substrate. The worm is imprisoned in its tube and depends on the production of a water current to obtain food.

▶   Carefully remove a *Chaetopterus* worm from its tube by cutting along the tube with a pair of scissors. The worm will usually move toward one end as you cut; when free of the tube, the worm can easily be slipped into a glass U-tube or piece of Tygon tubing. Place the tube with the enclosed worm in a larger vessel containing seawater, and make sure the entire tube is below the surface of the water. Observe the beating of the fanlike parapodia of segments 14, 15, and 16 which create the current (Figure 6.5C). Place a few drops of Congo-red-stained yeast suspension at one end of the tube and determine whether water enters anteriorly or posteriorly. Segment 12 bears a pair of winglike notopodia, which hold out a mucus bag spun by the worm. The bag acts as a sieve; small particles are trapped in it and periodically the notopodia of the twelfth segment release the open end of the bag; it is then rolled up in the cupule anterior to the fans and transferred to the mouth by a ciliated middorsal groove. Observe that all the notopodia of the middle section of the worm are modified and adapted for spinning the bag, maintaining the water current, and handling the bag of particles after collection.

d. Carnivores

One logically expects those animals that feed on flesh to be armed with teeth or clawing devices, and annelids that feed in this way do not disappoint us.

▶   Place a preserved specimen of *Nereis* in a wax dissecting pan with the dorsal surface up. Anchor the animal to the pan by placing a pin through the prostomium and, with a sharp pair of scissors, make a dorsal longitudinal incision. Continue the incision anteriorly and posteriorly, carefully tease the body wall from the septa, and pin out the body wall (Figure 6.2**D**). Cover the entire worm with tap water so that it does not dry out.

▶   Just posterior to the mouth is the muscular eversible **pharynx (proboscis),** containing a pair of **jaws.** The pharynx leads to a short **esophagus** into which open two **esophageal glands.** Posterior to the esophagus is the **stomach-intestine.** The **rectum** is the most posterior portion of the stomach-intestine. Contrast the organization of the gut of *Nereis* with that of *Lumbricus.*

▶   When protruded, the proboscis of *Nereis* shows two dark-colored chitinous jaws and a number of small denticles (Figure 6.2**A**). To observe the protrusion of the proboscis in living worms, introduce a specimen into a small glass tube and place in a bowl of seawater; allow the animal to relax and then offer it a small piece of clam or mussel. How is the protrusion of the pharynx accomplished? How is the proboscis retracted? What spreads the teeth apart? What brings them together again? What is the position of the pharyngeal jaws when the proboscis is retracted?

▶   A more dramatic demonstration of pharyngeal protrusion may be seen in the predaceous polychaete *Glycera*. This worm, often called the four-jawed worm, is commonly used for fishing bait. The anterior end of the worm is relatively nondescript and bears only two pairs of very small prostomial tentacles. If living specimens are offered meat or disturbed by handling, the pharynx may be everted; as in *Nereis* it is armed with jaws. How are protrusion and retraction accomplished?

Leeches provide good examples of predaceous carnivores. *Haemopis* has blunt teeth and food organisms such as earthworms and slugs are devoured whole. *Hirudo* feeds principally on blood, which it sucks from living animals.

▶ Obtain a preserved specimen of *Hirudo* (Figure 6.3A) and observe the simple anterior end. Open the worm by making a dorsal incision in a manner similar to that described for *Lumbricus* and *Nereis*. Stretch the leech out and pin it down in a dish under saline. Carefully separate the skin from the underlying botryoidal tissue with a needle and fine scalpel; observe your progress under a binocular microscope. Just within the mouth cavity locate three razor-like **teeth,** which are used for piercing the skin of the prey. Carefully remove the botryoidal tissue covering the gut. Blood from the prey is pumped into the large thin-walled **crop** by the muscular pharynx. Muscle strands radiate from the pharynx to the body wall; among these muscle strands are located numbers of diffuse **salivary glands** that produce an anticoagulant **hirudin.** Why is hirudin required by leeches? From the pharynx, a short **esophagus** leads into the **crop,** which consists of 11 pairs of expandable **caeca.** What is their function? Why is such extensive development not found in other annelids? The last pair of crop caeca turn back to the posterior end of the body. The small bilobed **stomach** leads into the thin-walled **intestine** and **rectum.** Attempt to locate the parts of the digestive system in your specimen, or in a prepared whole mount (Figure 6.3C). (Leech anatomy is often obscured by large amounts of botryoidal tissue and oblique muscles, and the organ systems are difficult to find. Much may be seen in a prepared stained whole mount). Compare the structure of the leech gut with that of *Lumbricus* and *Nereis*.

How does the structure of the gut differ between herbivores and carnivores? How do you correlate the modifications of the head with the habit of the organism?

6–5. COELOMIC FLUID, CIRCULATION, AND RESPIRATION

The coelomic cavity of annelids is fairly extensive and may be filled with coelomic fluid. One function of the coelomic fluid is as a component of the hydrostatic skeleton; it may take on additional functions in relation to increased size. As animals increase in size, the problems of distribution of gases, metabolites, and wastes multiply proportionally. The active movement of such materials in the body distributes them more efficiently than does diffusion alone. In the annelids the coelomic fluid, moved by the activity of the body wall musculature, may serve as a circulatory and respiratory fluid; in addition, there may be a true blood vascular system developed from the coelomic epithelium.

a. Circulatory System

▶ i. OLIGOCHAETA.   Oligochaetes such as *Lumbricus* have a fairly well-developed blood vascular system with main pulsating contractile vessels, and capillary networks supplying the gut, body wall, and other organs. Recall the location of the dorsal and ventral blood vessels or examine a cross section of *Lumbricus* and locate them now (Figure 6.1**B**). Blood is moved anteriorly in the contractile dorsal blood vessel; in the ventral vessel the blood flows posteriorly. Five pairs of **pseudohearts** connect the dorsal and ventral blood vessels, contain valves and contractile tissue, and act as accessory pumps.

▶ Return to your dissected specimen of *Lumbricus* or open the body cavity of a freshly chloretone-anesthetized earthworm as previously directed (see Section 6–4a) and observe the contractile vessels (Figure 6.4**B**). Obtain a small drop of coelomic fluid and examine under the microscope. Are cells present? Try to obtain a drop of blood from the hearts or the dorsal blood vessel with a hypodermic syringe fitted with a small needle (No. 27) or by cutting the vessels with a sharp razor blade. The red color of the blood is due to hemoglobin; this is not found in the cells, but in solution. Save some of this fluid for further analysis of its absorption spectrum. (See the following exercise.)

▶ ii. POLYCHAETA.   Cross sections of *Nereis* demonstrate a similar arrangement of the dorsal and ventral blood vessels, and there are lateral intestinal vessels in every segment except the first few. If you have not examined a cross section and located the components of the circulatory system, do so now (Figure 6.2**C**).

As in *Lumbricus* the dorsal blood vessel is the main propulsive element, and hemoglobin is dissolved in the circulatory fluids. In *Lumbricus,* which has no specialized respiratory organs, the transfer of gases occurs over the moist body surface, but in *Nereis* the parapodia serve as respiratory organs (gills). Blood flows from the ventral intestinal vessel to a capillary network in the parapodium and returns from the parapodium via dorsal parapodial vessels to flow into the dorsal blood vessel (Figure 6.2**E**). Blood in the parapodium is moved more by ebb and flow than by a true directed circulation, but even this system affords some advantage in respiration. In effect, the parapodium is the simplest sort of gill. Return to your opened specimen of *Nereis* (Section 6–4d) and in one segment, locate as many vessels of the blood system as you can.

▶ *Glycera* has no formed circulatory system, but the coelomic fluid is red and hemoglobin is contained in cells. Distribution of gases and nutrients is accomplished by body movements. Circulation of the coelomic fluid can be observed in a living specimen of *Glycera* under the dissecting microscope in the lateral projections from the body, called **branchiae** (gills). How is circulation in the branchiae accomplished? Obtain and examine a drop of coelomic fluid from a living *Glycera* by snipping off a small piece of the worm's posterior end and collecting the fluid. The red cells are erythrocytes and the larger, colorless cells are eggs.

▶ The gills of other polychaetes may be more arborescent than those of *Glycera*. Study whole mounts or preserved specimens of *Arenicola* (Figure 6.5**A**) and *Amphitrite* (Figure 6.5**B**) to see the form of these gills. In *Arenicola* the gills are associated with the notopodia and are located on segments 7–17, but in *Amphitrite* the gills are found on segments 4–6, are homologous with the dorsal cirri, and are hollow outgrowths of the body containing coelomic fluid. In living specimens of *Amphitrite,* the vessels will change in their intensity of red color as the blood ebbs and flows in the gills.

iii. HIRUDINEA.   The circulatory system of the leech is difficult to study, because most leeches have no true vascular system; instead they have a series of sinuses and scattered vessels. Details of this system in the leech will not be considered further.

b. Respiratory Pigments

To compare the respiratory pigments of annelids with those of other animals (e.g., humans), the absorption characteristics of the pigments will be examined. Solutions of the pigments must be prepared and then examined with an instrument that gives light of different wavelengths (colors) and indicates how much light is absorbed by the pigment under study.

▶   Take the blood of *Lumbricus* obtained previously and dilute it with some distilled water until you obtain a pink to light red solution. Do the same with a few drops of human blood (obtained by pricking your finger with a sterile lancet). To prepare chlorocruorin, a green iron-containing pigment found in the blood of some polychaetes, grind up some whole specimens of *Eudistylia* or *Hydroides* in a small volume (5 to 10 ml) of distilled water. Centrifuge all extracts to remove particles and transfer the clarified supernatants to separate labeled colorimeter tubes.

▶   You will be provided with a colorimeter or other similar apparatus. Turn the colorimeter on and allow the instrument to warm up. With no tube in the instrument, set the scale reading (percent transmittance) to zero; place a tube of distilled water (called the **blank**) in the instrument, turn the wavelength control to 340 nm and set the percent transmittance to 100% (absorbance = 0). Record the absorbance of the solution for 10 nm intervals between 340 and 700 nm for all pigments. Remember to recalibrate the instrument using the blank for each new wavelength measurement; the absorbance of the blank will vary with wavelength and compensation for this must be made. Plot percent transmittance (or absorbance) versus wavelength on a graph for each pigment.

Does earthworm hemoglobin differ from that of a human? Is chlorocruorin different from hemoglobin? Why? How do these pigments differ chemically and physiologically? How is respiration accomplished in annelids without respiratory pigments, e.g., *Chaetopterus*? Is there any correlation between annelid size, annelid habitat, and the presence or absence of respiratory pigment? What would be the disadvantage to the earthworm of having a large respiratory surface such as a parapodium? Why do you think *Ascaris* attains a maximum size of 30 cm and nematodes rarely reach a meter, whereas the largest annelids are 3½ m long?

6–6. EXCRETION

Increase in body size not only requires organs specialized to function in distribution, but organs for excretion. The excretory organ found in most annelids is the **nephridium** (Figure 6.6).

In *Lumbricus* almost every segment contains a pair of nephridia. Each one consists of a ciliated funnel or **nephrostome,** which communicates with the coelomic cavity and is situated posteriorly in the segment. A duct leads from the nephrostome, passes through the intersegmental septum posterior to the nephrostome, and continues as a fine ciliated coiled tubule surrounded by blood capillaries. The tubule leads to a bladder that discharges to the exterior via a **nephridiopore,** which opens near the ventral pair of setae. The cilia of the funnel and tubule beat and draw fluid from the coelom; wastes and other materials are exchanged with the blood in the capillaries surrounding the tubule as the fluid moves along it. By filtration, reabsorption, and tubular secretion, the nephridia perform their excretory function and maintain salt and water balance in the body.

▶   Carefully dissect a nephridium from an opened, chloretone-anesthetized earthworm. Do your dissection in saline for best results. Examine the nephridium in saline in a watch glass under the dissecting and compound microscopes, observing the activity of the nephrostome and the secretory tubule (Figure 6.6). If you have not already observed nephridiopores, they can be seen by careful examination of the ventrolateral exterior of an earthworm.

6–7. REPRODUCTION

i. OLIGOCHAETA.   Oligochaetes have permanent gonads and are hermaphroditic; however, cross-fertilization is the rule.

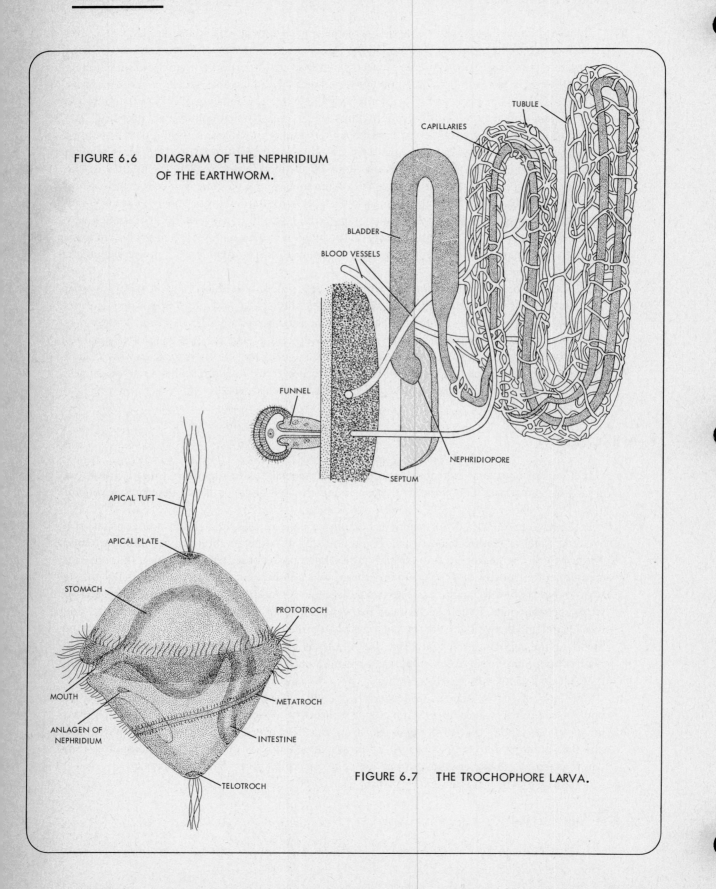

FIGURE 6.6   DIAGRAM OF THE NEPHRIDIUM
OF THE EARTHWORM.

TUBULE

CAPILLARIES

BLADDER

BLOOD VESSELS

FUNNEL

NEPHRIDIOPORE

SEPTUM

APICAL TUFT

APICAL PLATE

STOMACH

PROTOTROCH

MOUTH

METATROCH

ANLAGEN OF
NEPHRIDIUM

INTESTINE

TELOTROCH

FIGURE 6.7   THE TROCHOPHORE LARVA.

▶ Return to the specimen of the earthworm that you dissected to reveal the gut or open a specimen as described in Section 6–4a; carefully remove the gut. Ventrally in segments 9–13 will be found the trilobed **seminal vesicles** in which the sperm develop (Figure 6.4**B**). Tear off one of the lobes and make a smear on a microscope slide to observe the sperm morulae. If you remove the dorsal surface of the seminal vesicles and wash out the milky contents, the **sperm funnels** can be seen in segments 10 and 11, and in mature specimens, the small lobed **testes.** From the sperm funnels run the **vasa efferentia,** which join to form a **vas deferens** that opens to the outside in segment 15.

▶ The female reproductive organs may be obscured by the seminal vesicles. Once these are removed, the two tiny **ovaries** may be seen hanging from the anterior septum of segment 13. The **ovarian funnel** projects forward from the posterior septum in the same segment and leads to the short **oviduct** in segment 14. Remove the ovary and make a wet mount. Oocytes may be seen. Segments 9 and 10 contain paired yellow sperm receptacles **(spermathecae),** which store sperm from another worm. How is sperm transfer accomplished?

ii. HIRUDINEA.   Like oligochaetes, hirudineans have permanent gonads and are hermaphroditic, and cross-fertilization is the rule.

▶ Return to the specimen of the leech that you dissected to reveal the gut or open a specimen as described in Section 6–4d. Remove the gut completely, and attempt to identify the components of the reproductive system (Figure 6.3**D**). The small pair of coiled **ovaries** is found in segment 11. Short, paired **oviducts** run posteriorly to unite as a common duct leading into a median **vagina.** A bulbous **albumen gland** is present as an enlargement of the common duct, and the vagina opens to the exterior in segment 12. The ten pairs of **testes** are found in segments 12–21, and running from each testis is a coiled **vas deferens,** which leads to an anterior pair of **seminal vesicles.** Each seminal vesicle sends a branch **(ejaculatory duct)** into the muscular **penis,** which is enlarged basally in the tenth segment as a **prostate gland.** The **spermatozoa** are deposited in bundles or **spermatophores** on the body of another leech and penetrate the skin, making their way to the ovaries where fertilization occurs. The eggs pass through the vagina to the exterior, are laid in cocoons produced by the clitellum, and hatch as small leeches.

iii. POLYCHAETA.   Polychaetes do not have permanent gonads, and the coelomic epithelium buds off eggs and sperm, the sexes in most cases being separate.

▶ By examining the coelomic fluids of various living polychaetes available in the laboratory, it may be possible to observe living eggs, sperm, and sperm morulae. In many polychaetes the nephridia function as gonoducts and fertilization occurs in the sea.

In *Nereis,* the breeding season is mainly in spring and summer. At this time, the worms leave the sand or mud tubes in which they live and swim freely. They differentiate into a breeding form called the **heteronereis** (Figure 6.2**F**) with two dintinctly different body regions: an **atoke,** which is normal in structure, and a posterior reproductive **epitoke,** which bears the gametes. The epitoke has enlarged, leaflike parapodia and setae, an adaptation to swimming (Figure 6.2**G**). Release of the gametes takes place by rupture of the epitoke wall. Fertilization and development take place in the water.

▶ If available, examine a heteronereis and notice the large eyes. Compare the parapodia and setae of the epitoke with those of the atoke and of a nonreproductive specimen.

Why are copulation and internal fertilization the rule in oligochaetes and hirudineans and not in polychaetes?

▶ The larva characteristic of annelids is the **trochophore.** Study slides of this larva and look at Figure 6.7. Students interested in the development of annelids should consult the reference by Costello et al. (1957) for further details.

6–8. REGENERATION

Because of their simple anatomy and remarkable regenerative capacities, annelids provide excellent material for the analysis of mechanisms involved in regeneration and factors controlling growth and differentiation. There is, however, considerable variation in regenerative ability among the various species of annelids. For example, *Lumbricus* does not regenerate amputated segments, whereas *Eisenia* and *Nereis diversicolor* have high regenerative capacities. The fundamental problem to be explored in this section is the question of what determines the length of an annelid. Students undertaking this work should consult the references by Moment (1953) and Herlant-Meewis (1964).

▶ Obtain six specimens of the manure worm *Eisenia foetida* and lightly anesthetize the worms in 0.2% chloretone. Count and record the number of segments in each worm. Using a sharp razor blade, remove 5, 10, or 15 segments from the posterior end of each worm. Rinse operated worms in tapwater and return them to their containers, taking care to keep the groups separate, i.e., those with 5 segments amputated in one container, those with 10 removed in another, and those with 15 removed in another. In order for regeneration to occur successfully, the worms must be kept at 25°C. It will take approximately 2 weeks for regeneration to occur.

▶ Upon completion of the regenerative process, record and graph the number of segments amputated versus the number of segments regenerated. What is the relationship? Can you suggest experiments to test the reason why regeneration proceeds in the manner observed by you? Repeat these experiments using anterior-end amputation and follow regeneration. What happens? Why? What is the source of regenerative material?

6–9. NERVOUS SYSTEM

▶ Many of the functional aspects of the nervous system are dealt with in Section 6–3. If you have not already performed the studies in Section 6–3, and if time permits, do so after you have completed this section. You may find it helpful to read Section 6–3 through before you begin the dissection of the nervous system of *Lumbricus*.

▶ Return to your opened specimen of *Lumbricus* from which you removed the gut and reproductive structures, or open another specimen and remove these structures now (see Sections 6–4a and 6–7 for directions). Proceed to dissect a little deeper to locate the **supra-** and **subesophageal ganglia,** the **ventral nerve cord,** and the **segmental ganglia** (Figure 6.4C). The gross anatomy of the nervous system in *Nereis* and *Hirudo* is quite similar and need not be dissected.

6–10. BEHAVIOR

Although most oligochaetes and hirudineans do not have conspicuous or highly developed sense organs, they can respond to light, chemicals, touch, and water currents. Unfortunately, most of these responses are difficult to observe in the short time available in the laboratory. Conditioning of earthworms in the classical manner has been accomplished by some workers, but, as one might imagine from the anatomy of the nervous system, such studies require much patience and time. Students interested in this aspect of annelid biology should consult the papers by Krivanek (1956) and Ratner and Miller (1959).

Polychaetes generally have better developed sense organs, but behavioral studies require considerable attention to detail, and observations must be made over extended periods of time. The following problems are suggested for those individuals who have considerable time available or for use as reading material.

a. Activity Rhythms

Continuous observations of worm behavior are usually tedious and time-consuming because these animals perform their activities slowly and almost imperceptibly. In order to overcome some of the inherent difficulties and at the same time produce a satisfactory, reproducible, and permanent record of the behavior of annelids, Wells has devised a technique whereby the polychaete worm records its own activities, and the observer need not rely on his eye alone. Remember, however, that these experiments are extremely time-consuming and may, because of technical difficulties, yield nothing.

▶ i. ARENICOLA. Introduce a specimen of *Arenicola* into an apparatus that consists of two 35 cm square glass plates, held about 1 cm apart by pieces of modeling clay. One piece of clay forms a lopsided U and the other provides a semipartition inside the U (Figure 6.8A). The rest of the space between the plates is nearly filled with sand, and the whole apparatus is immersed in a tank of seawater.

▶ A worm placed on the surface of the sand soon burrows, and if the apparatus and worm are of the proper relative dimensions, the worm will position itself as shown in Figure 6.8A. After a time fecal castings will appear at one end of the burrow (tail shaft). On this side the exchange of water with the outer tank takes place via a glass capillary, whereas on the other (head) side water can flow freely from the tank into the apparatus and vice versa. A float placed above the tail shaft communicates fluctuations of the water level to a lever which in turn is connected to a stylus. The tip of the stylus communicates with a smoked kymograph drum, which should be set to turn very slowly (e.g., one revolution per day). The activities of the worm, reflected by changes in the water level above the tail shaft, are therefore scratched onto the smoked drum. Further details may be found in the references by Wells (1949).

▶ ii. AMPHITRITE. Tube-dwelling detritus feeders such as *Amphitrite* can be persuaded to accept glass U-tubes of appropriate bore, and will live in them for months when placed at the bottom of a seawater aquarium. After the worm is introduced into the U-tube, the rear end of the tube is sealed with a one-hole rubber stopper and glass tubing (Figure 6.8B). The glass tube runs to a flotation vessel and the latter communicates with the seawater aquarium via a capillary. The activity of the worm is recorded on a kymograph. Further details may be found in the reference by Dales (1955).

▶ iii. CHAETOPTERUS. This worm ordinarily lives in a parchment-like tube; it can be made to record its activities by cutting off the narrow ends of the tube and tying the remainder onto a pair of glass tubes, sleeved with rubber tubing (Figure 6.8C). One arm of the U-shaped tube runs to a flotation vessel and the other arm to a second cylinder. Further details concerning the construction of this apparatus may be found in the reference by Wells and Dales (1951).

▶ iv. SABELLA. The feather duster worms also live in parchment-like tubes; they can be made to record their activities by attaching the natural tube to a glass tube and then running this to a flotation vessel. As the worm irrigates the tube, feeds, or ejects waste material, the water current changes its pattern. Even headless *Sabella* carry out the characteristic patterns of activity.

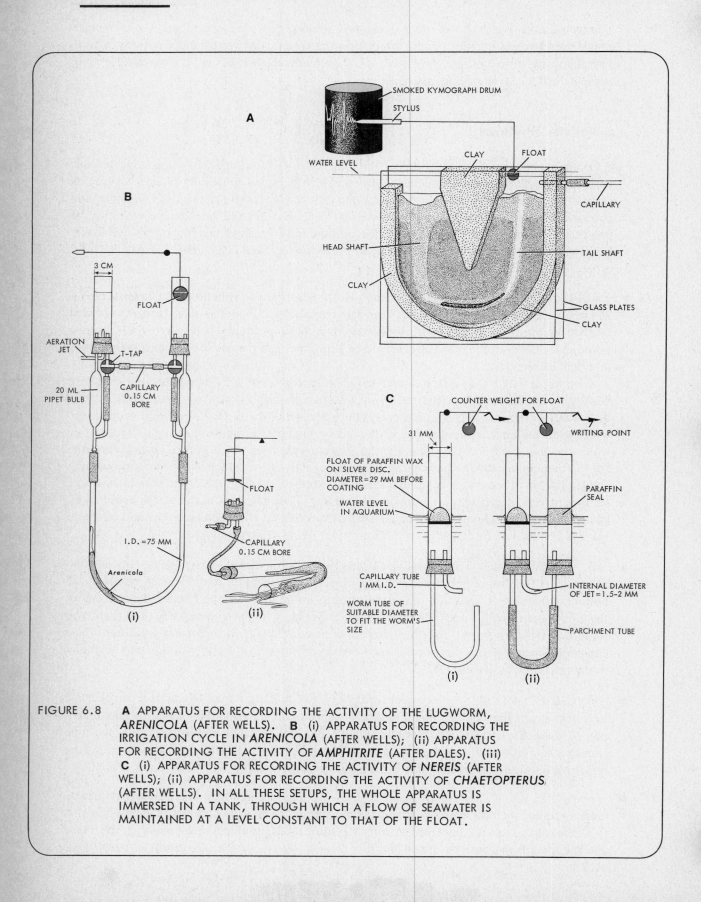

FIGURE 6.8   **A** APPARATUS FOR RECORDING THE ACTIVITY OF THE LUGWORM,
ARENICOLA (AFTER WELLS).   **B** (i) APPARATUS FOR RECORDING THE
IRRIGATION CYCLE IN *ARENICOLA* (AFTER WELLS); (ii) APPARATUS
FOR RECORDING THE ACTIVITY OF *AMPHITRITE* (AFTER DALES).   (iii)
C (i) APPARATUS FOR RECORDING THE ACTIVITY OF *NEREIS* (AFTER
WELLS); (ii) APPARATUS FOR RECORDING THE ACTIVITY OF *CHAETOPTERUS*
(AFTER WELLS).   IN ALL THESE SETUPS, THE WHOLE APPARATUS IS
IMMERSED IN A TANK, THROUGH WHICH A FLOW OF SEAWATER IS
MAINTAINED AT A LEVEL CONSTANT TO THAT OF THE FLOAT.

b. Tube Building

Examination of the construction of tubes by polychaetes is fascinating, for it shows the degree of coordination and selectivity that can be exercised by these worms.

▶ i. DIOPATRA CUPREA.   The tube-building activity of *Diopatra cuprea* can be studied in a period of less than 24 hours. Remove worms from their tubes and place the worms in large finger bowls containing bits of sand and a variety of materials, e.g., paper, glass, broken shells, pebbles, and string. Which materials are incorporated into the tube? How is selection effected?

▶ ii. PECTINARIA.   Another remarkable mason is the tube dweller *Pectinaria*. This worm is commonly called the ice-cream cone worm because it builds a conical tube of sand grains. Obtain a large number of worms in their tubes and remove 0.5 cm of the anterior end of the tubes. Supply worms, placed in individual containers, with sand grains of various dimensions (procured by using sieves of various mesh size). What sizes of sand grains are not used in building? Can the worms repair portions removed from the posterior end of the tube? Notice the soft posterior extension of the tube, often lost at collection, and renewable. Can the worm find its own tube after it is removed? Can you exchange the tube of one worm with that of another?

References

Barnes, R. D. *Invertebrate Zoology,* 3rd ed. Philadelphia: W. B. Saunders Co., 1974.

Berrill, N. J. Regeneration and budding in worms, *Biol. Rev.* **27**:401–38, 1952.

Brown, F. A., Jr. (ed.). *Selected Invertebrate Types.* New York: John Wiley & Sons, 1950.

Bullock, T. H. Functional organization of the giant fiber system of *Lumbricus, J. Neurophysiol.* **8**:55–71, 1945.

Chapman, G. The hydraulic skeleton in invertebrates, *Biol. Rev.* **33**:338–71, 1958.

Clark, R. B. The origin and formation of the heteronereis, *Biol. Rev.* **36**:199, 1961.

Clark, R. B. Structure and function of polychaete septa, *Proc. Zool. Soc. London.* **138**:543–78, 1962.

Costello, D. P., M. E. Davidson, A. Eggers, M. H. Fox, and C. Henley. *Methods of Handling Marine Eggs and Embryos.* Woods Hole, Mass.: The Marine Biological Laboratory, 1957.

Dales, R. P. *Annelids.* London: Hutchinson and Co., 1963.

Dales, R. P. Feeding and digestion in terebellid polychaetes, *J. Mar. Biol. Ass. U.K.* **34**:55–79, 1955.

Gray, J. Studies in animal locomotion, VIII. The kinetics of locomotion in *Nereis diversicolor, J. Exp. Biol.* **16**:9–17, 1939.

Gray, J., and H. W. Lissman. Studies in animal locomotion, VII. Locomotory reflexes in the earthworm, *J. Exp. Biol.* **15**:408–30, 1938.

Gray, J., H. W. Lissman, and H. J. Pumphrey. The mechanism of locomotion in the leech (*Hirudo medicinalis*), *J. Exp. Biol.* **15**:408–30, 1938.

Herlant-Meewis, H. Regeneration in annelids, *Advan. Morphol.* **4**:155–215, 1964.

Krivanek, J. Habit formation in the earthworm, *Lumbricus terrestris, Physiol. Zool.* **29**:241–50, 1956.

MacGinitie, G. E. The method of feeding of *Chaetopterus, Biol. Bull.* **77**:115–18, 1939.

Mann, H. K. *Leeches (Hirudinea). Their Structure, Physiology, Ecology and Embryology.* Oxford: Pergamon Press, 1962.

Meglitsch, P. *Invertebrate Zoology*. New York: Oxford University Press, 1967.

Mettam, C. Segmental musculature and parapodial movement of *Nereis diversicolor* and *Nepthys homberghi* (Annelida: Polychaeta), *J. Zool. London* **153**:245–75, 1967.

Moment, G. B. A theory of growth limitation, *Amer. Natur.* **87**:139–53, 1953.

Nicholls, J. G., and D. Van Essen. The nervous system of the leech, *Sci. Amer.* **230**(1):39–48, 1974.

Nicol, J. A. C. The feeding mechanism, formation of the tube and physiology of digestion in *Sabella pavonia, Trans. Roy. Soc. Edinburgh,* **56**:537–98, 1931.

Prosser, C. L. *Comparative Animal Physiology,* 3rd ed. Philadelphia: W. B. Saunders Co., 1973.

Ramsay, J. A. *Physiological Approach to the Lower Animals*. London: Cambridge University Press, 1952.

Ratner, S. C., and K. R. Miller. Effect of spacing and ganglionic removal on conditioning in earthworms, *J. Comp. Physiol. Psychol.* **52**:102–105, 1959.

Wells, G. P. Respiratory movements of *Arenicola marina L.:* Intermittent irrigation of the tube and intermittent aerial respiration, *J. Mar. Biol. Ass. U.K.* **28**:447–64, 1949.

Wells, G. P. The behavior of *Arenicola marina* L. in sand and the role of spontaneous activity cycles, *J. Mar. Biol. Ass. U.K.* **28**:465–77, 1949.

Wells, G. P., and R. P. Dales. Spontaneous activity patterns in animal behavior: The irrigation of the burrow in the polychaetes *Chaetopterus variopedatus* Renler and *Nereis diversicolor* O. F. Müller, *J. Mar. Biol. Ass. U.K.* **29**:661–80, 1951.

Welsh, J. H., R. I. Smith, and A. E. Kammer, *Laboratory Exercises in Invertebrate Physiology*. Minneapolis: Burgess Publishing Co., 1968.

Wilson, D. M. Nervous control of movement in annelids, *J. Exp. Biol.* **37**:46, 1969.

An Abbreviated Classification of the Phylum Annelida (After Various Authors)

Phylum **Annelida**

> *Class I* **Polychaeta**
>
> > Well segmented with setae that arise from special body projections called parapodia; spacious coelom; usually with distinct head and appendages; dioecious; primarily marine.
>
> *Subclass 1* **Polychaeta errantia,** *e.g., Nereis, Glycera.*
>
> *Subclass 2* **Polychaeta sedentaria,** *e.g., Chaetopterus, Arenicola, Pectinaria, Amphitrite, Sabella.*
>
> *Class II* **Oligochaeta**
>
> > Well segmented but with sparse setae, which are not found on the parapodia; distinct prostomium without appendages; spacious coelom; hermaphroditic; terrestrial and freshwater.
>
> *Order 1* **Plesiopora plesiothecata**
>
> > Spermathecae open in vicinity of male gonopore, e.g., *Tubifex.*
>
> *Order 2* **Plesiopora prosothecata**
>
> > Spermathecae open far anterior to male gonopore, e.g., *Enchytraeus.*

Order 3 **Prosopora**

Male gonopore opens on testes-bearing segment (first or last), e.g., *Lumbriculus*.

Order 4 **Opisthopora**

Male gonopore opens some distance posterior to the most posterior pair of testes, e.g., *Lumbricus, Eisenia*.

Class III **Hirudinea**

Body shortened, with a fixed number of segments; segments subdivided into a number of rings; no setae or parapodia; anterior and posterior ends bear suckers; coelomic space reduced; hermaphroditic; primarily freshwater.

Order 1 **Acanthobdellida**

Setae on first five anterior segments; no anterior sucker; coelom not obliterated

Order 2 **Rhyncobdellida**

Mouth a small pore; proboscis; no jaws; eyes, e.g., *Glossiphonia*.

Order 3 **Gnathobdellida**

Mouth large; no proboscis; jaws with serrate teeth; eyes, e.g., *Hirudo, Haemopis*.

Order 4 **Pharyngobdellida**

Pharynx with ridges but no teeth or jaws; gut without diverticula; carnivorous, not blood-sucking.

Equipment and Materials

Equipment

Dissecting and compound
  microscopes
Microscope slides
Cover slips
Razor blades
String or sewing thread
Paper towels
Soil
Glass tubing
Facial tissue
Dissection tray
Dissecting instruments
Tygon tubing
Colorimeter
Centrifuge and tubes
Pasteur pipets or droppers

Solutions and Chemicals

0.2% chloretone
Congo-red-stained yeast suspension

Preserved Material

Nereis
Lumbricus
Hirudo or *Haemopis*
Arenicola
Amphitrite

Prepared Slides

Nereis, whole mount (plastic)
Nereis, cross section of pharynx
Lumbricus, whole mount (plastic)
Lumbricus, cross section
Lumbricus, circulation in parapodium

Hirudo, whole mount (plastic)
Hirudo, cross section
Trochophores

Living Material*

Lumbricus (3,4,6,11)
Nereis (9,12)
Diopatra (7,12)
Sabella (7,12)
Hydroides or *Spirorbis* (7,12)

Eudistylia (9)
Arenicola (12)
Amphitrite or *Terebella* (9,12)
Chaetopterus (7,9)
Glycera (9,12)
Eisenia foetida (5)
Pectinarid (9,12)
Hirudo or *Macrobdella* (8)
Leeches (5,8,10,11)

*Numbers in parentheses identify sources listed in Appendix B.

ARTHROPODA

Contents

*Parts of these exercises should be started at the beginning of the laboratory period to allow time for completion.

Arthropods (*arthros,* joint; *podos,* foot; Gk.) are among the most numerous and successful of all animals. The phylum includes such familiar organisms as insects, crabs, lobsters, shrimp, barnacles, ticks, spiders, scorpions, centipedes, and millipedes. Representatives of this phylum exhibit adaptations necessary for life in water, on land, and in air.

Arthropods are **bilaterally symmetrical** and **metamerically segmented;** in their fundamental organization they are related to the annelid worms. Both groups show **segmental locomotor appendages,** for example, but only in the arthropods is the limb **jointed** and supported by a firm, usually chitinous **exoskeleton** or **cuticle.** The exoskeleton surrounds the body and serves for support and protection as well as in locomotion. Associated with the exoskeleton and jointed appendages is a well-developed musculature.

The arthropod nervous system consists of a ventral nerve cord having segmental ganglia and shows considerable cephalization. The circulatory system is of the open type, and in the adult the coelom is reduced and for the most part replaced by large blood sinuses **(hemocoel).** The excretory system is not nephridial and the respiratory organs are usually **gills** or **tracheae.** Cilia are generally absent from the body.

7–1. BODY ORGANIZATION

a. *Peripatus,* an Onychophoran

Onychophorans possess features characteristic of arthropods but they also show affinities with the annelids. They have been described by some authors as a "missing link" between the Arthropoda and the Annelida. They are sometimes included in the phylum Arthropoda and sometimes accorded separate phyletic status. (See Classification of Arthropoda, p. 220.)

▶ Examine a preserved specimen or a plastic mount of the subtropical onychophoran *Peripatus* (Figure 7.1**A**) and note its wormlike appearance. The soft chitinous **cuticle** is ridged and covered with rows of sensory **spines** set on **tubercles.** Can you detect obvious body segmentation? Turn the animal so that the ventral surface faces you and observe the anteriorly situated **preantennae,** behind which are paired **oral papillae.** The **mouth** is surrounded by ridged **lips** and contains paired toothed **jaws.** The trunk bears unsegmented lobelike **paired appendages.** Does each appendage bear claws? Although it is not apparent from your examination, *Peripatus* possesses ciliated **nephridia** and **gonoducts;** the external openings of the nephridia may be seen at the bases of the appendages, and the **gonopore** is subterminal. How many segments are there in your specimen? What is the basis for your calculation? Turn the animal so that the dorsal surface is visible. Can you see any sensory structures?

Internally, *Peripatus* has a system of **tracheal tubes,** which are respiratory in function, and a **hemocoelic body cavity.** The gut is a straight tube running from the subterminal mouth to the terminal anus. The sexes are separate and development is direct.

▶ Examine a prepared slide of a cross section through the body of *Peripatus* (Figure 7.1**B**) and compare its organization with that of the polychaete *Nereis* (Figure 6.2**C**). List the features that *Peripatus* has in common exclusively with annelids, exclusively with arthropods, and with both groups.

b. Body Regions—Segmentation and Tagmatosis

Arthropods are metamerically segmented animals; each segment or **metamere** is enclosed by an integumentary skeletal cylinder. The integument consists of hard cuticular plates (**sclerites**)

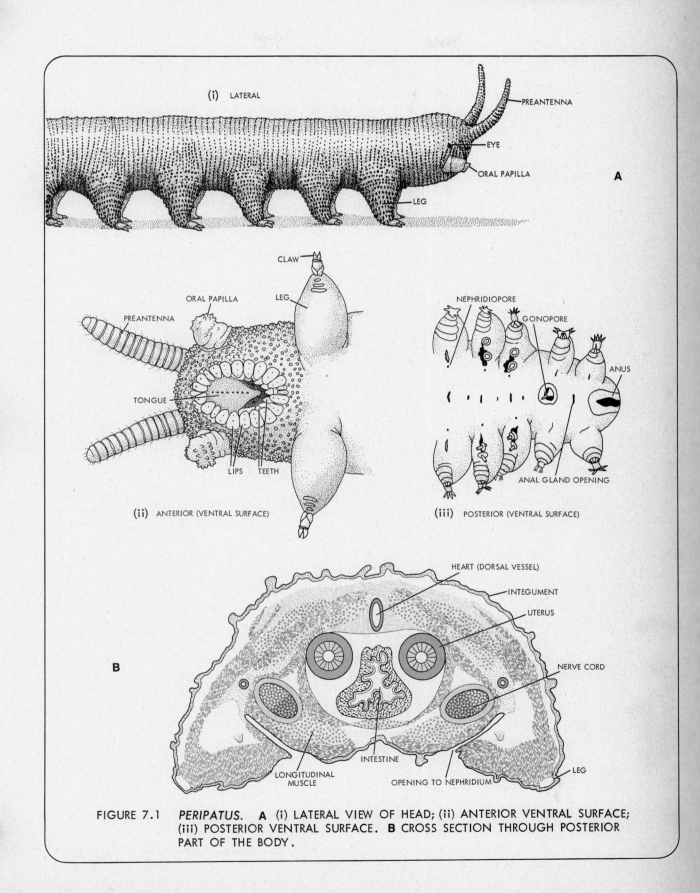

FIGURE 7.1   *PERIPATUS*.   **A** (i) LATERAL VIEW OF HEAD; (ii) ANTERIOR VENTRAL SURFACE;
(iii) POSTERIOR VENTRAL SURFACE.   **B** CROSS SECTION THROUGH POSTERIOR
PART OF THE BODY.

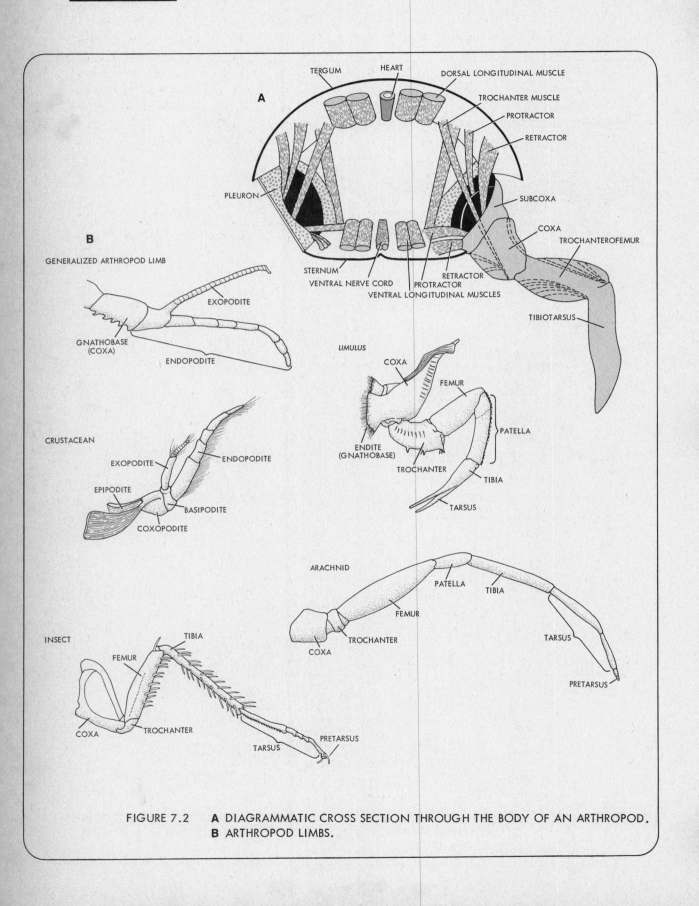

FIGURE 7.2 **A** DIAGRAMMATIC CROSS SECTION THROUGH THE BODY OF AN ARTHROPOD. **B** ARTHROPOD LIMBS.

and the underlying **epidermis.** The sclerites of each segment are basically arranged as a dorsal **tergum,** a ventral **sternum,** and two lateral **pleura** (Figure 7.2A). Metameric segmentation of the body permits regional specialization, and the arthropod body can be separated into regions consisting of fused segments called **tagmata.** The fusion and modification of originally separate and uniform segments makes it difficult to determine exactly how many segments comprise each tagma in the different arthropod classes. Because each segment is usually present in the embryo and generally bears a pair of appendages, some estimate of the number can be gained. However, in this connection it is important to remember that each tagma varies in size from

TABLE 7.1 Segments and Limbs of Arthropods
(Modified from Borradaile and Potts)

Somite	Body Region	Onychophora	Crustacea Malacostraca	Insecta	Chilopoda	Arachnida Scorpionida	Body Region	Somite
1[a]		Preantennae	Embryonic	Embryonic	Embryonic	Embryonic		1
2		Jaws	Antennules	Antennae	Antennae	Chelicerae		2
3	Head	Oral papillae	Antennae	Embryonic	Embryonic	Pedipalpi		3
4		1st pr. legs	Mandibles	Mandibles	Mandibles	1st pr. legs		4
5		2nd pr. legs	Maxillules	1st maxillae	1st maxillae	2nd pr. legs	Prosoma	5
6			Maxillae	Labium (fused 2nd maxillae)	2nd maxillae	3rd pr. legs		6
7		Many (17–43)	(1st) maxillipeds	1st pr. thoracic legs	Maxillipeds	4th pr. legs		7
8		somites, each	2nd thoracic limb	2nd pr. thoracic legs	1st pr. legs	Embryonic[b]		8
9		bearing a pair	3rd thoracic limb	3rd pr. thoracic legs	2nd pr. legs	Genital operc. ♀ ♂		9
10		of legs	4th thoracic limb	1st abd. seg.	3rd pr. legs	Pectines		10
11			5th thoracic limb	2nd abd. seg.	4th pr. legs	1st lung books		11
12			6th thoracic limb ♀	3rd abd. seg.	5th pr. legs	2nd lung books		12
13			7th thoracic limb	4th abd. seg.	6th pr. legs	3rd lung books	Opistho- soma	13
14			8th thoracic limb ♂	5th abd. seg.	7th pr. legs	4th lung books	(Meso- soma	14
15	Trunk (Thorax and abdomen)		1st. abd. limb	6th abd. seg.	8th pr. legs	No limbs	and	15
16			2nd abd. limb	7th abd. seg.	9th pr. legs	1st seg. metasoma	meta- soma)	16
17			3rd abd. limb	8th abd. seg. ♀	10th pr. legs	2nd seg. metasoma		17
18			4th abd. limb	9th abd. seg.	11th pr. legs	3rd seg. metasoma		18
19			5th abd. limb	10th abd. seg. (styles) ♂	12th pr. legs	4th seg. metasoma		19
20			6th abd. limb	11th abd. seg. (cerci)	13th pr. legs	5th seg. metasoma		20
21					14th pr. legs			...
22					15th pr. legs			
23					16th pr. legs			
24					17th pr. legs			
25					18th pr. legs			
26					19th pr. legs			
27					20th pr. legs			
28					21st pr. legs			
29					Genital limbs ♀ ♂			
...	Postsegmental region	Last pr. legs Embryonic	Telson	Embryonic	Telson	Telson	Postsegmental region	

[a]Eyes and frontal organs belong to a presegmental region.
[b]Chilaria in *Limulus.*
♀ ♂ indicates position of female and male openings, respectively.

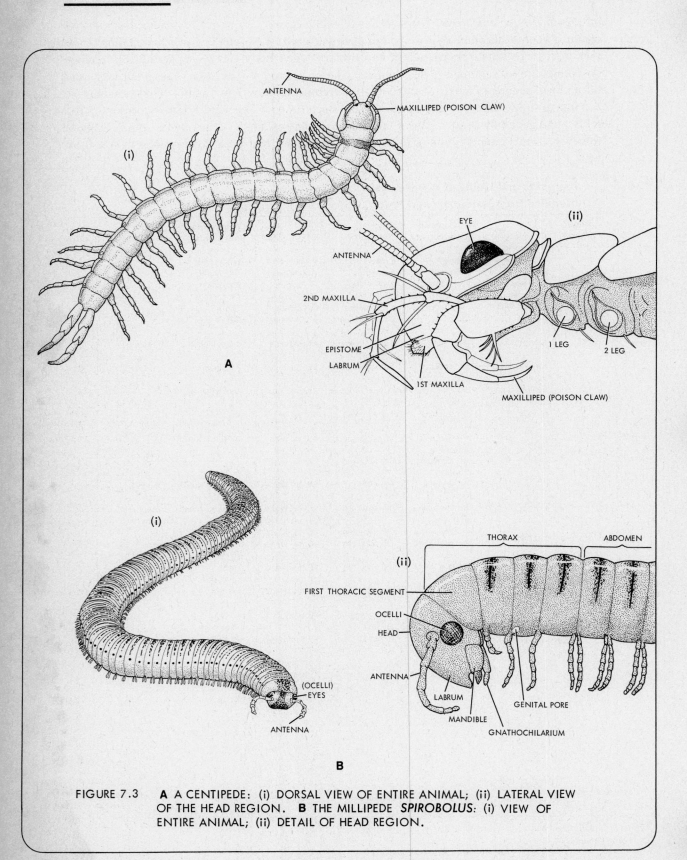

FIGURE 7.3   **A** A CENTIPEDE: (i) DORSAL VIEW OF ENTIRE ANIMAL; (ii) LATERAL VIEW
OF THE HEAD REGION.   **B** THE MILLIPEDE *SPIROBOLUS*: (i) VIEW OF
ENTIRE ANIMAL; (ii) DETAIL OF HEAD REGION.

group to group, as a consequence of which no tagma consists of the same number of embryonic segments (somites) in all arthropods. Therefore, tagmata may not be homologous between groups but only functionally analogous. Table 7.1 summarizes the arrangement of segments and appendages for some arthropods.

Two basic kinds of body organization are found among the arthropods. (1) In the Onychophora, Myriapoda, Insecta, and Crustacea the body is divided into an anterior tagma, the **head,** which consists of fused anterior segments, and the remainder of the body is the **trunk.** In insects the trunk is further divided into the **thorax** and **abdomen,** whereas in crustaceans the head and thorax are usually fused to form a **cephalothorax** and the remainder of the body is the abdomen. (2) In the Arachnida the most anterior tagma is the **prosoma,** which also consists of fused anterior segments. Although the prosoma is functionally comparable to the crustacean cephalothorax, the two body regions are not homologous. The hinder part of the arachnid body is called the **opisthosoma** and is functionally analogous but not homologous to the abdomen of other arthropods.

i. HEAD, PROSOMA.   The chief functions of the most anterior arthropod tagma are concerned with sensation, food selection and ingestion, and coordination.

▶   Return to the specimen of *Peripatus* (Figure 7.1A) and recall that the head bears three pairs of appendages: the preantennae, oral papillae, and jaws. How many segments comprise the onychophoran head? Examine specimens of myriapods (e.g., the centipede *Scolopendra,* Figure 7.3A, and the millipede *Spirobolus,* Figure 7.3B), an insect (e.g., the grasshopper *Romalea,* Figure 7.4A, or the cockroach *Periplaneta,* Figure 7.16A) and a crustacean (e.g., crab, lobster, or crayfish, Figure 7.5A). Identify the head and briefly examine its form. You will study the arrangement, anatomy, and functional role of the appendages later in this exercise. In insects, myriapods, and crustaceans the head is usually composed of six fused segments, and the first segment bears no appendages. Confirm the fact that the head is primarily concerned with sensory and alimentary functions by locating antennae, eyes, jaws, and mouth.

▶   Briefly examine the prosoma of arachnids (e.g., the horseshoe crab *Limulus,* Figure 7.6; a spider such as *Argiope,* Figure 7.7; and a scorpion, Figure 7.8). Note that the arachnid prosoma usually consists of eight segments and carries ambulatory limbs as well as appendages used in feeding. Locate sensory structures and the mouth. Are antennae present on the arachnid prosoma?

ii. THORAX.   The thorax may bear legs, wings, accessory feeding devices, and respiratory organs.

▶   Examine the thorax of your representative arthropods. In the myriapods the thorax is without distinctive character. In the insects it is divided from the head by a flexible neck and consists of three segments, the **prothorax, mesothorax,** and **metathorax.** In crustaceans and arachnids the thorax is fused to the head to form a **cephalothorax** or **prosoma.** In some crustaceans (e.g., the decapods) the head and thorax are enclosed in a cephalothoracic shield or **carapace.** The head is the region anterior to a fold of the carapace known as the **cephalothoracic groove.** The presence of an exoskeleton with such a structure is generally a diagnostic feature of crustaceans, but it is absent in copepods, isopods, amphipods, and anostracans.

iii. ABDOMEN, OPISTHOSOMA.   This body region may be concerned in locomotion, reproduction, and/or respiration. It may consist of fused segments in some arthropods, but very often the degree of fusion is less and segmentation is more obvious than in the head and thorax.

A

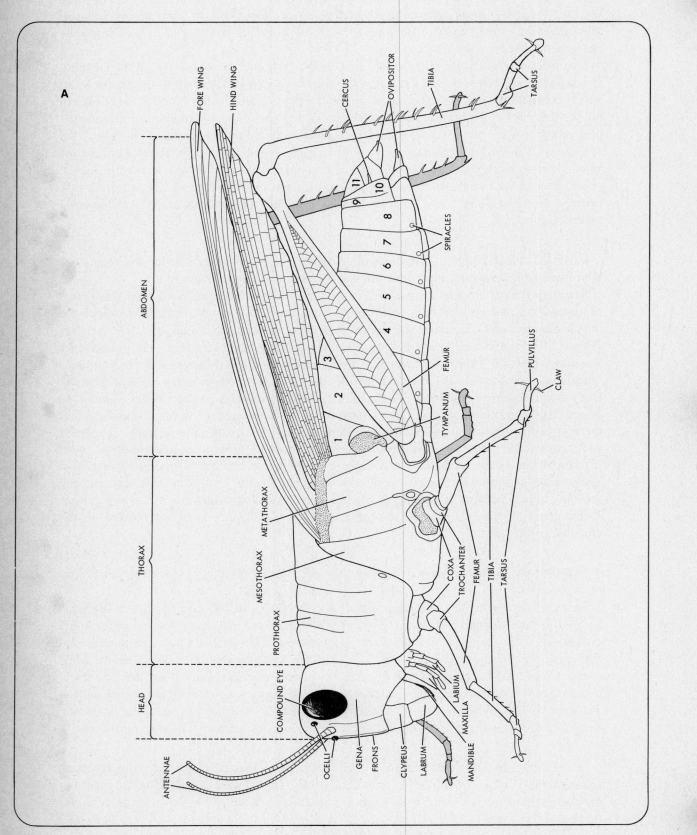

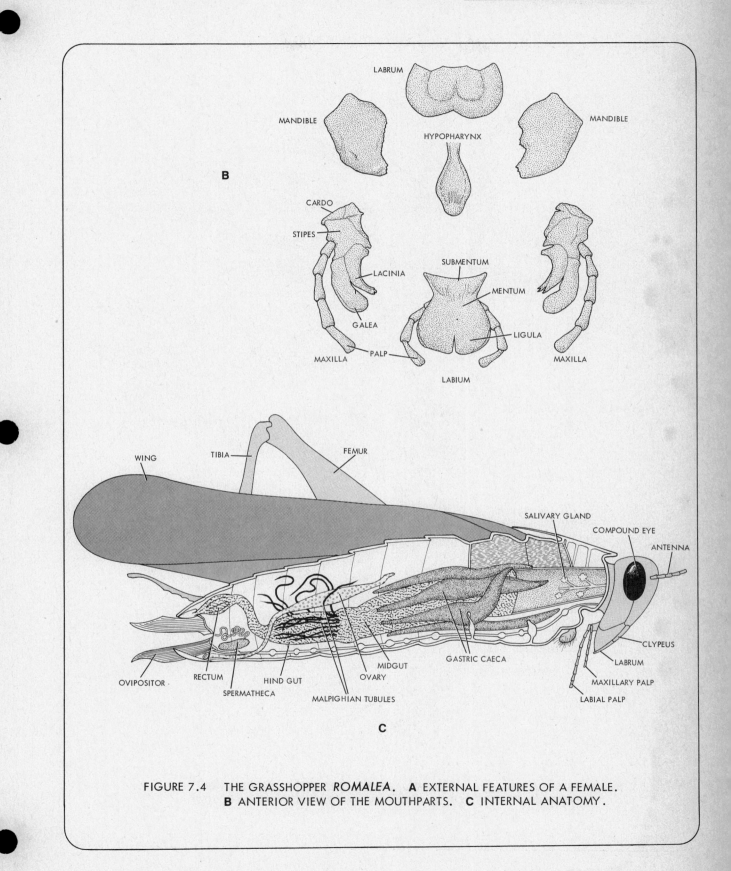

FIGURE 7.4   THE GRASSHOPPER *ROMALEA*. **A** EXTERNAL FEATURES OF A FEMALE.
B ANTERIOR VIEW OF THE MOUTHPARTS. **C** INTERNAL ANATOMY.

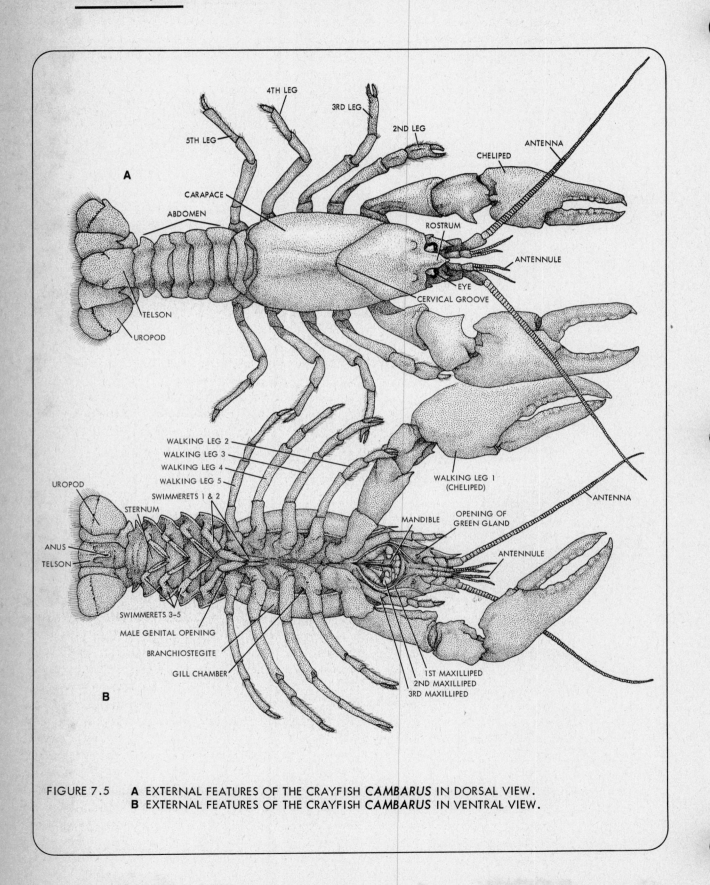

FIGURE 7.5   **A** EXTERNAL FEATURES OF THE CRAYFISH *CAMBARUS* IN DORSAL VIEW.
B EXTERNAL FEATURES OF THE CRAYFISH *CAMBARUS* IN VENTRAL VIEW.

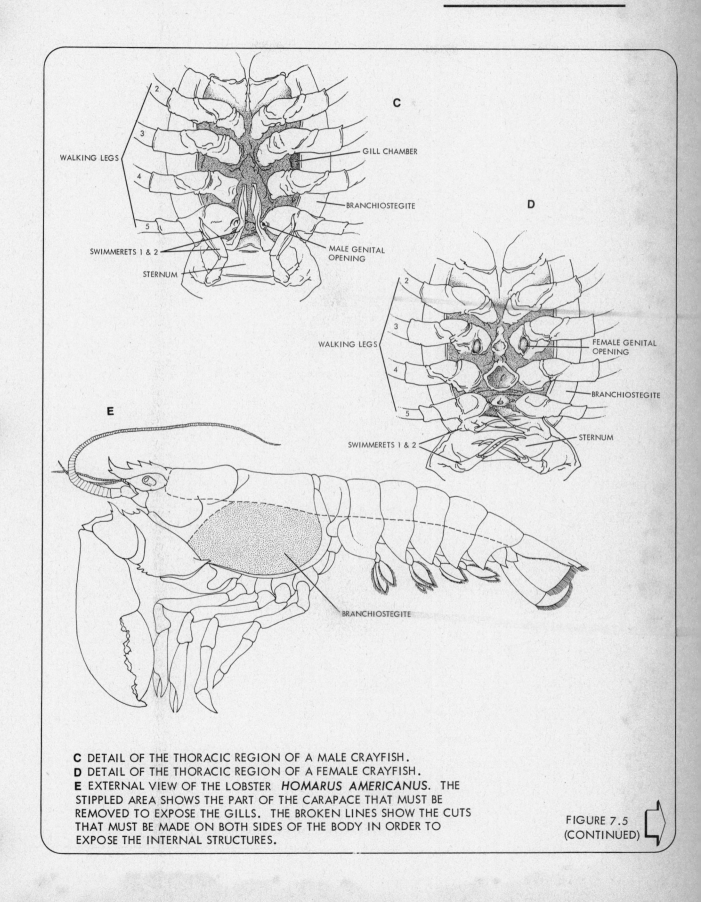

C DETAIL OF THE THORACIC REGION OF A MALE CRAYFISH.
D DETAIL OF THE THORACIC REGION OF A FEMALE CRAYFISH.
E EXTERNAL VIEW OF THE LOBSTER *HOMARUS AMERICANUS*. THE
STIPPLED AREA SHOWS THE PART OF THE CARAPACE THAT MUST BE
REMOVED TO EXPOSE THE GILLS. THE BROKEN LINES SHOW THE CUTS
THAT MUST BE MADE ON BOTH SIDES OF THE BODY IN ORDER TO
EXPOSE THE INTERNAL STRUCTURES.

FIGURE 7.5
(CONTINUED)

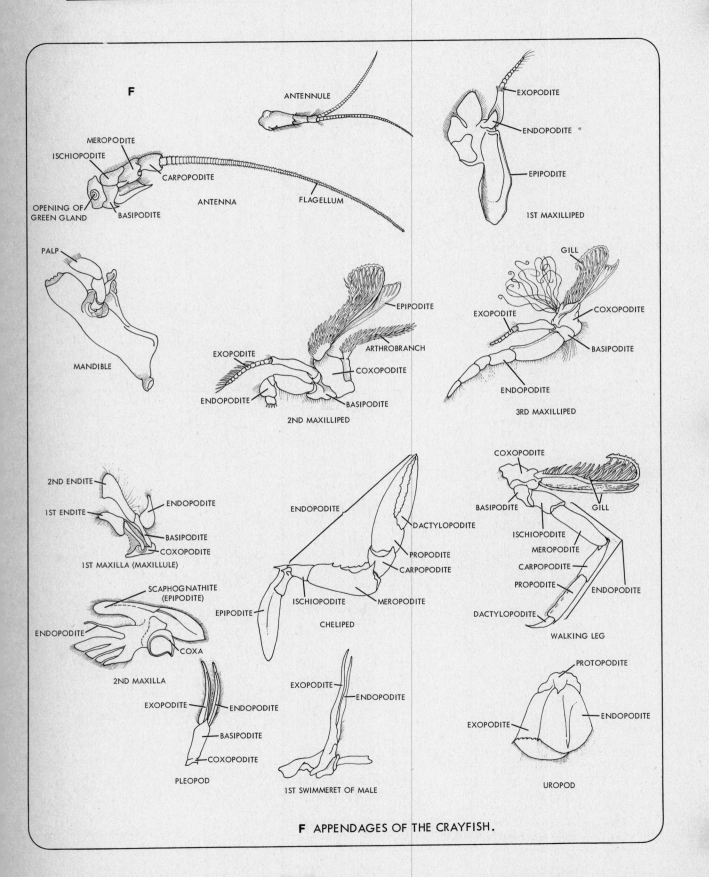

F APPENDAGES OF THE CRAYFISH.

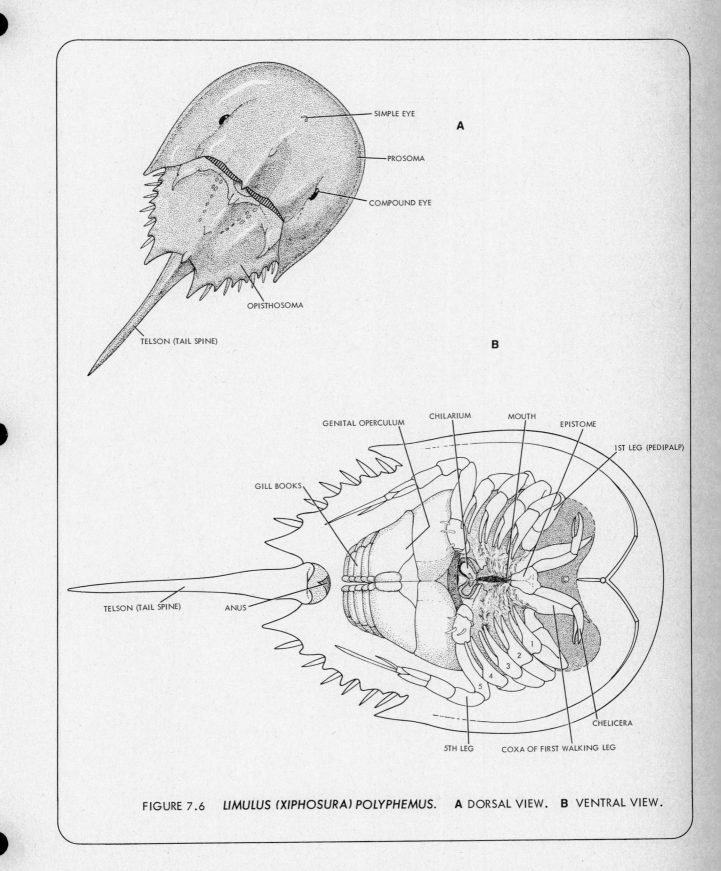

FIGURE 7.6   *LIMULUS (XIPHOSURA) POLYPHEMUS.*   **A** DORSAL VIEW.   **B** VENTRAL VIEW.

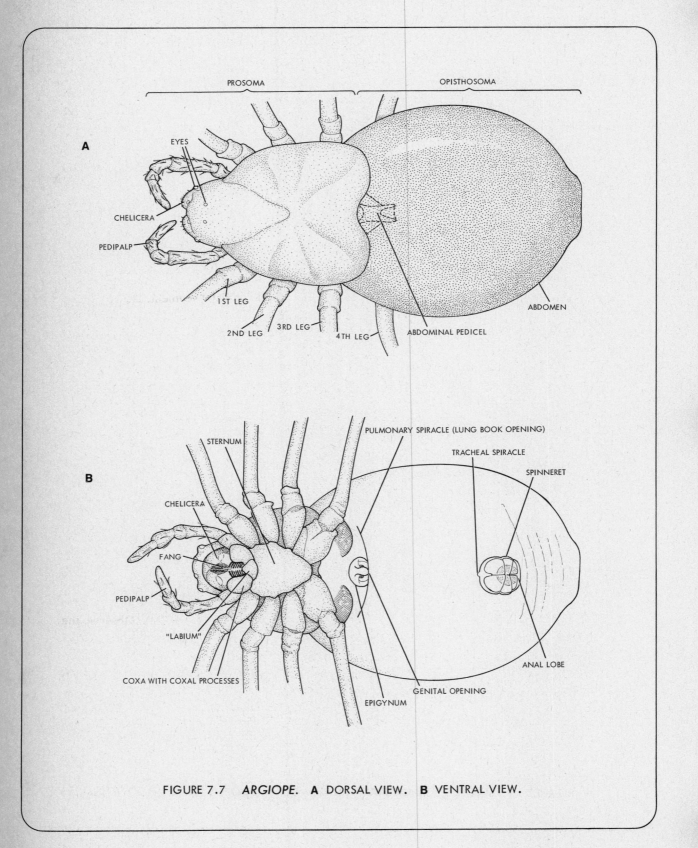

FIGURE 7.7 *ARGIOPE*. **A** DORSAL VIEW. **B** VENTRAL VIEW.

▶ Observe that the separation of thorax and abdomen is indistinct in centipedes and millipedes and that there are functional locomotor appendages on almost all segments; this body region is referred to as the **trunk.** The abdomen of insects bears no legs; in the crab, lobster, and crayfish it bears pleopods (swimmerets); in the crab the abdomen is generally tucked under the cephalothorax.

▶ Examine the opisthosoma of the horseshoe crab, spider, and scorpion. Note that in the scorpion two opisthosomal regions can be distinguished: the **mesosoma** and the **metasoma** (Figure 7.8). Are there external appendages on the opisthosoma of any of your specimens?

c. Appendages

Arthropod appendages are hollow integumentary evaginations formed just above the outer edges of the sternites (ventral sclerites). The appendages are sclerotized cylinders made flexible by unsclerotized **arthrodial membranes** (joints). Attempts have been made to reduce arthropod appendages to a common type form. Some authors have suggested an archetype with a main axis composed of nine segments (**endopodite**) that bears a biting process on the first segment (**gnathobase**) and an outer branch (**exopodite**) on a more distal segment. However, the metameric and jointed nature of the arthropod limb has allowed specialization of the appendages of the various body regions for different functions and this, together with the enormous adaptive radiation displayed by arthropods, makes the concept of a basic or archetypal limb difficult to apply in all cases. Nevertheless, within the various classes some uniformity of structure can be discerned. Compare the ''archetypal limb'' with the basic crustacean, insect, and arachnid appendages (Figure 7.2**B**). Some of the regional differentiation of arthropod appendages was seen in the study of body regions. In the following survey refer to Table 7.1 and your representative specimens so that you can relate each limb examined to its correct body segment.

i. HEAD, PROSOMA.   The Arthropoda can be separated into two major groups, the Mandibulata and the Chelicerata, on the basis of the kind of head appendages they bear. **Mandibles** are jawlike appendages associated with the mouth, and are found in crustaceans, insects, and myriapods. **Chelicerae** are clawlike appendages located in the second prosomal segment and are found in arachnids.

Mandibulates.   Group the mandible-bearing specimens and orient them so you have a clear view of the oral region.

▶ Examine a preserved specimen of the grasshopper *Romalea* (Figure 7.4). Observe the anteriorly situated **antennae** and note that they carry sensory hairs. The mouthparts (Figure 7.4**B**) are covered by a bilobed flap called the **labrum.** Lift this with a pair of forceps and expose the **mandibles.** Observe that they are heavily sclerotized and move in a lateral plane. In what plane does your own mandible operate? Use your forceps to remove the labrum, mandibles, and accessory appendages (**maxillae**). The lower lip or **labium** consists of the fused **second maxillae.** Both sets of maxillae are provided with conspicuous sensory **palps.** *Romalea* possesses the typical basic mandibulate arrangement of mouthparts adapted for biting and chewing. The mouthparts may be modified in other insects according to their method of feeding (see Section 7–2).

▶ The head appendages of myriapods (centipedes and millipedes) are arranged in a similar manner to those of insects. Using preserved specimens of a centipede and a millipede, locate the **antennae, mandibles,** and **maxillae** (Figures 7.3**A** and 7.3**B**). The **labrum** is a toothed plate continuous with the front of the head. Behind the mandible of the millipede is the **gnathochilarium,** reminiscent of the labium of insects in structure and position. What therefore is its relationship to the maxillae? Does the millipede have a second maxillary segment?

▶ In contrast to the insects and myriapods the second segment of the crustacean head typically bears a pair of sensory **antennules** and the third segment bears the antennae. Examine a specimen of the lobster or crayfish and locate the antennules, antennae, and the mandibles (Figure 7.5**A** and 7.5**B**). Closely associated with the mandibles are two pairs of accessory feeding appendages, the **maxillules** and **maxillae,** corresponding in their segmental arrangement to the first and second maxillae of the insects and myriapods. They may be difficult to see without dissection, as they are obscured by the most anterior thoracic appendages **(maxillipeds).**

▶ Examine a preserved specimen of the gooseneck barnacle, *Lepas* (Figure 7.9**A**), a highly modified crustacean. Observe that the bilaterally compressed body is encased in a **carapace** composed of two leathery **mantle folds** or **valves,** in which are embedded five **calcareous plates.** It is mounted on a stalk or **peduncle,** by which the animal was attached in life. The peduncle is pre-oral and represents the second segment. The distal processes (cirri) of the thoracic limbs may be seen projecting from a gap between the edges of the carapace, curled toward the mouth. Remove one valve (Figure 7.9**B**). The tiny vestigial antennules are located at the adhesive end of the peduncle; the antennae are present only in the larva. The mouth is overhung by the labrum. Locate the mandibles at the sides of the mouth. The maxillules are single structures with a fringe of setae on the notched, median edge. The maxillae are a pair of hairy lobes, fused to form a hind lip.

Chelicerates.   The body of chelicerates is divided into two tagmata, an anterior prosoma and a posterior opisthosoma. The clawlike **chelicerae** are found on the second prosomal segment, in a similar segmental position to the antennae of mandibulates. The third segment bears a pair of **pedipalps** and segments four through seven bear ambulatory limbs. The eighth segment and its appendages may or may not be present (see Table 7.1).

▶ Examine the ventral surface of a small (8 to 10 cm) preserved specimen of the horseshoe crab *Limulus* (Figure 7.6**B**). Locate the chelicerae just anterior to the mouth. Posterior to the chelicerae are the pedipalps; these function in locomotion in *Limulus* and resemble the four pairs of ambulatory limbs found on segments four through seven. On the base of each of the walking legs, which are also used in feeding, observe the presence of masticatory processes or **gnathobases.** The seventh pair of appendages are nonchelate and are used in digging. They also bear a spatulate process, which functions in cleaning the appendages of the opisthosoma. On the eighth prosomal segment, locate a pair of flat processes called **chilaria,** of unknown function.

▶ Now turn your attention to some of the more familiar chelicerates, spiders and scorpions. Locate the prosoma of a preserved specimen of the spider *Argiope,* and turn the animal so that the ventral surface faces you (Figure 7.7**B**); identify the fanglike chelicerae, which contain a poison secreted by a venom gland. The pedipalps are used for gripping prey and function as an intromittent organ in the male. There are four pairs of walking legs. If specimens of the scorpion (Figure 7.8) are available, compare the structure of the chelicerae and pedipalps with that of the chelicerae and pedipalps of *Limulus* and the spider.

ii. THORAX, ABDOMEN, AND OPISTHOSOMA

▶ *Mandibulates.*   Study the appendages of the thorax and abdomen in your mandibulate arthropod specimens.

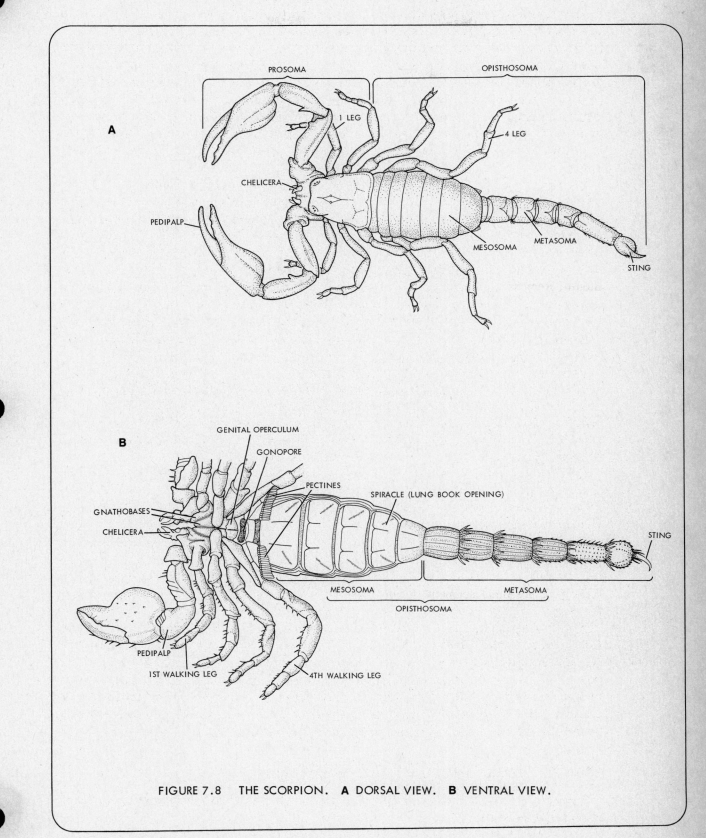

FIGURE 7.8   THE SCORPION. **A** DORSAL VIEW. **B** VENTRAL VIEW.

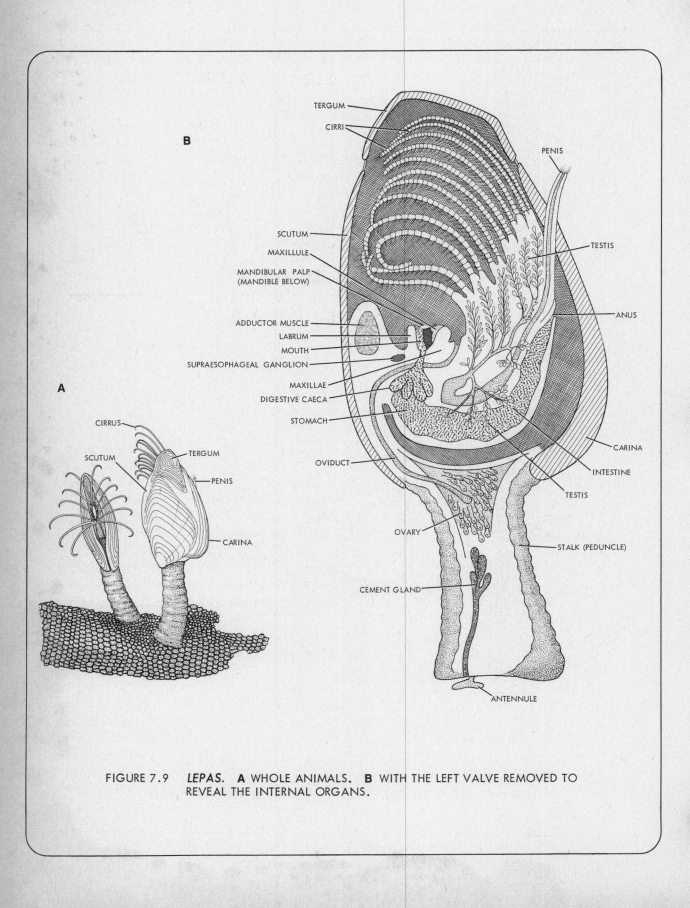

TERGUM

CIRRI

PENIS

B

SCUTUM

MAXILLULE

TESTIS

MANDIBULAR PALP
(MANDIBLE BELOW)

ADDUCTOR MUSCLE

LABRUM

ANUS

MOUTH

SUPRAESOPHAGEAL GANGLION

MAXILLAE

DIGESTIVE CAECA

STOMACH

CARINA

INTESTINE

A

OVIDUCT

TESTIS

CIRRUS

SCUTUM

TERGUM

PENIS

CARINA

OVARY

STALK (PEDUNCLE)

CEMENT GLAND

ANTENNULE

FIGURE 7.9   *LEPAS.* **A** WHOLE ANIMALS. **B** WITH THE LEFT VALVE REMOVED TO
REVEAL THE INTERNAL ORGANS.

▶ In the insects there are three pairs of thoracic legs, used for walking and jumping, and wings may be borne on the meso- and metathoracic segments (Figure 7.4A). The few abdominal appendages are generally represented by genitalia and posterior jointed **cerci.**

▶ The thoracic and most of the abdominal segments of myriapods bear appendages similar in structure to the walking legs of insects that are principally ambulatory in function (Figure 7.3). In centipedes the first pair of thoracic appendages or **maxillipeds** are modified to form poison claws (Figure 7.3A).

▶ The thoracic and abdominal limbs of the Crustacea, a group that is principally aquatic, function in locomotion, feeding, and respiration. Return to the preserved specimen of the lobster or crayfish (Figure 7.5A). Identify the first three pairs of thoracic appendages (**maxillipeds),** the paired **chelipeds,** and the four pairs of **pereiopods** (walking legs). The chelipeds and walking legs are uniramous, having lost the exopodite, and the chelipeds and first two pairs of walking legs have their distal joints modified to form **chelae** for gripping and tearing the food. Observe that the carapace, as the **branchiostegite,** hangs down on each side of the thorax to enclose a **gill chamber.** Insert one point of a pair of scissors under the left posterodorsal edge of the carapace and cut anteriorly until the branchiostegite can be lifted off (Figure 7.5C). Move the various appendages and observe a corresponding movement of the gills in the gill chamber. Carefully remove the thoracic appendages from the left side of the body and examine their structural modifications (Figure 7.5F). Do all the thoracic appendages bear gills?

▶ Turn the animal so that the ventral surface faces you (Figure 7.5B). In the abdominal region identify the five pairs of **pleopods (swimmerets),** the **uropods,** and the **telson.** Identify the male genital openings on the base of the fifth pair of thoracic pereiopods if your specimen is a male, or the female openings on the third pair of pereiopods if it is a female. Notice the structural modifications of the first and second pairs of abdominal pleopods in a male specimen. These are important in reproductive activity. The uropods and the telson form the tail fan, used in making rapid backward swimming movements.

▶ Return to your opened specimen of the barnacle *Lepas* (Figure 7.9B) and locate the six pairs of feathery thoracic limbs or **cirri.** Each has two long jointed bristly **rami,** which increase in length posteriorly. Behind the cirri is a long median-ventral **penis,** and behind this is the **anus** with a pair of vestigial **caudal rami.** The abdomen is absent in the adult barnacle.

Chelicerates. The opisthosoma of chelicerates bears reproductive, respiratory and/or swimming appendages.

▶ Return to the specimen of *Limulus,* and examine the ventral surface of the opisthosoma. The opisthosomal limbs (six pairs in all) are fused in the midline to form flattened platelike flaps, and each has a narrow inner endopodite and a broad outer exopodite (Figure 7.6B). The first of these flaps is the **genital operculum** with paired genital openings on its posterior surface. On the posterior surface of the exopodites of the remaining five limbs are 150 to 200 leaflike **gills,** together called a **gill book.** The last two opisthosomal segments are limbless. The posterior hinged spine or **telson** is probably not a modified appendage and therefore not homologous to the telson of other arthropods. It is used to right the body when it is overturned and to push it forward during burrowing.

▶ Terrestrial chelicerates, e.g., scorpions and spiders, have respiratory organs (**lung books)** on the opisthosoma, considered by some authorities to have been evolutionarily derived from an insinking of gill books. Locate the openings of these on one of your specimens (Figure 7.7**B** and 7.8**B**). Strip the integument from the area covering the lung books, and observe the numerous vertically arranged white lamellae of the gills. In the spider locate the **genital opening** on the ventral surface between the lung books near the anterior end of the opisthosoma. The opening is protected by a pair of fused appendages that form a plate called the **epigynum.** Many spiders spin complex **webs** by means of silk glands and organs called **spinnerets,** located behind the lung books on the opisthosoma. Does your specimen possess spinnerets? In the scorpion find the **genital operculum,** a small plate covering the openings of the genital ducts, consisting of the fused rudiments of two appendages, and the **pectines,** which are spiny flaplike structures, sensory (tactile) in function. The most posterior portion of the opisthosoma of the scorpion is used for manipulating the poisonous **sting.** How does the opisthosoma of the scorpion differ from that of the spider? The opisthosoma of the spider is separated from the prosoma by a narrow waist or **pedicel.** Is this true of the scorpion?

7–2. FEEDING, DIGESTION, AND THE STRUCTURE OF THE GUT

The head or prosoma, the mouthparts and other appendages, and the gut itself may show various modifications among the arthropods which reflect the feeding habits.

a. Predators

Predators are found among both the Mandibulata and the Chelicerata, and they employ a variety of methods to catch, subdue, and handle their prey. Recall the chelipeds of lobsters, crabs, and crayfish, the poison tail of scorpions, the poison fangs of chilopods, and the chelicerae of spiders.

i. CHELICERATES.   Chelicerates feed by means of a suctorial pharynx or stomach, and digestion is largely external. Digestive enzymes are injected into the prey after it has been caught and subdued; some species crunch up their prey with their chelicerae prior to digestion. The products of digestion are sucked up and the undigested parts remain. An exception to this is the aquatic bottom dweller *Limulus,* which ingests solid food and utilizes the gnathobases of the walking legs for crushing, shredding, and nipping other invertebrates, usually worms and clams.

▶   If living *Limulus* are available (small specimens are best), place bits of clam tissue or small annelids in a marine aquarium with *Limulus* and observe the action of gnathobases and chelicerae in manipulating food and feeding.

ii. MANDIBULATES.   Mandibulate mouthparts are designed for biting and chewing, thereby reducing the size of food materials so that they can be handled by the oral opening and conveniently acted upon by digestive enzymes.

▶   To observe the basic mandibulate feeding mechanism in action, introduce a few water fleas *(Daphnia)* into a syracuse watch glass containing a damselfly nymph. Observe the feeding behavior under the dissecting microscope. On seeing the *Daphnia* the nymph stalks it and then quickly lashes out the labium to effect capture. The prey is manipulated by the labial palps as the dark mandibles rip into the flesh and push it into the digestive tract.

▶   If living crayfish are available, hold some bits of clam with a pair of forceps and offer them to the animals. How are the appendages of the head and thorax used?

b. Modifications of the Mandibulate Plan—Insect Mouthparts

The mouthparts of insects are all evolved from the chewing mandibulate types, as described for the grasshopper *Romalea* in section 7–1; they offer particularly fine material for studies of functional adaptation, adaptive radiation, and homology.

▶   i. HYMENOPTERA.   The honeybee bites and manipulates wax with its jaws, but sucks or paps its food (pollen and nectar). Examine prepared slides of the mouthparts (Figure 7.10**A**) and note the small **mandibles** projecting laterally from the vicinity of the **labrum;** the **maxillae** forming two long pointed sheaths that in life protect the actual sucking parts; the sucking parts, consisting of a medial ''tongue'' or glossa, which is a modified **labium,** and laterally the projecting **labial palps.**

▶   ii. LEPIDOPTERA.   The mouthparts of butterflies are modified solely for sucking and are reduced in number. Examine a prepared mount (Figure 7.10**B**), and observe the sucking tube or **proboscis.** It consists of the elongated **maxillae.** These are grooved on their medial surfaces and fit together in life to form a tightly constructed coilable tube. They may be separated on your slide and should not be confused with the jointed **antennae.** Locate the sensory **labial palps** at the base of the proboscis. When not in use, the proboscis is coiled like a watch spring between the labial palps. The proboscis is used for sucking nectar from flowers.

▶   iii. HEMIPTERA.   All true bugs have piercing and sucking mouthparts; the majority feed on the juices of plants or animals. Examine a prepared slide of the mouthparts of the squash bug (Figure 7.10**C**) and note the short, fleshy **labrum,** which in life closes the upper part of the groove in the longer, fleshy, jointed **labium.** The elongated hairlike **stylets** may or may not be completely separated in your mount. They represent the highly modified **mandibles** and **maxillae** and therefore total four in number. In life the stylets are united to form the sucking parts. The maxillae fit together to form two enclosed tubes; the upper one is the food tube and the lower the salivary channel. The outer mandibles closely surround the maxillae. When not in use, the stylets are protected within the labium and held in place by the labrum. They alone enter the tissues of the food organisms; the other mouthparts stay outside. In addition to these structures, you may be able to discern the muscles lying within the head, which function in moving the stylets in and out of the tissue to which they have been applied. Do not confuse the **antennae** with the mouthparts.

iv. DIPTERA.   In the order Diptera belong many flies with lapping mouthparts (e.g., the housefly) as well as the very important bloodsucking flies (e.g., the horsefly and the mosquito). The several kinds of mouthparts found in members of this group have departed widely from the primitive biting type. There is always a **proboscis,** formed principally from the elongated **labium,** ending in a pair of lobes, the **labella.** The labium may function as a support and guide to the remaining mouthparts, which are enclosed within it.

▶   Examine a prepared mount of the mouthparts of the mosquito (Figure 7.10**D**). Observe the median, dagger-like **labrum-epipharynx,** the edges of which are curved to form the food tube; the **hypopharynx,** which carries the salivary duct; the sharp, stilleto-like **maxillae** and **mandibles,** which are also designed for piercing. Identify the **labium,** which acts as a protective sheath for the other mouthparts when not in use, and the jointed **maxillary palps,** which are sensory in function.

▶   Anesthetize some houseflies *(Musca)* or blowflies *(Calliphora)* with carbon dioxide gas, cut off the head of one specimen, and examine the mouthparts (Figure 7.10**E**). The mandibles are absent and the feeding apparatus consists of a **proboscis,** which is hinged to the head. The basal section or **rostrum** is really part of the head and the **clypeus** is situated on its anterior section. Laterally are found the paired **maxillary palps,** which are all that remain of the maxillae. The middle region of the proboscis, the **haustellum,** is formed from the highly modified **labium** and is hinged to the rostellum. Over the anterior face of the haustellum is hinged the **labrum-epipharynx;** this covers a deep labial groove in which lies the bladelike **hypopharynx.** The hinged arrangement of the proboscis allows it to be folded up under the head when not in use.

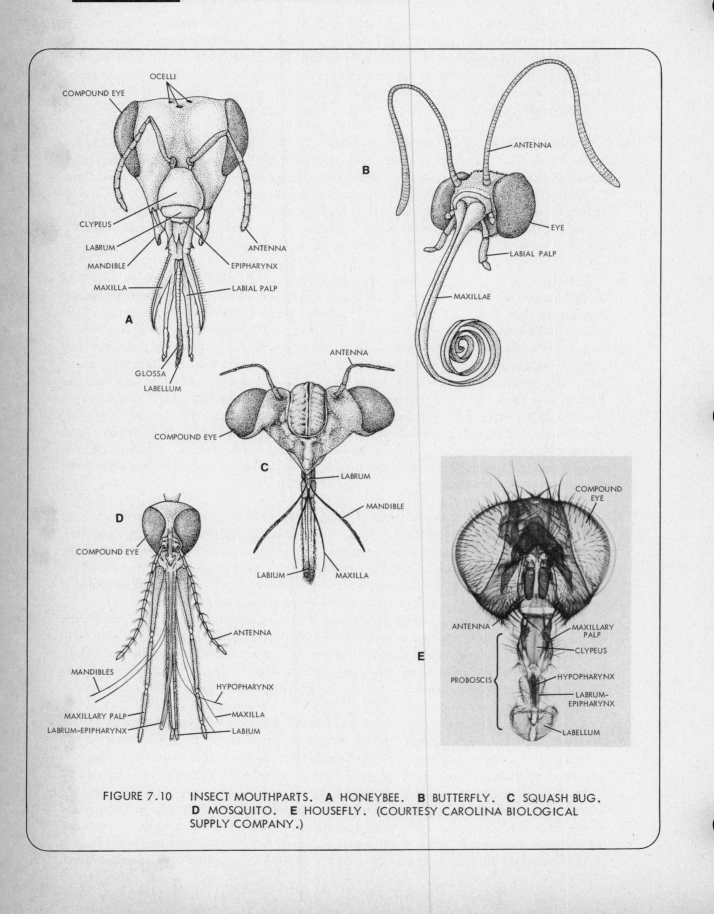

FIGURE 7.10 INSECT MOUTHPARTS. **A** HONEYBEE. **B** BUTTERFLY. **C** SQUASH BUG. **D** MOSQUITO. **E** HOUSEFLY. (COURTESY CAROLINA BIOLOGICAL SUPPLY COMPANY.)

Distally are two fleshy lobes, the **labella,** formed of the distal parts of the labium. They contain a series of fine food channels, strengthened by rings of chitin and referred to as **pseudotracheae.** These open externally via tiny pores through which liquid food and fine food particles can enter. In the groove between the labella are a series of **prestomal teeth,** which can be everted and are used for rasping.

Musca and *Calliphora* feed largely on fluid, but can effect solution of soluble food by regurgitating alimentary fluids onto it. In addition, the prestomal teeth can abrade dried surfaces, bringing regurgitated fluids more effectively into contact with the food material.

▶ Attach a glass rod or piece of wire (e.g., a paper clip) to the thorax of an anesthetized fly by means of melted paraffin wax (Figure 7.11). Allow the fly to recover from anesthetization and place on a sugar cube that has been lightly moistened with a drop of water. Examine under the dissecting microscope and attempt to observe feeding and the operation of the mouthparts.

c. Filter Feeders

Crustaceans, like all other arthropods, generally lack true body cilia, yet many are microphagic feeders. Instead of employing cilia in feeding, the crustacean may use the limb and its **setae** (hairlike extensions of the exoskeleton) as filtering devices.

▶ Obtain a live adult specimen of the brine shrimp *Artemia* or the freshwater fairy shrimp *Eubranchipus* in a syracuse watch glass containing water. Observe that under normal conditions the animal swims on its back by the metachronal beat of its 11 foliaceous thoracic appendages **(phyllopodia).** The inner edge of each appendage bears a fringe of setae used in filtering the water. Because of the anatomical arrangement and metachronal rhythm of the limbs, water is brought into a space between the limbs. As the water passes into the interlimb space, food particles are trapped on the setae; from here the food passes to the ventral food groove and then to the mouth. Place your specimen on a paper towel and allow it to dry briefly. Fix the abdomen ventral side uppermost to a piece of plasticine in a dry depression slide. Add a small amount of fluid. Study the process of feeding by adding a small drop of Congo-red-stained yeast suspension. Do changes in gut pH occur after feeding?

▶ *Daphnia* can also be conveniently used for studying filter-feeding mechanisms. Place a small dab of petroleum jelly in a dry depression slide. Take up a *Daphnia* in a pipet, and place the drop containing the animal on paper toweling so that the animal becomes dry. With forceps transfer the animal to the depression slide and press it so that the dorsal region of the carapace is firmly held by the petroleum jelly. Cover with fresh water and observe under low magnification (Figure 7.12). The bilaterally compressed body of *Daphnia* consists of a head bearing large antennae and a thorax entirely encased in the large carapace. There are only five pairs of thoracic appendages, and the third and fourth pairs act as the principal filters. As in *Artemia,* water is drawn into the space between the limbs and particles are filtered by the setae. After adding a small drop of Congro-red-stained yeast suspension, trace the current of water and the movement of food particles.

▶ If living barnacles are available, place some in a finger bowl containing seawater. Observe that the feeding activity of barnacles occurs in two phases. During the slow initial phase the valves open, the limbs **(cirri)** are protruded and unroll, water enters the mantle cavity and passes over the thoracic appendages and mouthparts. The second phase is rapid: the cirri are driven forward and withdrawn as water is sieved through the bristles on the limbs, the mantle cavity contracts, and water is forced out. This rhythmic activity functions in feeding and respiration. Test the feeding response as follows: make a suspension of one part milk in three parts seawater and gently deliver a drop of this suspension by means of a Pasteur pipet in the

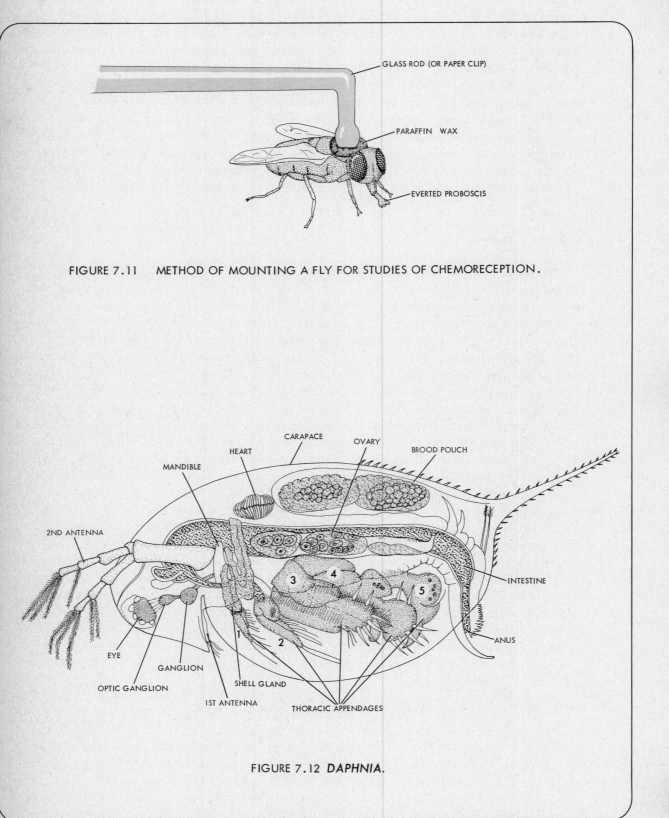

GLASS ROD (OR PAPER CLIP)

PARAFFIN WAX

EVERTED PROBOSCIS

FIGURE 7.11   METHOD OF MOUNTING A FLY FOR STUDIES OF CHEMORECEPTION.

HEART   CARAPACE   OVARY   BROOD POUCH

MANDIBLE

2ND ANTENNA

INTESTINE

3   4   5

ANUS

EYE

GANGLION   1   2

OPTIC GANGLION   SHELL GLAND

1ST ANTENNA   THORACIC APPENDAGES

FIGURE 7.12 *DAPHNIA*.

vicinity of the protruded cirri. Follow the response in another barnacle using seawater alone. Attempt to feed the barnacles small pieces of mussel *(Mytilus)* or some brine shrimp *(Artemia)*.

d. Structure of the Gut

Modifications in the structure of the gut in the different groups of arthropods also reflect variations in the dietary habits. There are normally three regions to the gut, and involutions of ectoderm at the mouth and anus delineate the **foregut** and **hindgut** from the **midgut.** The lining of the foregut may give rise to bristles or teeth, which function in straining or breaking up the food. In many arthropods the length of the gut and area of absorptive surface may be increased by coiling. The spiders are an exception; increased gut area is achieved through development of many diverticula. Digestion is extracellular, except in a few mites.

▶ Gut structure, feeding, and the digestive processes may be studied in the larvae of the fruit fly *Drosophila.* Sprinkle an indicator dye (e.g., neutral red, methylene blue, litmus) into a culture bottle containing *Drosophila* larvae and allow the larvae to feed for about 1 hour.* Remove a larva or two from the food and place on a microscope slide together with a drop of water. Gently press on a cover slip that has been ringed with petroleum jelly so that the larva flattens out sufficiently (when the finger-like spiracles of the anterior end flare out). Identify the various gut regions (Figure 7.14). Trace the course of food in the gut and observe any changes in pH along the length of the gut.

▶ The structure of the gut as well as chewing and digestion can also be studied in the cockroach *Periplaneta* (Figure 7.16A). Adult roaches and nymphs in the last instar are most suitable; they should be deprived of food, but not water, for 48 hours before beginning feeding tests. Prepare a paste of Pablum or mashed banana to which Congo red or litmus powder has been added and offer it to an animal placed in a covered glass jar or in a glass vial. Observe the activity of the animal. How is food detected? After feeding is complete, remove the animal and enclose it between two glass slides, separated by walls of plasticine. The size of the glass-and-plasticine chamber should be adapted to the size of the insect. If enough light is provided, the activity of various parts of the gut can be seen with the dissecting microscope.

Gross dissection of the gut of *Periplaneta* (Figure 7.13) and the crustacean should be delayed until after the circulatory and respiratory systems have been studied unless more than one specimen is available, in which case complete the dissections at this time following the directions on page 200. Dissection of the chelicerate gut will be omitted.

7–3. CIRCULATION

a. Vascular System

In contrast to the spacious coelom of annelids, the coelom of arthropods is obliterated by blood-filled **perivisceral sinuses,** which constitute a **hemocoel.** During development the **blasto-coel** becomes the cavity into which blood is discharged, and the coelom is restricted to the cavities of the gonads and excretory organs. The vascular system is described as being of the open type; that is, it consists of arteries that discharge blood into the perivisceral sinuses, rather than into capillaries in the tissues. The blood surrounds and bathes various tissues and

*This exercise should be started at the beginning of the laboratory period to allow enough time for completion.

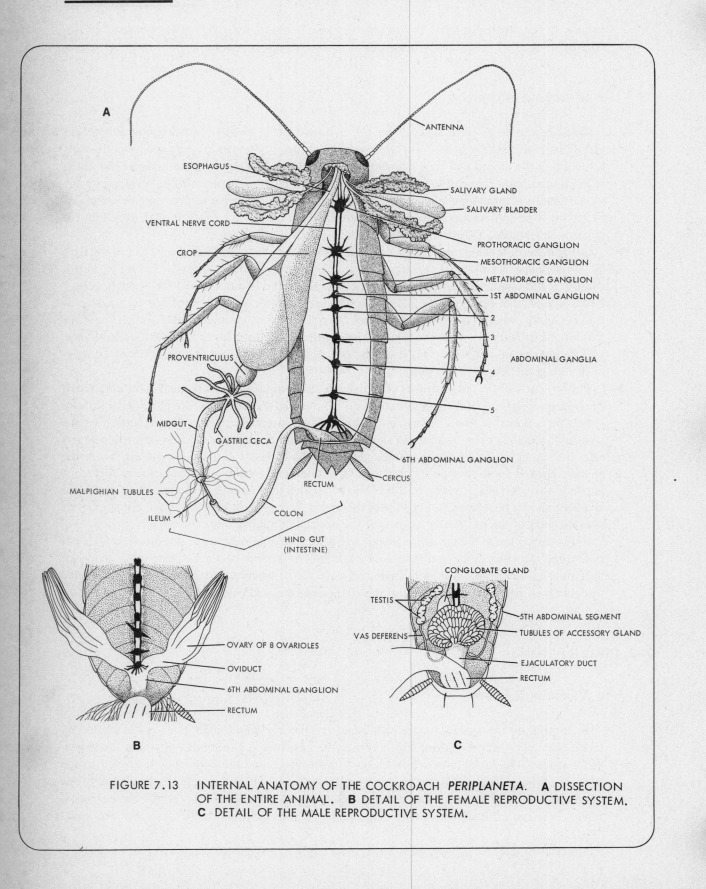

A

ANTENNA

ESOPHAGUS

SALIVARY GLAND

SALIVARY BLADDER

VENTRAL NERVE CORD

PROTHORACIC GANGLION

CROP

MESOTHORACIC GANGLION

METATHORACIC GANGLION

1ST ABDOMINAL GANGLION

2

3

ABDOMINAL GANGLIA

PROVENTRICULUS

4

5

MIDGUT

GASTRIC CECA

6TH ABDOMINAL GANGLION

CERCUS

RECTUM

MALPIGHIAN TUBULES

ILEUM

COLON

HIND GUT
(INTESTINE)

OVARY OF 8 OVARIOLES

OVIDUCT

6TH ABDOMINAL GANGLION

RECTUM

B

CONGLOBATE GLAND

TESTIS

5TH ABDOMINAL SEGMENT

VAS DEFERENS

TUBULES OF ACCESSORY GLAND

EJACULATORY DUCT

RECTUM

C

FIGURE 7.13   INTERNAL ANATOMY OF THE COCKROACH *PERIPLANETA*.   **A** DISSECTION
OF THE ENTIRE ANIMAL.   **B** DETAIL OF THE FEMALE REPRODUCTIVE SYSTEM.
C DETAIL OF THE MALE REPRODUCTIVE SYSTEM.

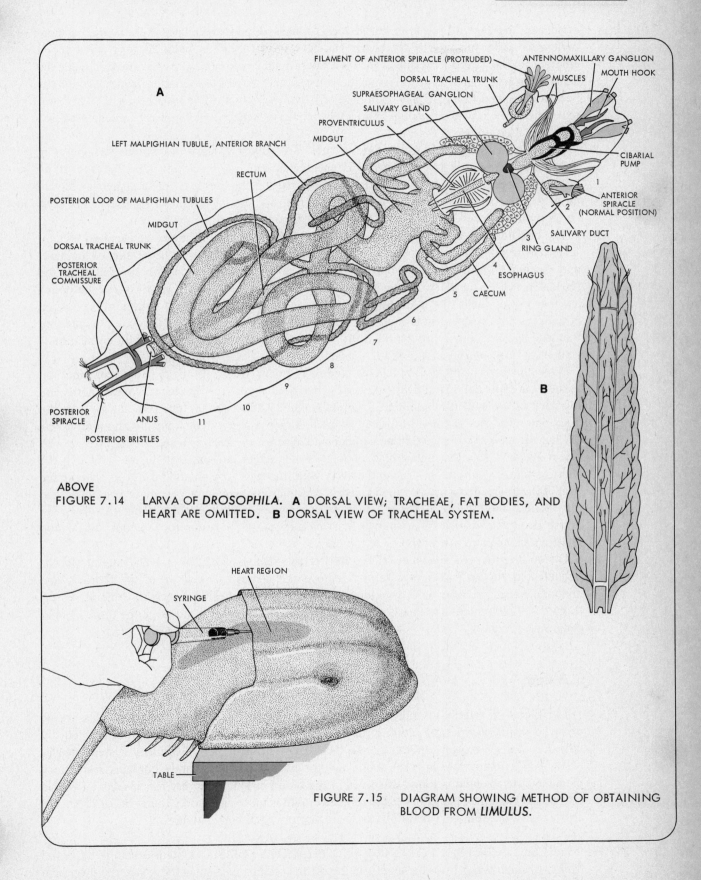

FILAMENT OF ANTERIOR SPIRACLE (PROTRUDED)
ANTENNOMAXILLARY GANGLION
DORSAL TRACHEAL TRUNK
MUSCLES
MOUTH HOOK
SUPRAESOPHAGEAL GANGLION
SALIVARY GLAND
PROVENTRICULUS
MIDGUT
A
LEFT MALPIGHIAN TUBULE, ANTERIOR BRANCH
RECTUM
CIBARIAL PUMP
POSTERIOR LOOP OF MALPIGHIAN TUBULES
1
MIDGUT
2
ANTERIOR SPIRACLE (NORMAL POSITION)
DORSAL TRACHEAL TRUNK
3
SALIVARY DUCT
POSTERIOR TRACHEAL COMMISSURE
RING GLAND
4
ESOPHAGUS
5
CAECUM
6
7
8
9
10
11
POSTERIOR SPIRACLE
ANUS
POSTERIOR BRISTLES

B

ABOVE
FIGURE 7.14   LARVA OF **DROSOPHILA**. **A** DORSAL VIEW; TRACHEAE, FAT BODIES, AND HEART ARE OMITTED. **B** DORSAL VIEW OF TRACHEAL SYSTEM.

HEART REGION
SYRINGE

TABLE

FIGURE 7.15   DIAGRAM SHOWING METHOD OF OBTAINING BLOOD FROM **LIMULUS**.

organs of the body. From the perivisceral sinuses, blood eventually collects in the large pericardial sinus surrounding the **heart.** The heart is perforated by **ostia,** through which blood enters from the perivisceral sinus. The ostia are provided with one-way valves so that when the heart contracts, blood is forced into the arteries and there is little or no backflow into the pericardial sinus. Compare this open type of system with that of the molluscs, and note that it resembles the circulatory system of annelids only in that the main propulsive force is provided by a dorsal contractile vessel.

▶ Insert a fine hypodermic needle into the heart of a living specimen of *Limulus* as shown in Figure 7.15. Small, relatively transparent specimens are best. Withdraw approximately 10 ml of blood into a syringe and save for use in a later part of this exercise. Inject approximately 10 ml of a saline suspension of India ink or carmine-in-saline into the heart. Observe the path taken by the colored solution into the sinuses of the body and the gill books.

▶ The heart can be seen *in situ* in wet mounts of whole living specimens of *Daphnia* (Figure 7.12). *Drosophila* larvae or the damselfly nymph may be substituted. The beating of the heart may be more easily observed if its activity is slowed down by cooling specimens in the refrigerator (4°C) for 15 to 20 minutes. In *Daphnia* the heart lies in front of the brood pouch and has muscle cells and a single pair of ostia in its wall. You should be able to see blood corpuscles passing forward and into the heart via the ostia, from where they are pumped into the head and then back through the body ventral to the midgut. The pathway of circulation is similar in *Drosophila* larvae and the damselfly nymph.

▶ Anesthetize an adult cockroach by holding it under water. Without penetrating the soft tissues, carefully cut up the right edges of the terga and across the first tergum of the thorax and the back of the seventh abdominal segment (Figure 7.16A). Dissect the terga away from the body and turn aside the flap so formed. Attach the specimen to a dish lined with paraffin wax (by means of pins through the dissected flap and edges of the abdomen and thorax). Cover with insect Ringer's solution and observe under the dissecting microscope (Figure 7.16C). If the heart remains *in situ* and has not come away with the terga, you should be able to observe the beat of the long narrow heart, which consists of numerous segmentally arranged chambers, and the opening and closing of the ostia. There is a pair of ostia in each segment. At this point notice also the glistening silvery network of fine **tracheae** of the respiratory system (see Section 7–4) that ramifies over the organs of the body.

▶ Obtain a preserved specimen of the lobster or crayfish. Make dorsal and lateral cuts as indicated in Figure 7.5E and carefully remove the dorsal exoskeleton of the thoracic and abdominal regions. After you have identified the polygonal heart and principal blood vessels (Figure 7.17A), remove the heart, place in tap water, and study the valves of the three pairs of ostia more closely.

b. Blood

The blood of arthropods may function in transport of food and gases, as well as being involved in phagocytic and clotting reactions. Insects have a system of air tubes which, by diffusion, carries oxygen directly to the tissues. As a result, the circulatory system usually plays only a minor role in respiration, and the blood of insects rarely contains respiratory pigments. **Hemoglobin** is found dissolved in the blood plasma of a very few insects and some crustacea. The blue respiratory pigment **hemocyanin** is more commonly found in the plasma of crustaceans, as well as in that of *Limulus,* scorpions, and some spiders.

▶ i. BLOOD CELLS, CLOTTING, AND PHAGOCYTOSIS.   Obtain a sample of blood from a living crayfish by snipping off the tip of the antenna or a small leg. Place the drop on a

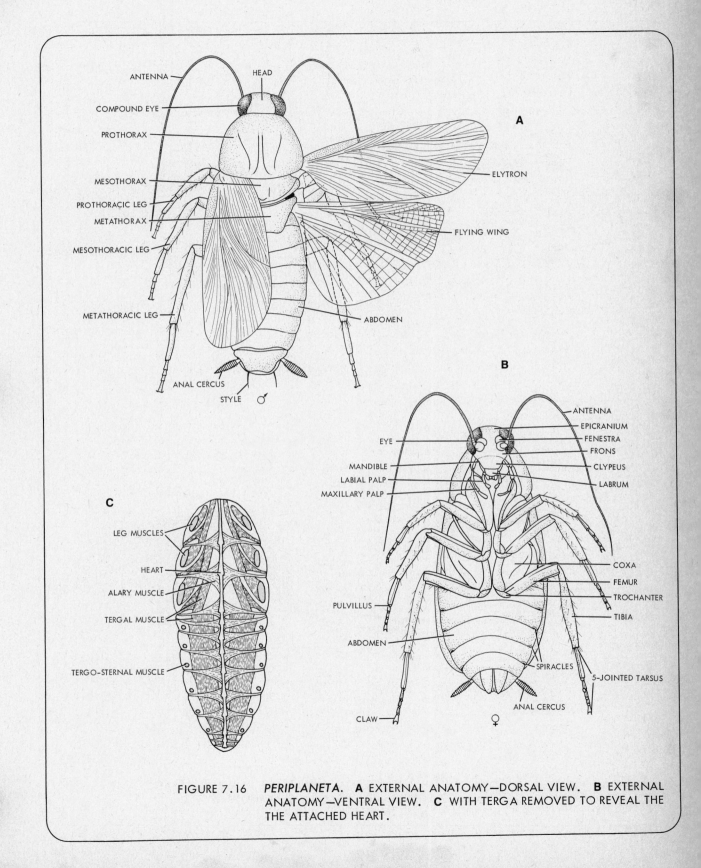

FIGURE 7.16   *PERIPLANETA.*   **A** EXTERNAL ANATOMY—DORSAL VIEW.   **B** EXTERNAL ANATOMY—VENTRAL VIEW.   **C** WITH TERGA REMOVED TO REVEAL THE THE ATTACHED HEART.

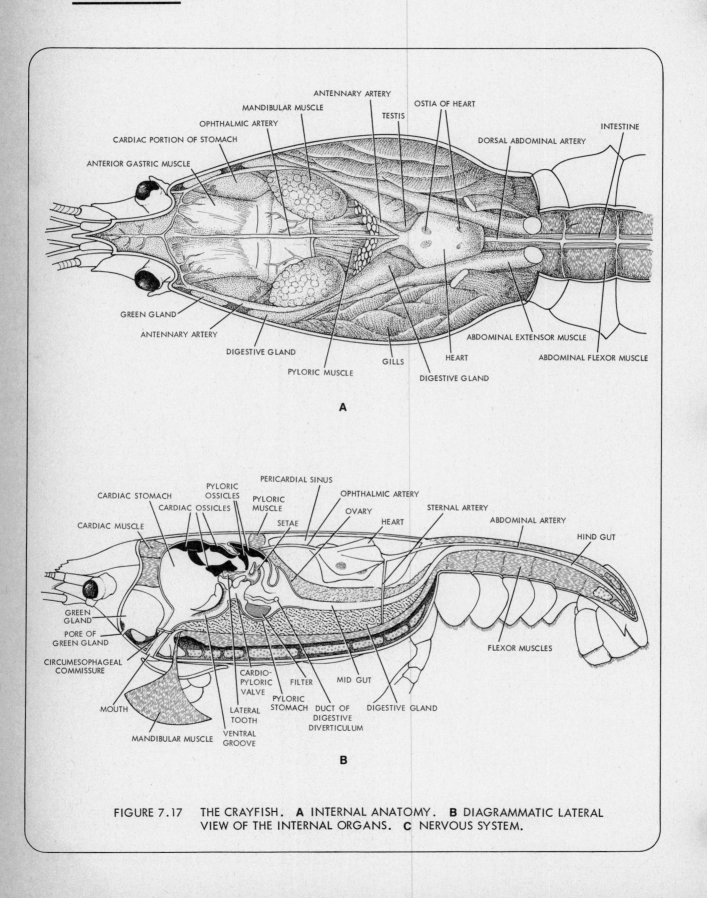

FIGURE 7.17   THE CRAYFISH. **A** INTERNAL ANATOMY. **B** DIAGRAMMATIC LATERAL
VIEW OF THE INTERNAL ORGANS. **C** NERVOUS SYSTEM.

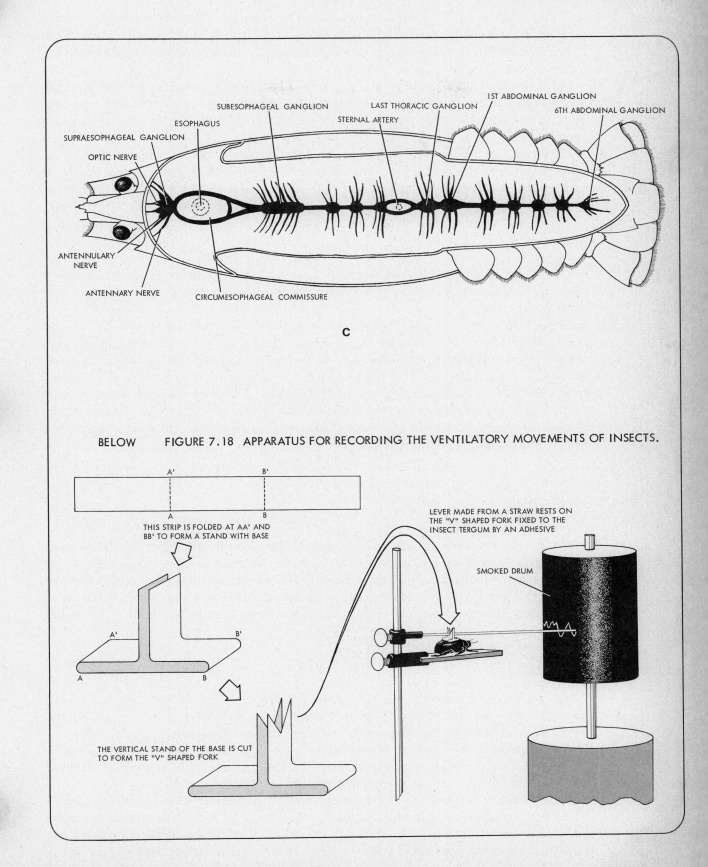

OPTIC NERVE

SUPRAESOPHAGEAL GANGLION

ESOPHAGUS

SUBESOPHAGEAL GANGLION

STERNAL ARTERY

LAST THORACIC GANGLION

1ST ABDOMINAL GANGLION

6TH ABDOMINAL GANGLION

ANTENNULARY NERVE

ANTENNARY NERVE

CIRCUMESOPHAGEAL COMMISSURE

c

BELOW   FIGURE 7.18 APPARATUS FOR RECORDING THE VENTILATORY MOVEMENTS OF INSECTS.

A'   B'

A   B

THIS STRIP IS FOLDED AT AA' AND BB' TO FORM A STAND WITH BASE

A'   B'

A   B

LEVER MADE FROM A STRAW RESTS ON THE "V" SHAPED FORK FIXED TO THE INSECT TERGUM BY AN ADHESIVE

SMOKED DRUM

THE VERTICAL STAND OF THE BASE IS CUT TO FORM THE "V" SHAPED FORK

slide and cover with a cover slip. Observe the cells under the microscope. Do they disintegrate? Add 1% iodine solution to the edge of the cover slip and allow it to stain the fibers. What is their function? In order to delay changes in cell morphology, place a drop of crayfish blood in a drop of mineral oil.

▶ For easier observation of cell types, make a film of crustacean or cockroach blood. Keep an animal immersed in water at 60°C for 10 minutes, take a drop of blood, and then smear the drop on a slide with another slide that has its edge coated with paraffin. Dry in air and stain with Giemsa stain.

▶ The phagocytic properties of crustacean and insect blood may be demonstrated by injecting some India ink or carmine into the limb joints or directly into the heart of a living specimen with a fine hypodermic needle and syringe. Allow the animal to rest for 1 hour, and then obtain a drop of blood as directed previously and observe under the microscope. Alternatively, withdraw and observe some blood from the previously injected specimen of *Limulus* (see Section 7–3).

ii. DISSOCIATION OF HEMOCYANIN. Hemocyanin is a copper-containing respiratory pigment characteristically present in the blood of crustaceans, *Limulus,* some arachnids, and some molluscs.

▶ Hemocyanin may be obtained from a specimen of *Limulus* by inserting a needle into the heart (Figure 7.15) and withdrawing approximately 10 ml of blood into a syringe, or you may use the blood obtained previously before you injected a specimen with ink or carmine (see Section 7–3). The blood can be defibrinated if it is shaken in an Erlenmeyer flask with glass beads, filtered through cheesecloth, and centrifuged. Place 5 ml of the blood into a test tube and gently bubble oxygen through it. Observe the reaction. Fit the test tube with a one-hole rubber stopper and evacuate the tube by means of a water faucet aspirator. What do you observe? What is the explanation of these results?

7–4. RESPIRATION

The circulatory system of arthropods is closely linked to the respiratory system. In most aquatic arthropods blood circulates through **gills** and contains a respiratory pigment such as hemocyanin or hemoglobin. Small crustaceans and midge (chironomid) larvae have no special respiratory devices. The air-breathing arthropods have a **tracheal** respiratory system; oxygen is brought directly to the tissues and respiratory pigments are the exception rather than the rule.

a. Gills

Gills are enlarged respiratory surfaces, formed as thin evaginations of the integument and invested with a network of blood vessels. They increase the body surface contact with the surrounding medium, facilitating gaseous exchange. In their simplest form gills are vascularized platelike or saclike processes usually associated with the limbs; the surface area available for gaseous exchange may be greatly increased by the development of filaments or lamellae along a central axis containing afferent and efferent blood vessels.

▶ In the lobster and crayfish the gills are associated with the limbs. The respiratory chamber, in which the gills are found, is formed by the dorsal overhang of the carapace (**branchiostegite)** and is really outside the body proper. Obtain a preserved specimen of a lobster or crayfish and remove a branchiostegite (Figure 7.5E), or return to the specimen from which you previously removed a branchiostegite (Section 7–1) and remove the branchiostegite of the other side. Examine the gills. How many gills are there? Is there any pattern to their distribution? What are the origins of the gills aside from those originating on the legs?

▶ Obtain a living crayfish, remove the anterolateral wall of the gill cover, and return the animal to a freshwater aquarium. Observe the action of the "bailer" or **scaphognathite** (second maxilla). The movement of water through the gill chamber can be demonstrated by adding a few drops of carmine suspension to the water in the vicinity of the animal. What is the direction of water movement? How do the appendages function in production of the water current?

▶ If live *Limulus* are available, study the beating of the gill books. With every forward beat of the covering segmental flap, the gill lamellae fill with blood; they empty each time the flap moves backward. The beating of the opisthosomal gill covers is also used during swimming and burrowing into the sand. In a demonstration injected specimen observe the ramifying network of blood vessels in each "leaflet" of the gill book.

b. Tracheae

Tracheae are ectodermal structures formed by invaginations of the integument. Composed of epithelial cells and lined with cuticle, they branch throughout the body and end in extremely fine **tracheoles.** The tracheal system communicates with the exterior via openings called **spiracles.**

▶ Observe the nature and distribution of spiracles in the grasshopper *(Romalea)* and a larva of the meal worm *(Tenebrio).*

▶ Return to the specimen of the cockroach in which you observed the heart, or kill a fresh cockroach with chloroform and open the body according to the directions given in Section 7–3; cover your specimen with insect Ringer's solution. Under the dissecting microscope observe the ramifying network of tracheal tubes over the surface of the gut and in the fat body (Figure 7.13). Remove a small piece of the fat body and place it on a slide with a drop of insect Ringer's solution. Add a drop of methylene blue and cover with a cover slip. Observe under high power. How does the structure of the tracheal tube relate to its function? Is there any evidence of nuclei associated with the tube?

▶ Anesthetize a cockroach nymph with carbon dioxide gas and mount it in a culture dish in such a way that a thoracic spiracle may be clearly seen under a dissecting microscope (Figure 7.16**B**). Use plasticine to restrain the limbs. Disturb the animal and observe the effect on the spiracles. After recovery, gently fill the dish with carbon dioxide gas and observe the spiracles.

▶ Large active insects show specialized movements of the body to produce mechanical ventilation of the tracheal system. Obtain a live cockroach or grasshopper and fix the insect to a glass slide by means of waterproof cement. Fold a strip of aluminum foil so that it forms a T and then cut a V in the vertical portion. Attach the base of the stand to the abdomen with adhesive. Rest a long straw or other suitable material in the V-shaped fork. The free end of this lever can be made to record on a kymograph (Figure 7.18). Observe the effects of low and high temperature (using dry ice or a hot plate and measuring the air temperature near the insect) and of carbon dioxide gas on the ventilatory rhythm.

c. Tracheal Gills

Some aquatic insects have gills that are thin-walled evaginations of the integument or rectum, invested with fine tracheoles. These gills expose a large surface to the surrounding medium and function in respiration.

▶   Obtain a damselfly nymph, a mayfly nymph, or a dragonfly nymph and examine them, or observe demonstration specimens. Notice the tracheal gills (Figure 7.19). Mayfly nymphs have tracheal gills along the sides and back of the abdomen; dragonfly nymphs have tracheal gills inside the rectum; and damselfly nymphs respire by means of three leaflike tracheal gills projecting from the abdomen in the position of anal cerci. Do not confuse the three "tails" of the mayfly nymph with tracheal gills; these are anal cerci and a posterior caudal appendage.

7–5. DISSECTION OF THE GUT

Dissection of the gut of the cockroach and crustacean can conveniently be completed at this time.

▶   i. PERIPLANETA.   Carefully remove the tracheae, muscles, heart, and blood vessels from the thorax of your opened specimen. Gently dissect out the cockroach gut and pull it to your left (Figure 7.13A). Anteriorly, locate the narrow **esophagus** leading dorsally from the pharynx in the head region. The pharynx will be seen following dissection of the nervous system. The esophagus widens into a thin-walled **crop;** this narrows posteriorly into a small globular **gizzard** or **proventriculus** with muscular walls, and internal **teeth** and **setae** for grinding and straining the food. The **midgut** leads from the proventriculus and coils counterclockwise. At the anterior end of the midgut, posterior to the proventriculus, arise eight diverticula, the **gastric caeca,** which increase the absorptive surface. The midgut leads into the **hindgut,** differentiated into three regions, the narrow anterior **ileum,** the wider **colon,** and the broad posterior **rectum** terminating at the **anus.** At the junction of midgut and hindgut lie 70 to 80 threadlike diverticula, the **Malpighian tubules,** arranged in six groups. They function in excretion.

▶   Anteriorly, locate a pair of white lobulate salivary glands and paired salivary bladders. The ducts from the gland and bladder on each side of the esophagus run anteroventrally and join the corresponding duct from the opposite side before emptying into the pharynx.

▶   Dissect the outer walls away from the proventriculus and cut longitudinally through it to observe the teeth of the grinding mill and the filtering setae. The structure of the proventriculus is easier to observe if it is removed completely and examined under water.

▶   ii. THE CRUSTACEAN GUT.   Examine the gut in your opened specimen of the lobster or crayfish, and identify the various organs (Figure 7.17B). Surrounding the midgut are two large, greenish midgut diverticula, the **digestive glands (hepatopancreas),** which function in the secretion of digestive enzymes and in absorption. Their cells also store reserve food materials and may have an excretory function. Carefully remove these glands and any gonadal tissue that obscures the digestive tract (white **testis** in males, yellow, salmon pink, or greenish **ovary** in females). Locate and identify the powerful **mandibular muscles.** From the mouth, a short vertical **esophagus** ascends and opens into a large **stomach,** which consists of an anterior **cardiac chamber** and a smaller posterior **pyloric chamber.** The **midgut** leads from the stomach and gives rise to a blind-ending **caecum** projecting forward over, and closely applied to, the pyloric stomach. The midgut is short in crayfish and long in lobsters, and leads to the **hindgut,** terminating in the **anus.**

▶   Remove the stomach and examine it under water with a dissecting microscope. Slit open the stomach to identify the grinding chitinous **teeth** of the **gastric mill** in the cardiac chamber and the rows of chitinous **setae,** which act as a sieve in the pyloric chamber. Compare the structure

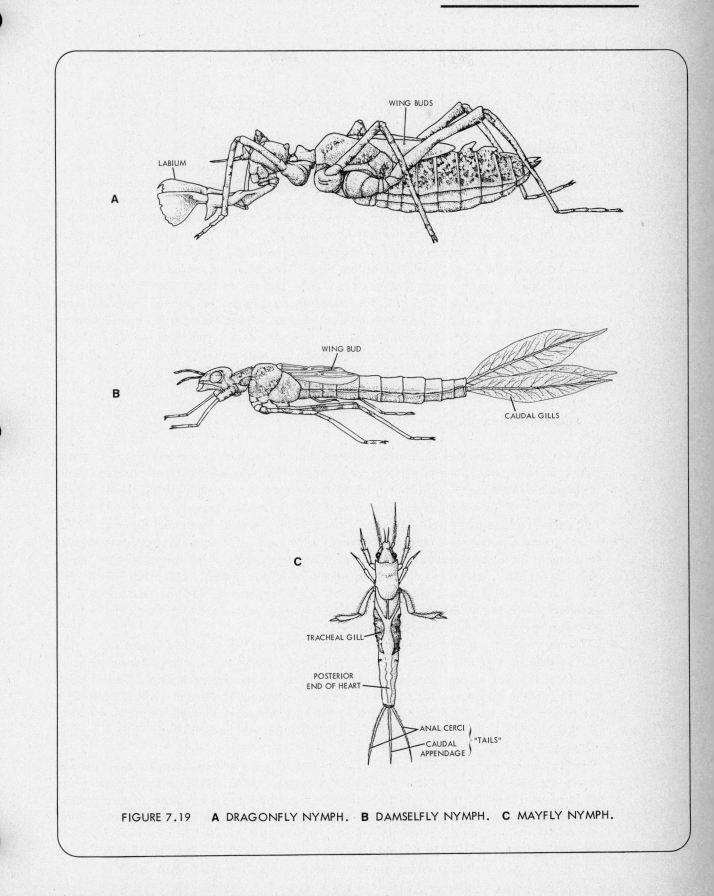

FIGURE 7.19   **A** DRAGONFLY NYMPH.   **B** DAMSELFLY NYMPH.   **C** MAYFLY NYMPH.

of the gastric mill with the proventriculus of *Periplaneta*. In what other ways is the gut of *Periplaneta* comparable to the crustacean gut?

7–6. EXCRETION, OSMOREGULATION, AND WATER RELATIONS

Maintenance of the internal milieu is one of the central problems facing all organisms. Both marine and freshwater environments pose problems of ion regulation and elimination of nitrogenous wastes. Because the body fluids of most arthropods living in the marine habitat are not exactly identical to seawater in ionic composition, some regulation does occur in these organisms. This problem becomes more acute with the invasion of fresh water, where there is a greater influx of water into the tissues and dilution of body fluids. This regulation is achieved in a variety of ways: (1) impermeability of the cuticle to salts and water, (2) resorption of salts from the urine, and (3) active uptake of ions from the surrounding medium. Invasion of land poses a further problem—control of water loss. In such cases the cuticle may be impermeable and the excretory products may be concentrated or solid. In general, aquatic forms eliminate ammonia and urea as the principal nitrogenous wastes, whereas terrestrial forms ordinarily excrete uric acid. In both aquatic and terrestrial forms the excretory organs are blind tubules lying in the hemocoel that discharge their concentrated waste products to the outside directly or via the gut.

a. Aquatic Arthropods

A variety of adaptations to marine and freshwater environments are found in the crustaceans, including variations in surface permeability, renal excretion, active extrarenal (gill) absorption of salts, and salt storage in the tissues. Only a few examples of this range will be studied here. Students interested in this aspect of arthropod biology should read in the references by Waterman (1960), Potts and Parry (1964), or Welsh, Smith, and Kammer (1968).

▶ The paired **maxillary** and **antennary (green) glands** are conspicuous excretory-osmoregulatory organs found in many crustaceans; they open at the bases of the appendages for which they are named. Locate these openings in specimens of the lobster or crayfish (Figure 7.5**B**). The glands usually consist of three principal parts: an internal **end sac (coelomosac),** an **excretory canal,** and an **exit duct** (Figure 7.20). The lower part of the excretory canal and the exit duct may be enlarged to form a **bladder.**

▶ During your dissection of the lobster or crayfish (Section 7–3), you may have observed the pair of antennary glands lying in the ventral region of the head near the mouth-esophagus (Figure 7.17). The various parts of the gland may be stained by injecting 0.25 ml of 0.5% aqueous solution of cyanol and Congo red into the hemocoel of a live crayfish. Twenty-four hours later remove a gland and observe it under the microscope. The cells of the coelomosac are red, and the multiple channels of the highly folded upper portion of the excretory canal (labyrinth) are blue.

When the antennary gland is in place in the crayfish, the labyrinth makes up the outer **cortical** region; the lower part of the excretory canal or **convoluted tubule** fills the central **medullary** part of the gland (Figure 7.20). The labyrinth and convoluted tubule together make up the glandular part of the organs. The convoluted tubule opens into the bladder. It has been postulated that the fluid in the end sac and labyrinth is a blood filtrate that is isosmotic with the blood, and chloride and other solutes are absorbed in the convoluted tubule. Is there any significance to the fact that the convoluted tubule is absent from the antennary gland of the lobster?

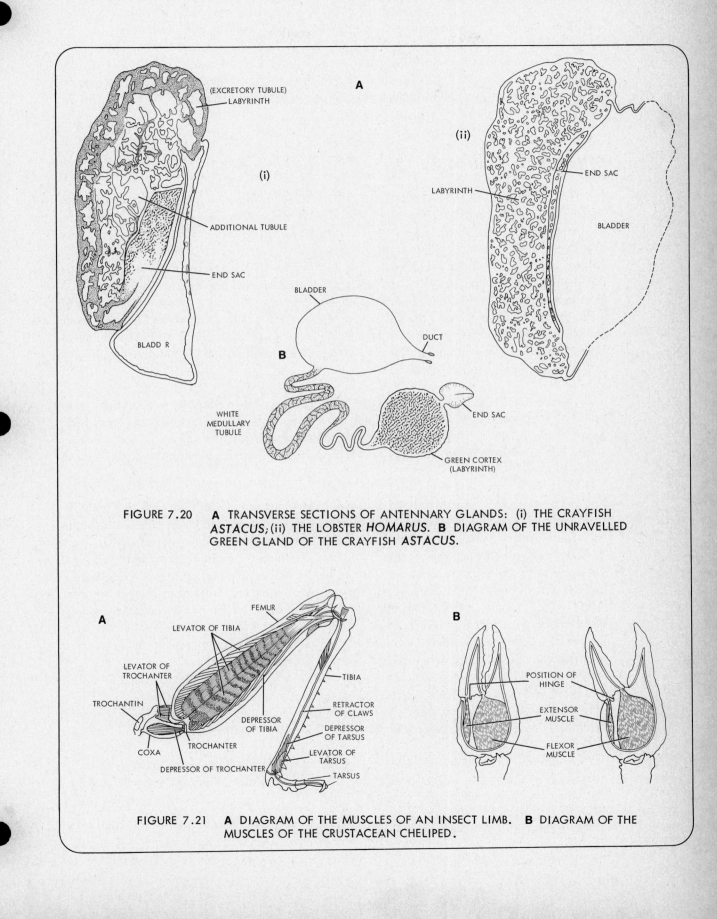

FIGURE 7.20 **A** TRANSVERSE SECTIONS OF ANTENNARY GLANDS: (i) THE CRAYFISH *ASTACUS*; (ii) THE LOBSTER *HOMARUS*. **B** DIAGRAM OF THE UNRAVELLED GREEN GLAND OF THE CRAYFISH *ASTACUS*.

FIGURE 7.21 **A** DIAGRAM OF THE MUSCLES OF AN INSECT LIMB. **B** DIAGRAM OF THE MUSCLES OF THE CRUSTACEAN CHELIPED.

Osmotic and ionic regulation may be effected by organs other than the antennary and maxillary glands; in some organisms the gills play a significant role in salt exchange.

▶ To localize areas of salt exchange in the brine shrimp *Artemia*, wash some living adult animals in several changes of distilled water (1 to 2 hours). Place the animals in 0.01 M $AgNO_3$ solution for 2 to 5 minutes, wash again in several changes of distilled water (1 to 2 hours) and then place in photographic developer. Where are silver grains deposited? Why? For further details refer to the paper by Croghan (1958).

b. Terrestrial Arthropods

i. EXCRETION AND MALPIGHIAN TUBULES.   The principal excretory organs of insects, myriapods, and arachnids are **Malpighian tubules,** ectodermal evaginations of the gut.

▶ Anesthetize a live grasshopper or meal worm *(Tenebrio)* larva. Under saline or insect Ringer's solution, carefully remove the dorsal body wall. Alternatively, return to the specimen of the cockroach in which you dissected the gut (Section 7–5). Observe the fine, filamentous Malpighian tubules that run from the posterior portion of the gut (Figure 7.13). Remove a portion of a tubule and place it on a slide with a drop of insect Ringer's solution. Add a drop of neutral red and observe under the microscope to see if the tubule accumulates the dye. Under higher magnification look for uric acid crystals in the lumen of the tubules. Alternatively, inject some *Tenebrio* larvae with indigo carmine; 30 minutes later dissect out the Malpighian tubules and observe the dye in the lumen.

ii. WATER RELATIONS.   Terrestrial arthropods may lose water by evaporation, transpiration, and excretion. They gain water by feeding, through the cuticle, and from their metabolism. Terrestrial arthropods show adaptations of the excretory and respiratory systems which enable them to conserve water, and the cuticle is important in cutting down the rate of water loss.

Cuticle structure.   The arthropod cuticle is a nonliving secretion of the cells that underlie it. Its principal constituent is **chitin,** a polymerized polysaccharide composed of repeating units of n-acetyl glucosamine. Often it has layers of calcium carbonate deposited within it (e.g., crustaceans). In many arthropods the cuticle becomes **tanned** by protein polymerization.

▶ Obtain pieces of exoskeleton from a fresh lobster or crayfish and an insect. Observe the cuticle surface under the dissecting microscope. Are pores and projections (setae) present? What is their function? What differences are apparent between the two exoskeletons? Drop some water on a piece of insect cuticle. What happens? Why? If the cuticle is relatively impervious, how does the animal receive stimuli from the external environment? What reaction occurs if the pieces of cuticle are placed in dilute hydrochloric acid?

▶ Observe the differences in cuticle strength and color in newly emerged *Drosophila* larvae and those that are older. What is the explanation for the observed differences?

In view of the fact that the cuticle is relatively rigid, how does growth of the animal occur? What functional role does the cuticle play aside from water regulation?

▶ *Cuticular and spiracular water loss.**   Label and weigh nine glass vials. Anesthetize three cockroach nymphs with carbon dioxide gas. Locate the spiracles (Figure 7.16**B**). Place some nail polish in a fine capillary and carefully introduce a small drop of the nail polish into each

*This exercise should be started at the beginning of the laboratory period to allow time for completion.

spiracle to block the spiracular opening. Allow the polish to dry. Place each animal in a vial and reweigh the vials. Place these three vials in a sealed desiccator containing desiccant and air. Place three more cockroach nymphs in separate vials and reweigh the vials. Place these in a sealed desiccator containing desiccant and an atmosphere of at least 10% carbon dioxide. Weigh three more animals in vials and place these in a sealed desiccator containing desiccant and air. Keep all desiccators at 30°C. After 2 hours open the desiccators and weigh the insects. Express your results as per cent weight lost. Explain your results. (For the effect of carbon dioxide on spiracles, refer to Section 7–4b).

▶ *Site of the water barrier.** Anesthetize 12 meal worm larvae and abrade six lightly with fine sandpaper. Weigh the animals in groups of three and dry for 2 hours at 30°C as just described. Reweigh the animals and express your results as per cent weight lost; explain your results.

▶ *Comparative rates of transpiration.** Weigh four pill bugs and four meal worm larvae. Hold at room temperature for 4 hours. Reweigh, express the results as per cent weight lost, and explain these results.

7–7. LOCOMOTION

One of the key features contributing to the success of arthropods is the presence of an exoskeleton, which resists deformation and is capable of being formed into levers. The metameric arrangement and jointed nature of the arthropod limb permits local contractions and movements. The hydrostatic skeleton is relatively unimportant in arthropod locomotion and plays a role only in those forms, having a flexible cuticle, e.g., branchiopods and centipedes, spiders, barnacles, and other forms just after molting.

a. Swimming

Swimming movements in arthropods are accomplished by a variety of appendages, e.g., the antennae of *Daphnia,* copepods, and of the nauplius larva of *Artemia;* the trunk phyllopodia of adult *Artemia;* and the abdominal pleopods (swimmerets) of lobsters and crayfish. Crabs do not do much swimming, but in those that do, such as *Ovalipes* and *Callinectes,* the fifth pair of thoracic appendages are modified into paddles. *Limulus* is basically a bottom-dwelling species, but swimming can be accomplished by undulations of the body and movements of the gill books.

▶ Study swimming activity in as many of the previously named specimens as are available in the laboratory.

b. Walking and Jumping

▶ Arthropods accomplish walking movements by the use of thoracic and/or abdominal appendages. Examine the terrestrial arthropods available in the laboratory and notice the length and

*These exercises should be started at the beginning of the laboratory period to allow time for completion.

thickness of the walking limbs. Where feasible, observe locomotion and the presence and location of joints and the various other modifications in structure that reflect different patterns of locomotion. How is movement related to number and arrangement of limbs?

▶ Obtain a living cockroach and place the animal in a dish over a sheet of graph paper. Study the normal pattern of locomotion. Diagram the movements of the legs. Remove the second pair of legs from another animal and observe the ambulatory pattern. Cut off the right anterior and left posterior legs from a third animal. How is the ambulatory pattern affected? What do these experiments tell you about coordination of leg movement? See the paper by Hughes (1952) for further details.

c. Flying

Movement through the air has been achieved with considerable success in the arthropods, and insects are truly aerial animals. The thoracic exoskeleton provides the structural framework of the wings, which are coupled to thoracic muscles for flying movements. Nervous control of flying is exercised by the thoracic ganglia (see Section 7–7), although there is some cerebral inhibition of the thoracic flight control centers. Flight movements may also be influenced by stimuli which impinge on other parts of the body.

▶ Grasp a living adult cockroach on the posterior edge of the large pronotum (first thoracic tergum, Figure 7.16A) with a pair of forceps held closed by a rubber band. Attach the animal and the forceps to a ring stand and screen the animal and apparatus to exclude air drafts. Cut off the head of another animal and mount the animal in the same way. Perform the following experiments on both animals, and record and compare the results.

▶ Stimulate various parts of the body by blowing on the animals through a small glass tube or a soda straw, noting in which regions stimulation elicits flying. Turn the animals upside down and observe flight behavior. Turn right side up again, place a small piece of paper or a pencil in contact with the legs while the wings are moving, and note the response. What is the minimum number of legs that must come into contact with the object to elicit a response? Remove the object from the animal's grasp and note the duration of the following flying activity. While the animal is flying, pull on one and then both of the metathoracic legs. What happens? Try pulling on other legs while the animal is flying; is the same response elicited? (Blowing on the cockroach from a posterior direction is a reliable way of inducing flying when the animal is beginning to fatigue.)

From the results of these experiments what can you conclude about the location of nervous flight control centers and of sensory receptors that influence flying, and about the effects of sensory stimuli on the initiation and maintenance of flight movements?

7–8. NEUROMUSCULAR SYSTEM

Striated muscle is characteristic of arthropods, which lack smooth muscle altogether. The muscular system of arthropods differs from that of most vertebrate striated muscle in that each muscle is innervated by more than one nerve, and in addition there are many nerve terminals on each muscle fiber. Basically, each muscle is innervated by two **axons** (nerve cell processes): a "fast" axon, which produces a rapid twitch when the frequency of stimulation is high, and a "slow" axon, which produces a tonic contraction when the stimulation frequency is low. Crustacean muscles are innervated by a third axon which inhibits the responses of the fast and slow axons. Neurophysiological studies are beyond the scope of this course, but students interested in this aspect of arthropod biology should consult the references by Waterman (1960), Wigglesworth (1953), and Welsh, Smith, and Kammer (1968).

a. Muscular System

▶ Return to the specimen of the crayfish or lobster that you opened to examine the vascular system and gut, or open a fresh or preserved specimen as previously described (Section 7–3). If you have not already removed the heart and alimentary canal from your specimen, do so now. If the gonads are large and obscure your view, they should also be removed (Figure 7.17). Notice the arrangement of the muscles in the thorax and abdomen and attempt to locate their origins and insertions. Pick up your specimen and flex it, and attempt to work out how these muscles operate.

▶ Place a living crayfish in a large finger bowl containing fresh water. Disturb the anterior end of the animal with a pencil and observe the **escape response.** What muscles and other parts of the body are involved in quick withdrawal?

▶ Acquaint yourself with the arrangement of muscles in the limbs of arthropods (Figure 7.21) so that you are fully aware of the mechanism by which the limb is extended and retracted. If there is time, remove a leg from your opened specimen of the crayfish or lobster, or from the grasshopper *Romalea,* and dissect it to reveal the muscles. Work out which are the extensors and which the flexors. Dissect a crustacean cheliped and attempt to work out the muscles responsible for opening and closing the claw (Figure 7.21).

b. Nervous System

The organization of the arthropod nervous system is basically similar to that of the annelids. It consists of paired **supraesophageal ganglia** ("brain"), **circumesophageal connectives,** a ventral **nerve cord,** and a series of **segmental ganglia.**

▶ Trim the base of the rostrum of your opened specimen of the lobster or crayfish to fully expose the supraesophageal ganglia (Figure 7.17**B**). Cut through the connective tissue on either side of the midline of the thorax to expose the circumesophageal connectives, the subesophageal ganglion, and the nerve cord with its segmentally arranged ganglia. Cut through the abdominal muscles close to their origins from the abdominal sterna, separate the muscle blocks carefully, and remove them. Watch for the abdominal part of the nerve cord, taking care not to pull it out with the muscles. Can you detect or trace any nerves running from the nerve cord or the ganglia? Variations in the amount of fusion of ganglia occur in certain groups, e.g., crabs and barnacles.

▶ Return to your opened specimens of the cockroach and/or grasshopper, or open new specimens (see Sections 7–3 and 7–5 for directions). Remove the gut and attempt to trace the components of the nervous system (Figure 7.13**A**). Compare the arrangement of the nerve cord and ganglia with that in the dissected crustacean.

7–9. SENSE ORGANS

Survival of an organism depends on its ability to receive signals from the environment and then to adjust its activities accordingly. Sense organs or receptors obtain information from the environment and the responses made are integrated by the nervous system. Arthropods possess a variety of sense organs. Determination of their functions and pathways of integration of information received by the nervous system can best be appreciated from studies of the behavior of the animal when the sense organs are stimulated.

a. Mechanoreception

Mechanoreceptors are sense organs that are sensitive to mechanical deformation. Sense organs for audioreception and detection of air currents are characteristically found in insects, whereas these organs are absent in crustaceans. The statocyst, an organ sensitive to gravity and movement, is characteristic of crustaceans and is not found in insects. Proprioreceptors, which are involved in the transfer of information concerning body position, muscle tension, etc., are common to all groups.

i. SENSILLAE.   Hairlike sensillae are located on the cerci of *Periplaneta* (Figure 7.16A), some of which are mechanoreceptors (phonoreceptors) responsive to air currents and/or sound vibrations. When stimulated, a reflex startle response is produced involving a sudden jump forward, followed by running activity, which carries the animal away from the source of stimulation.

▶   Place a cockroach nymph in a dishpan with a thin film of petroleum jelly smeared around the sides of the dish to prevent the animal from climbing out. Blow on the cerci through a small glass tube or a soda straw. Note the sudden leap forward and sustained running activity. Continue to stimulate the cerci. Is the response decreased? Why? Cut off the head and wait a few minutes for the animal to "recover." Again blow on the cerci. What is the response? Which response is diminished? Pinch the leg of this decapitated roach with a pair of forceps and observe the reflex retraction of the leg. Anesthetize the roach with carbon dioxide gas and carefully remove the thoracic sterna. Cut the neural connectives between the thoracic ganglia (Figure 7.13A). When the animal "recovers," test for the retraction reflex. What is the result? What do you conclude about the location of the center responsible for mediating this reflex?

▶   Continued cercal stimulation sometimes produces a cercal cleaning response in which the metathoracic legs stroke the abdomen or the cerci. Cut the nerve cord of a headless roach between the prothoracic and mesothoracic ganglia and observe the activity of the metathoracic legs. Can you elicit the cercal cleaning response from this roach? Now cut the neural connectives between the mesothoracic and metathoracic ganglia. Blow on the cerci and observe behavior. Cut off or pinch one of the cercal hairs and observe the reaction. What do you conclude about the location of the center for mediating this leg activity?

ii. TYMPANIC ORGANS.   Tympanic organs consist of a cuticular membrane and attached sensillae, which are involved in the reception of sound waves.

▶   Locate the **tympanum,** an audioreceptive membrane, on the thorax of a preserved specimen of the grasshopper *Romalea* (Figure 7.4A).

iii. STATOCYSTS.   Statocysts occur on the dorsal side of the proximal segment of the antennules of a number of crustaceans. These sense organs function in response to gravity and enable the animal to orient itself.

▶   Remove an antennule from a preserved specimen of a lobster or crayfish (Figure 7.5); carefully remove the lateral portion of the exoskeleton of the proximal segment of the antennule. Locate the statocyst and observe the structure under the dissecting microscope (Figure 7.22).

▶ Allow a small crayfish to swim in a dish of water. Note that the longitudinal and transverse axes are kept horizontal and the appendages are extended in a symmetrical fashion. Remove the animal from the water, dry the dorsal part of the carapace, and attach a vertical glass rod to it by means of paraffin wax or waterproof cement. When the rod is firmly attached, return the animal to the aquarium. Incline the animal to one side and observe the reactions of the limbs, then incline the animal in the other direction. Now destroy one statocyst by cutting off an antennule at its base. What is the orientation of this animal's appendages when the rod is vertical? Tilt the animal and compare the response of the appendages. Now destroy the other statocyst and test the response of the animal to tilting. Explain your results. Turn another crayfish on its back and observe the righting response. Repeat the experimental destruction of statocysts in another animal, or remove the glass rod from the first animal and turn it on its back. Is righting delayed?

▶ How may **eyes** play a role in the righting response? Can you devise an experiment that will enable you to determine the relative importance of eyes and statocysts in the righting response of crustaceans?

b. Chemoreception

Chemoreceptive sense organs respond to chemical changes in the environment. They may be located at various points on the body, e.g., antennae, pedipalps, tarsi.

▶ Anesthetize some blowflies with carbon dioxide gas and attach a glass rod or a piece of wire (e.g., paper clip) to the thorax of each by means of melted paraffin wax (Figure 7.11). The flies should have been fasted for the previous 24 hours. Flies may be held in position by inserting the wire into a piece of plasticine. Do not use injured flies. Arrange the flies under a dissecting microscope so that the proboscis can be readily observed (Figure 7.10E). Cover the tips of the tarsi (distal parts of the legs) one at a time with a fine glass capillary and observe the response of the proboscis. Now charge the capillary with water and again cover one tarsus at a time. Observe the reaction. Repeat, using a solution of 5% sucrose in water and then a solution of 10% NaCl in water. Use a piece of filter paper soaked with these substances and touch the tarsi. What is the action of the mouthparts?

c. Photoreception

Photoreceptors are sense organs capable of discriminating light energy. They are usually one of two kinds: the simple **ocellus** or the **compound eye.**

▶ Examine the specimens available to you in the laboratory, and determine the locations of the compound eyes and/or ocelli.

▶ Remove the compound eye of a lobster or crayfish and examine with the dissecting microscope. Bisect the eye longitudinally with a sharp razor, and by means of the accompanying diagram (Figure 7.23) and the figures in your text, identify the regions seen in your specimen. What are the functional differences between compound eyes and ocelli?

▶ Pill bugs generally prefer dark places. Set up a two-choice chamber, making one compartment dark and the other light. Put a few animals in the chamber, and allow them to move and settle. Which chamber do they seem to prefer? As a control, note the behavior of animals that have had their eyes covered with black lacquer. Explain your results.

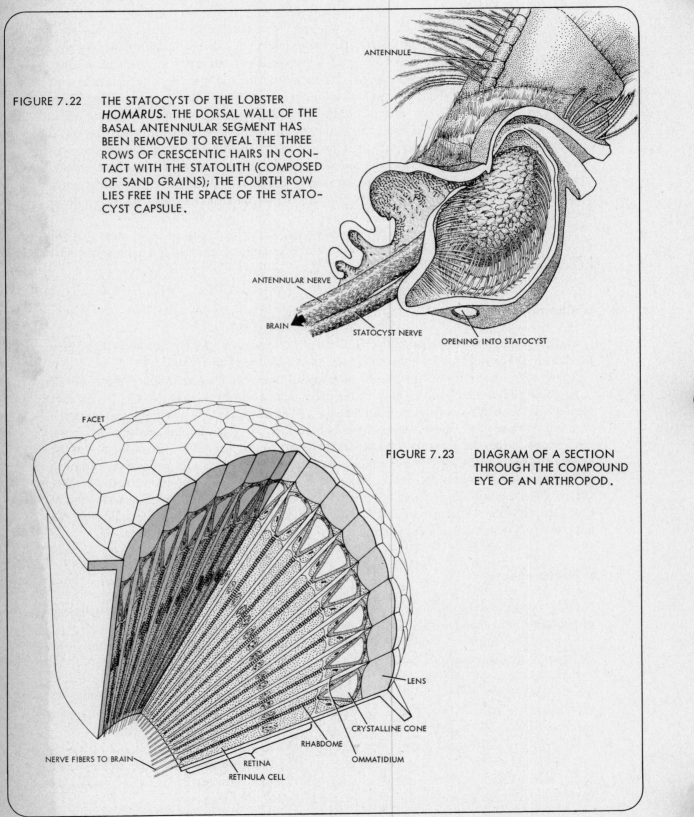

FIGURE 7.22   THE STATOCYST OF THE LOBSTER *HOMARUS*. THE DORSAL WALL OF THE BASAL ANTENNULAR SEGMENT HAS BEEN REMOVED TO REVEAL THE THREE ROWS OF CRESCENTIC HAIRS IN CONTACT WITH THE STATOLITH (COMPOSED OF SAND GRAINS); THE FOURTH ROW LIES FREE IN THE SPACE OF THE STATOCYST CAPSULE.

ANTENNULE

ANTENNULAR NERVE

BRAIN

STATOCYST NERVE

OPENING INTO STATOCYST

FACET

FIGURE 7.23   DIAGRAM OF A SECTION THROUGH THE COMPOUND EYE OF AN ARTHROPOD.

LENS

CRYSTALLINE CONE

RHABDOME

OMMATIDIUM

NERVE FIBERS TO BRAIN

RETINA

RETINULA CELL

7–10. REPRODUCTION AND DEVELOPMENT

In most arthropod species the sexes are separate. The barnacles and parasitic isopods present exceptional cases of hermaphroditism. Life cycles are extremely varied and complex, and only a few examples will be studied here.

i. CRUSTACEA.   The gonads of crustaceans are typically paired organs lying in the dorsal region of the thorax and/or abdomen. The oviducts and sperm ducts are simple tubes that open at the bases of the trunk appendages.

▶   You may have observed the gonads of the lobster or the crayfish during dissection of the digestive and nervous systems (Section 7–5 and 7–8). Review the location of these organs (Figure 7.17), or open a preserved specimen as described in Section 7–3 and locate the gonads. The male genital opening is found on the base of the fifth pereiopod, the female genital opening on the third pereiopod. Determine the sexes of specimens available in the laboratory from the modifications of their external appendages (Figure 7.5). What is the significance of the location of the openings of the genital ducts and the structural modifications of the first two pairs of pleopods?

▶   During development, many crustaceans spend some time as a free-swimming **nauplius larva,** which undergoes metamorphosis into the adult form. Using wet mounts of living material or prepared slides, study the structure of the nauplius larva (Figure 7.24**A**).

▶   In the cirripede crustaceans the late nauplius develops into a **cypris larva** (Figure 7.24**B**), having many of the features of the adult. Study demonstration slides of the cypris.

▶   If time and material are available, study the larval stages of the crab and the lobster (Figures 7.24**C,** 7.24**D**).

▶   ii. INSECTA.   Return to the specimen of the cockroach that you opened previously, or open another specimen according to the directions given in Section 7–3a, and locate the genital organs (Figure 7.13**C**).

▶   The reproductive organs can also be seen easily in male and female *Drosophila*. Etherize some flies, and from their external characteristics and with the aid of Figure 7.25**A**, sort out some females from some males. Place a female fly in a drop of insect Ringer's solution on a microscope slide. With the aid of the dissecting microscope, place a dissecting needle on the thorax, and, with a pair of watchmaker's forceps, pull the external genitalia away from the abdomen; the reproductive organs should accompany the genitalia. Place them in a drop of insect Ringer's solution on a slide and identify the component parts of the reproductive system (Figure 7.25**B**). Cover with a cover slip and observe under high power; careful examination of the ventral receptacle and spermatheca of the female will show sperm. What is the explanation for this? Make and examine a similar preparation of the reproductive organs of a male.

Development of insects may or may not involve metamorphosis from a larval or juvenile stage. **Metabolous** insects are those that undergo a distinct metamorphosis and they can be divided into two groups: the **Heterometabola** are those insects whose young stages (nymphs) closely resemble the adult, and generally metamorphosis involves increase in size, growth of wings, and development of the reproductive system. The **Holometabola** have young stages (larvae and pupae) in which the body form differs markedly from that of the adult.

▶   If available in the laboratory, observe all the stages in the life cycles of the cockroach *(Periplaneta),* grasshopper *(Romalea),* meal worm *(Tenebrio),* fruit fly *(Drosophila),* and the butterfly. What are the major differences between nymphs and adults? Larvae and pupae? Pupae and adults? What are the methods of locomotion in each of the stages? How is each

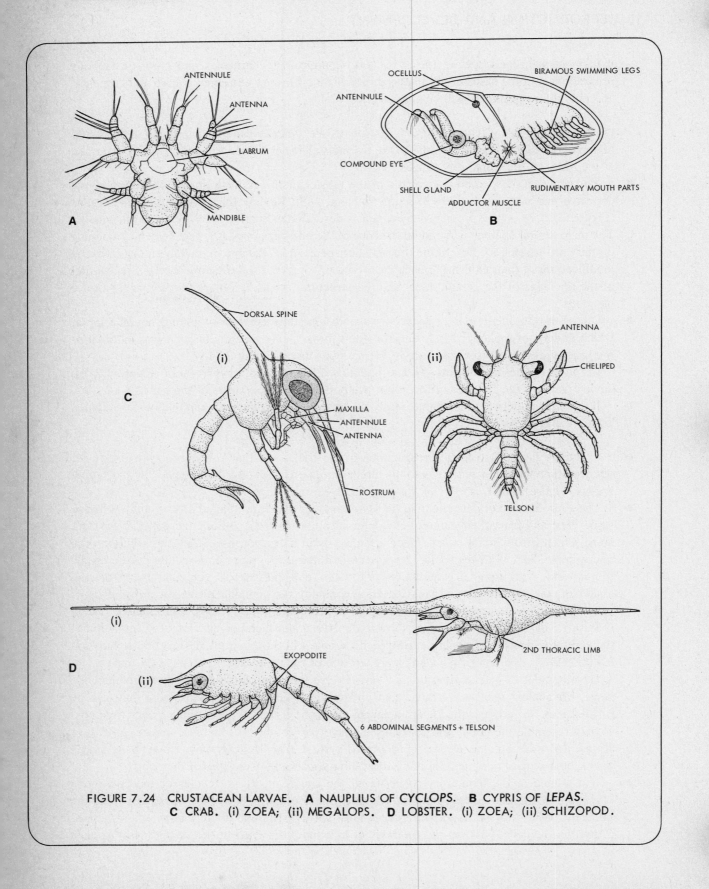

FIGURE 7.24 CRUSTACEAN LARVAE. **A** NAUPLIUS OF *CYCLOPS*. **B** CYPRIS OF *LEPAS*.
C CRAB. (i) ZOEA; (ii) MEGALOPS. **D** LOBSTER. (i) ZOEA; (ii) SCHIZOPOD.

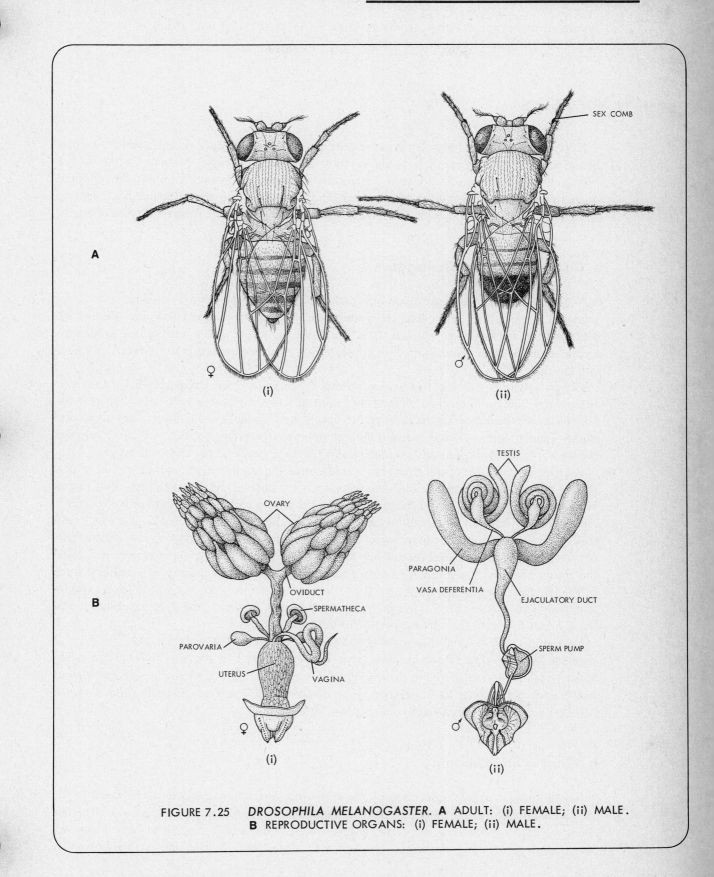

FIGURE 7.25 *DROSOPHILA MELANOGASTER.* **A** ADULT: (i) FEMALE; (ii) MALE. **B** REPRODUCTIVE ORGANS: (i) FEMALE; (ii) MALE.

developmental stage morphologically adapted to its particular habit? What is the adaptive significance of larval forms in arthropods?

7–11. HORMONES

Nerves are made up of bundles of nerve cells and their processes, or **neurons,** which function in the transfer and integration of information and facilitate rapid communication by the propagation of impulses. In addition, these cells may function in long-term coordination and regulation through the secretion of **hormones.** These are chemical substances, active in minute amounts, that produce a specific effect on the activity of cells remote from their point of origin.

a. Color Changes in Crustaceans

Many crustaceans undergo coloration changes related to the degree of dispersion of the pigments in specialized cells called **chromatophores.** The pigment migration is controlled by the secretions of the neurosecretory cells of the **X-organ,** whose axons (nerve cell processes) terminate in the **sinus gland.** Both X-organ and sinus gland are located in the eyestalk of higher crustaceans.

▶ Obtain specimens of the freshwater shrimp *Palaemonetes* or young crayfish *(Astacus).* Place a few animals in black containers and some in white containers for 1 hour. Examine the animals under the dissecting microscope before and after treatment, and observe the dispersion of the pigment within various types of chromatophores in each group (Figure 7.26). Exchange the backgrounds and again observe changes in the appearance of the chromatophores.

▶ Remove the eyestalks from five fresh specimens of *Palaemonetes* or young *Astacus* with a pair of fine scissors; to reduce bleeding, ligature at the base of the eye with a piece of thread before making your cut. Grind the eyestalks with a mortar and pestle in 1 ml of 0.85% NaCl. After thorough trituration transfer to a centrifuge tube, heat the mixture in a boiling water bath for 2 to 3 minutes, and centrifuge. Inject 0.02 to 0.04 ml of the clear liquid into the ventral part of the abdomen of each specimen. Place each animal in a black or white container. Examine the chromatophores under the microscope and record the degree of pigment dispersion (Figure 7.26). The first changes should appear 2 to 3 minutes after injection, and responses should be maximal in 10 minutes for *Palaemonetes,* 15 minutes for *Astacus.* Compare the coloration of these organisms with that of uninjected eyestalkless animals.

▶ Repeat this series of experiments with the fiddler crab *Uca* sp. (Changes in the dispersion of pigment in *Uca* chromatophores are most easily observed in the legs.) Note that when eyestalkless *Uca* are injected in the leg joints with eyestalk extract, the melanophore pigment becomes dispersed. Are these differing responses due to different hormones in *Palaemonetes* (or *Astacus*) and *Uca* or to different receptor systems?

b. Molting and Growth

The arthropod cuticle is secreted by the epidermis. Because the epidermis is bound to the relatively nondeformable cuticle, growth can take place only when the cellular layer detaches from the cuticle prior to laying down a new cuticle. This dissolution of the old cuticle—its loosening, rupture, and shedding—is referred to as **molting.** Molting and growth are under

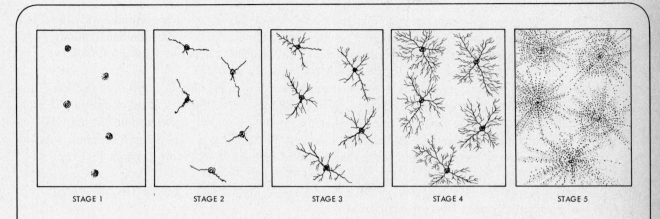

STAGE 1 STAGE 2 STAGE 3 STAGE 4 STAGE 5

FIGURE 7.26 AN ARBITRARY SCALE FOR SCORING THE DEGREE OF PIGMENT DISPERSION IN CHROMATOPHORES.

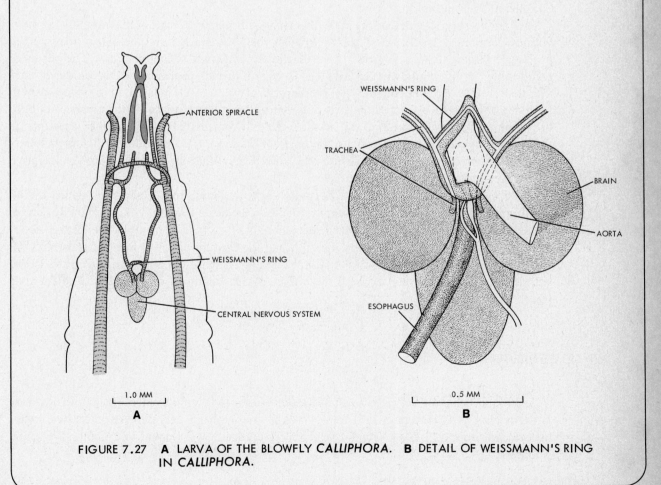

FIGURE 7.27 **A** LARVA OF THE BLOWFLY *CALLIPHORA*. **B** DETAIL OF WEISSMANN'S RING IN *CALLIPHORA*.

endocrine control; the hormones crustecdysone (crustaceans) and ecdysone (insects) are mutually effective.

▶ Obtain some healthy crayfish *(Procambarus clarki)* and remove the eyestalks with a scalpel; cauterize the stubs with a hot nail or needle. Leave the animals out of water for 10 minutes and then return them to the freshwater aquarium. Molting should occur in 6 to 8 days. Observe how the skeleton is shed. How does the eyestalk influence molting?

▶ If time permits, prepare an eyestalk extract according to the directions given in Section 7–11. Remove the eyestalks in two groups of *Procambarus clarki.* Inject 0.05 ml of eyestalk extract into the abdominal region of one group and a similar amount of seawater into the abdomen of the other group. Reinject animals every second day. Keep each animal in a separate, labeled container and record the number of individuals in each group that molt. Explain your results.

c. Molting and Metamorphosis

Metamorphosis, as well as growth, is controlled by hormonal substances released from the neurosecretory cells of the "brain." The number of molts and developmental stages of the life cycle is usually characteristic for the species. Arthropods are basically dimorphic organisms, existing as juvenile (larva) and adult, although the holometabolous insects have a more complex life cycle during which they may undergo a well-defined pupal stage, and these are considered to be trimorphic.

The interpolation of extra stages in the life cycle of insects is a result of the interplay between various hormones. The onset of metamorphosis may be retarded and the insect undergoes a prolonged period of somatic growth prior to the attainment of sexual maturity. The hormone responsible for inhibiting metamorphosis is called **juvenile hormone.** It is produced and secreted by the **corpora allata.** The **prothoracic glands** are stimulated to produce growth hormone or **ecdysone** by secretions from the brain and **corpora cardiaca.** In the presence of both ecdysone and juvenile hormone the insect molts but remains larval; diminished amounts or absence of juvenile hormone leads to the onset of maturation (to pupation in holometabolous insects) followed by the final molt into the adult form. What do you think are the adaptive advantages of delayed metamorphosis?

▶ In the higher Diptera the corpora cardiaca, corpora allata, and prothoracic glands are all fused in a ring called **Weismann's ring,** surrounding the aorta (Figure 7.27). In blowfly larvae, Weismann's ring is located in the fourth segment. Obtain early third instar larvae that are about to pupate,* immobilize with carbon dioxide gas, and ligature either in front of or behind the fourth segment, **not both.** A few days after treatment compare the appearance of these larvae with untreated animals, and observe the effects of this treatment on pupation. Explain your results.

7–12. BEHAVIOR

Arthropod behavior varies from simple reflex reactions to complicated instinctive or innate patterns of activity. Only a few examples will be studied here, and students interested in this aspect of arthropod biology should read the references by Fraenkel and Gunn (1961) and Thorpe (1963).

*This can be determined by examining the larvae under the dissecting microscope. The crop should be empty.

a. Orthokinesis

In orthokinetic behavior, average speed of movement or frequency of activity is related to the intensity of the stimulus; when it ceases, animals come to rest and tend to aggregate where their rate of movement is least.

▶ Pill bugs prefer moist, cool places and show orthokinetic behavior. Prepare three finger bowls as follows: place some moist paper toweling in one, dry toweling in another, and wet toweling in the third. Obtain 15 pill bugs and place 5 in each container. Cover each bowl with some aluminum foil to exclude the possibility of light acting as a stimulus. After 5 to 10 minutes remove the foil and observe the degree of activity in each bowl. What is the source of stimulus?

b. Hygrotaxis

Hygrotaxis involves orientation toward or away from humidity.

▶ Line a flat glass dish with two pieces of paper toweling, separated in the middle and overlapped by a short piece of aluminum foil. Secure the foil to the paper with tape. Make sure that the entire bottom of the dish is covered with toweling; moisten one of the towels and leave the other dry. Place 10 pill bugs on the foil and observe the position of the isopods. After 10 to 15 minutes remove the foil and observe the position of the isopods. How many are on the moist paper? On the dry paper? What is the state of activity of the animals on each?

c. Klinotaxis

Klinotactic behavior involves movement away from or toward a stimulus, with the animal orienting itself by comparing the intensity of stimulation on the two sides of the body.

▶ Blowfly larvae are negatively phototactic. Inscribe a circle on a piece of paper, and divide the circle into eight equal sections. Mark an x in the center. Place a blowfly larva on the x in subdued light. Record the path of movement of the animal on the paper with a pencil. Repeat several times with different larvae. Are the movements random? Does the head of the animal move from side to side? Now shine a narrow beam of light horizontally so that it strikes the side of the animal's body. The source should be 25 to 35 cm from the larva. Does the animal make directed movements? Move the beam of light so that the larva is illuminated from different directions. Record the movement of the head and the path of movement, relating these to the position of the light beam.

▶ Arrange two lamps so that their beams shine on the larva from the horizontal and form an angle of 90° where they cross. Place a larva equidistant (25 to 35 cm) from both light sources so that the intensity of both lamps is equal. Observe and record the path of locomotion. Observe the response of the larva when one light is switched off at the same time as the other is switched on. Where are the photoreceptors located?

▶ Try similar experiments with pill bugs. Cover one of the eyes with black lacquer and repeat. What does this tell you?

d. Thigmotaxis

Thigmotaxis involves orientation toward or away from sources of stimulation that touch the body of the animal. Contact, especially with a solid body, is the directive factor.

▶ Prepare sections of rubber tubing of various lengths (½ to 5 cm) and, using a cork borer heated in a flame, cut holes in the center of five or six plastic petri dishes. The holes should be large enough to just fit over a wooden dowel. Place sections of rubber tubing over the wooden dowel, followed by a petri dish, building a series of platforms of different heights (Figure 7.28A). Place the apparatus in a jar that has its top ringed with petroleum jelly. Introduce cockroaches into the jar, place in a subdued light, and after 30 minutes observe their location on the platforms. Repeat in bright light. What is the relationship of animal size to the size of the space occupied? Where are the sense organs for thigmotaxis located?

e. Rhythmic Behavior

Many animals show rhythmic physiological patterns that may persist even under constant conditions. These may be manifested by rhythmic changes in activity, changes of color, changes in metabolism, and so on. Some of the rhythms show a periodicity of about 24 hours and are therefore called **circadian rhythms** (*circa,* about; *dian,* day; L.). Others may show an annual periodicity, which can be referred to as **circannian rhythms** (*circa,* about; *annal,* annual; L.). The mechanism through which the animal is capable of timing its activity remains unknown; some workers believe endogenous factors are important, whereas others support an exogenous factor as the timer.

▶ i. ACTIVITY RHYTHMS. Construct a simple actograph such as the one illustrated (Figure 7.28B). This can be used to measure the locomotor activity of *Periplaneta* or *Uca.* Place the animal in the running wheel (with some seawater in the case of *Uca*). Record the activity of the animal on a slow-moving smoked kymograph drum for 48 hours under normal day-night conditions. Now place the apparatus in constant darkness. Does the rhythmic locomotor activity persist?

▶ ii. COLOR CHANGES. The chromatophores of *Uca,* particularly the melanophores, show quite a precise 24-hour period in the degree of dispersion of the pigment. Obtain some *Uca* and place them in individual beakers containing a small amount of seawater. Place one group in constant light and subject the other to normal day-night conditions. Using the dissecting microscope, observe and record the degree of dispersion of pigment (Figure 7.26) in the melanophores of the walking legs of individuals in each group over a period of 24 hours. Plot the degree of dispersion versus time.

References

Barnes, R. D. *Invertebrate Zoology,* 3rd ed. Philadelphia: W. B. Saunders Co., 1974.

Borradaile, L. A., and F. A. Potts. *The Invertebrata.* London: Cambridge University Press, 1961.

Brown, F. A., Jr. (ed.). *Selected Invertebrate Types.* New York: John Wiley & Sons, 1950.

Bullough, W. S. *Practical Invertebrate Anatomy.* New York: St. Martin's Press, 1962.

Carlisle, D. B., and F. Knowles. *Endocrine Control in Crustaceans.* London: Cambridge University Press, 1959.

Carthy, J. D. *An Introduction to the Behavior of Invertebrates.* London: Allen and Unwin, 1958.

Clements, A. N. *The Physiology of Mosquitoes.* New York: Macmillan Publishing Co., 1963.

Croghan, P. C. The mechanism of osmotic regulation in *Artemia salina* L.: The physiology of the branchiae, *J. Exp. Biol.* **35**:234–42, 1958.

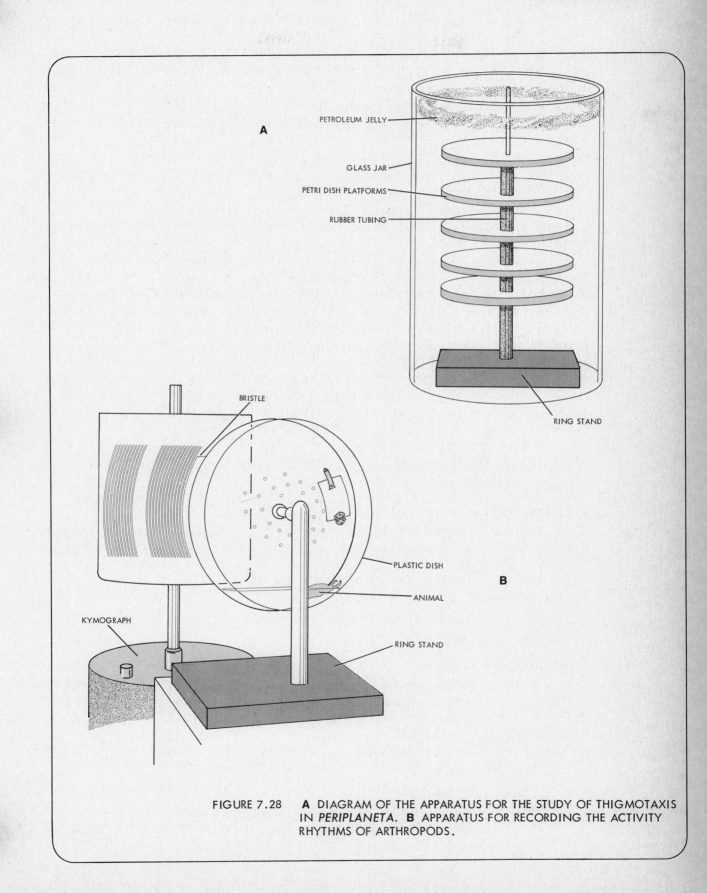

FIGURE 7.28   **A** DIAGRAM OF THE APPARATUS FOR THE STUDY OF THIGMOTAXIS
IN *PERIPLANETA*. **B** APPARATUS FOR RECORDING THE ACTIVITY
RHYTHMS OF ARTHROPODS.

Edney, E. B. *The Water Relations of Terrestrial Arthropods*. London: Cambridge University Press, 1957.

Fingerman, M. *The Control of Chromatophores*. Oxford: Pergamon Press, 1963.

Fraenkel, G. S., and D. L. Gunn. *The Orientation of Animals*. New York: Dover Publications, 1961.

Green, J. *A Biology of Crustacea*. London: H. F. and G. Witherby, 1961.

Hughes, G. M. Coordination of insect movement, I. The walking movements of insects, *J. Exp. Biol.* **29**:267–84, 1952.

Meglitsch, P. *Invertebrate Zoology*. New York: Oxford University Press, 1967.

Potts, W. T. W., and G. Parry. *Osmotic and Ionic Regulation in Animals*. Oxford: Pergamon Press, 1964.

Pringle, J. W. S. *Insect Flight*. London: Cambridge University Press, 1957.

Richards, A. G. *The Integument of Arthropods*. Minneapolis: University of Minnesota Press, 1951.

Rockstein, M. (ed.). *The Physiology of Insecta*. New York: Academic Press, 1964.

Roeder, K. D. *Insect Physiology*. New York: John Wiley & Sons, 1953.

Savory, T. H. *The Biology of Spiders*. New York: Macmillan Publishing Co., 1928.

Snodgrass, R. W. *A Textbook of Arthropod Anatomy*. Ithaca, N.Y.: Comstock Publishing Associates, Cornell University Press, 1938.

Snodgrass, R. W. *Principles of Insect Morphology*. New York: McGraw-Hill Book Co., 1952.

Waterman, T. H. (ed.). *The Physiology of Crustacea*. New York: Academic Press, 1960.

Welsh, J. H., R. I. Smith, and A. E. Kammer. *Laboratory Exercises in Invertebrate Physiology*. Minneapolis: Burgess Publishing Co., 1968.

Wigglesworth, V. B. *The Principles of Insect Physiology*. London: Methuen and Co., 1953.

An Abbreviated Classification of the Phylum Arthropoda and Its Allies (After Various Authors)

Phylum Onychophora

Vermiform, tracheate, soft cuticle, indistinct head bearing eyes and antennae, parapodia-like limbs bearing claws, e.g., *Peripatus, Peripatopsis*.

Phylum Arthropoda

Metameric animals with a chitinous exoskeleton and paired, jointed appendages.

Subphylum I Trilobita

Extinct aquatic arthropods with body molded into three longitudinal lobes, one pair of antennae, appendages of body all similar.

Subphylum II Chelicerata

Arthropods lacking antennae and mandibles, mouthparts consist of chelicerae, body divided into anterior prosoma and posterior opisthosoma.

Class 1 Merostomata

Aquatic chelicerates with broad prosoma divided by hinge from opisthosoma, pedipalps similar to five pairs of ambulatory appendages, gnathobases on walking legs, gill books, e.g., *Limulus*.

Class 2 **Arachnida**

Terrestrial chelicerates, four pairs of ambulatory appendages, gill books, and/or tracheae.

Order 1 **Scorpionida**

Prosoma covered by carapace, opisthosoma divided into mesosoma and metasoma, chelicerae and pedipalps both chelate, four pairs of lung books, e.g., scorpions.

Order 2 **Araneida**

Prosoma separated from opisthosoma by narrow pedicel or "waist," prosoma covered by single tergal shield, head marked off by groove, spinning glands, e.g., spiders.

Order 3 **Acarina**

Rounded body, no boundary between prosoma and opisthosoma, base of pedipalps united behind mouth, tracheae, e.g., ticks and mites.

Subphylum III **Mandibulata**

Arthropods with antennae and mandibles.

Class 1 **Crustacea**

Aquatic mandibulates bearing two pairs of antennae.

Subclass 1 **Cirripedia**

Sessile forms lacking compound eyes, carapace encloses trunk, six thoracic limbs, barnacles, e.g., *Lepas, Balanus*.

Subclass 2 **Branchiopoda**

Free swimming with compound eyes, carapace usually present, at least four pairs of trunk limbs, which are broad (phyllopods) and have setae.

Order 1 **Anostraca**

No carapace, stalked eyes, trunk limbs numerous and similar, fairy shrimps, e.g., *Artemia*.

Order 2 **Notostraca**

Carapace a broad trunk shield, sessile eyes, trunk limbs numerous with first pair different, tadpole shrimps, e.g., *Apus*.

Order 3 **Diplostraca**

Compressed carapace encloses trunk and limbs.

Suborder 1 **Cladocera**

Four to six trunk limbs, e.g., *Daphnia*

Suborder 2 **Chonostraca**

Ten to twenty-seven trunk limbs

Subclass 3 **Ostracoda**

Bivalve shell with adductor muscle, two pairs of trunk limbs, e.g., *Cypris, Cypridina*.

Subclass 4 **Copepoda**

No compound eyes or carapace, typically six pairs of limbs.

Subclass 5 **Malacostraca**

Compound eyes usually stalked, carapace typically over thorax, thorax of eight segments, abdomen usually of six segments.

Series 1 **Leptostraca**

Seven abdominal segments, phyllopodia, carapace present or absent.

Series 2 **Eumalacostraca**

Six abdominal segments or less.

Superorder 1 **Pericarida**

Carapace absent or when present not fused.

Order 1 **Isopoda**

No carapace, body flattened dorsoventrally, sowbugs, pill bugs, e.g., *Oniscus, Porcellio.*

Order 2 **Amphipoda**

No carapace, body flattened laterally.

Order 3 **Mysidacea**

Carapace over most of thorax.

Order 4 **Cumacea**

Carapace over three to four segments.

Superorder 2 **Eucarida**

Carapace fused to all thoracic segments.

Order 1 **Decapoda**

Big scaphognathite, statocyst, five pairs of ambulatory appendages, e.g., crabs, lobsters, shrimps.

Order 2 **Euphausiacea**

Small scaphognathite, no statocyst.

Class 2 **Labiata**

First pair of antennae present, bases of second maxillae fused to form labium.

Subclass 1 **Chilopoda**

Body flattened, segments all similar and typically each abdominal one bearing a pair of appendages, first segment bears poison claws, genital opening posterior, e.g., centipedes.

Subclass 2 **Diplopoda**

Body cylindrical, segments all similar and typically each abdominal segment bearing two pairs of appendages, genital opening anterior, e.g., millipedes.

Subclass 3 **Insecta**

Body divided into three regions, three pairs of legs, genital opening near anus.

Order 1 **Thysanura**

Chewing mouthparts, wingless, no metamorphosis, e.g., silverfish, fire brats.

Order 2 **Collembola**

Chewing mouthparts, wingless, no metamorphosis, most species leap, e.g., springtails.

Order 3 **Ephemeroptera**

Vestigial chewing mouthparts, winged, incomplete metamorphosis, e.g., mayflies.

Order 4 **Odonata**

Chewing mouthparts, membranous wings, huge eyes, metamorphosis without pupa, e.g., dragonflies, damselflies.

Order 5 **Orthoptera**

Chewing mouthparts, hind wings membranous, metamorphosis gradual, e.g., cockroaches, grasshoppers, crickets, mantids.

Order 6 **Isoptera**

Chewing mouthparts, metamorphosis gradual, e.g., termites.

Order 7 **Dermaptera**

Chewing mouthparts, metamorphosis gradual, e.g., earwigs.

Order 8 **Mallophaga**

Chewing or biting mouthparts, wingless, slight metamorphosis, e.g., chewing lice.

Order 9 **Anoplura**

Sucking mouthparts, wingless, slight metamorphosis, e.g., sucking lice.

Order 10 **Hemiptera**

Piercing-sucking mouthparts, winged, e.g., true bugs.

Order 11 **Homoptera**

Piercing-sucking mouthparts, winged or wingless, e.g., cicadas, aphids, leafhoppers.

Order 12 **Neuroptera**

Chewing mouthparts, membranous wings, complete metamorphosis, e.g., dobson flies, ant lions.

Order 13 **Coleoptera**

Chewing mouthparts, thick forewings (elytra); complete metamorphosis, e.g., beetles.

Order 14 **Lepidoptera**

Sucking mouthparts in adult, complete metamorphosis, mandibles may be absent, e.g., butterflies and moths.

Order 15 **Diptera**

Piercing-sucking or lapping mouthparts, pair of membranous wings, complete metamorphosis, e.g., flies, mosquitoes.

Order 16 **Siphonaptera**

Piercing-sucking mouthparts, body laterally compressed, wingless, complete metamorphosis, e.g., fleas.

Order 17 **Hymenoptera**

Chewing or lapping mouthparts, four wings, complete metamorphosis, e.g., wasps, bees, ants.

Equipment and Materials

Equipment

Dissecting instruments
Dissecting microscope
Compound microscope
Microscope slides
Lamp
Glass vials and jars
Plasticine
Pasteur pipets
Paraffin
Hypodermic syringes and needles
Tank of oxygen
Tank of carbon dioxide
Aspirator (water faucet)
Aluminum foil
Kymograph
Plastic petri dishes
Cork borer
Bunsen burner
Rubber tubing
Dry ice
Hot plate
Sandpaper
Nail polish
Balance
Desiccators
Thread
Ring stand
Glass rod
Syracuse watch glasses

Solutions and Chemicals

Pablum or mashed banana
Carmine powder
Congo red
Congo-red-stained yeast suspension
Litmus powder
Neutral red (powder and 1% solution)
India ink
Milk
0.5% cyanol–Congo red
1% iodine solution
Giemsa stain
10% sodium chloride
Insect Ringer's solution
Chloroform
Methylene blue
Waterproof cement
0.01 M $AgNO_3$
0.1 N HCl
1% indigo carmine
Photographic developer
Petroleum jelly
5% sucrose

Prepared Slides

Peripatus, whole mount (plastic)
Peripatus, cross section
Honeybee mouthparts
Butterfly mouthparts

Squash bug mouthparts
Mosquito mouthparts

Preserved Materials

Lobster *(Homarus)* or crayfish
   (Astacus or *Cambarus)*
Crab
Centipede *(Scolopendra)*
Millipede *(Spirobolus)*
Grasshopper *(Romalea)*
Spider *(Argiope)*
Scorpion
Limulus
Lepas

Living Material*

Limulus (7,12)
Astacus or *Cambarus* (3,4,6,10)
Balanus or *Mitella* (7,9,12)
Daphnia (3,4,6)

Damselfly nymph (3,5)
Eubranchipus (5)
Mayfly nymph (5)
Dragonfly nymph (3,5)
Meal worm *(Tenebrio)* larvae
   (3,4,5,6,10)
Periplaneta, nymphs and adults
   (3,4,6)
Artemia,† adults (3,4,6,11), or
   Eubranchipus, adults (5)
Drosophila, larvae and adults (4,5,11)
Clams or mussels (7,9,12)
Pill bugs *(Armadillidium),* (or
   Porcellio or *Oniscus)* (3,5)
Blowfly‡ *(Calliphora)* larvae (11)
Housefly‡ *(Musca)*
Procambarus clarki (small) (3)
Palaemonetes (7,9,12)
Uca (4,7,9)

*Numbers in parentheses identify sources listed in Appendix B.
†*Artemia* are obtained as eggs and may be reared according to the supplier's directions.
‡Blowflies and houseflies must be reared in the laboratory. Good directions for doing this in large numbers may be found in Welsh, Smith, and Kammer (1968).

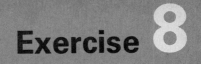
MOLLUSCA

Contents

Molluscs (*molluscus,* soft; L.), as their name implies, are soft-bodied animals, more familiarly known as clams, octopods, and snails. They are bilaterally symmetrical with well-developed digestive, circulatory, excretory, and respiratory systems. A calcareous shell may or may not be present.

Molluscs are closely related to annelids in their mode of development and the presence of a **trochophore larva,** but they are distinguished from the latter by a lack of segmentation. In addition, the extensive coelom, so important in annelid locomotion, is reduced in size in most molluscs; it is usually restricted to the area surrounding the heart and to spaces within the gonads and kidneys. The metameric arrangement of appendages and body organs, allowing streamlined locomotion and specialization of body regions in the annelids, is absent in the majority of molluscs. Most molluscs, as one might suspect, are slow-moving, creeping forms, but in some instances the organization of the body regions has been modified so that swift movements are possible. In the evolutionary sense they are extremely plastic, and even without segmentation they demonstrate highly successful adaptations to a variety of habitats.

The considerable adaptive radiation exhibited by members of this phylum is reflected by the varied functional morphology of the group. In numbers of species they are among the most abundant of all organisms. Although primarily marine, representatives are found in fresh water and on land.

8–1. BODY ORGANIZATION

All molluscs can theoretically be derived from a generalized plan consisting of three main body regions: **head-foot, visceral mass,** and **mantle** or **pallium** (Figure 8.1). The head-foot region is the locomotory and sensory portion of the body, upon which rides the visceral mass containing the excretory, digestive, and circulatory organs. Whereas the head-foot region is operated by muscles, the visceral mass works by means of cilia and mucus. The third component of the body, the mantle, forms a fleshy cover over the visceral mass and a skirt around the foot. It secretes the **shell,** which lies on its outer surface, contains the gills or **ctenidia,** and itself functions in respiration. The mantle alone might be considered the hallmark of the molluscs, for the plasticity of the mantle in both form and function has contributed greatly to the success of the group.

Each of the five classes of molluscs reflects the elaboration or suppression of one or more of the three body regions. (See Classification of Mollusca, p. 270.)

i. AMPHINEURA.   The class Amphineura contains two distinct subclasses: the Polyplacophora or chitons, and the specialized Aplacophora, wormlike molluscs that have lost the shell. The chitons are in some ways the most primitive of molluscs, showing traces of segmentation that hark back to their ancestors of preannelid days. They spend most of their lives tightly compressed to flat rocks. The body is compressed in the dorsoventral axis and elongated in the anteroposterior axis (Figure 8.1).

▶   Obtain a specimen of *Chaetopleura, Ischnochiton* or *Mopalia,* and examine the external structure (Figure 8.8A). The dorsal surface is occupied by the **shell,** which is composed of eight plates. Round the edge of these runs a muscular **girdle** of tissue, an extension of the mantle, into which the plates are anchored. Underlying these plates is the **mantle,** which in turn covers the **visceral mass.** Turn the chiton so that the ventral surface faces you. The muscular **foot** occupies a major portion of the ventral surface. Between the mantle edge and the foot is a groove (the **pallial groove**) in which the **ctenidia** lie. The ctenidia are not paired but are arranged

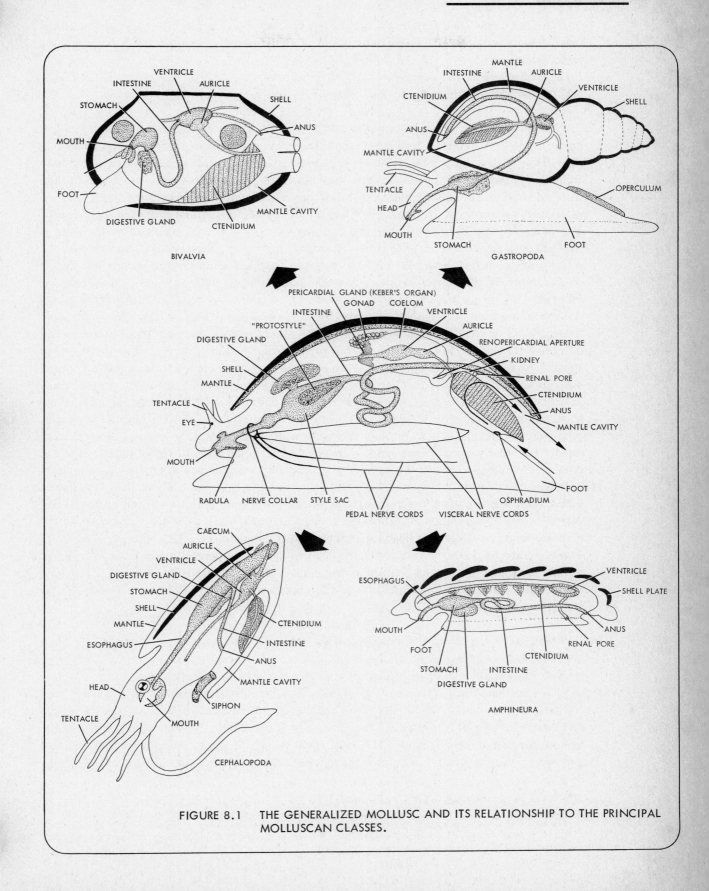

FIGURE 8.1   THE GENERALIZED MOLLUSC AND ITS RELATIONSHIP TO THE PRINCIPAL MOLLUSCAN CLASSES.

in a series that runs the entire length of the body on each side. Identify the **mouth** and **anus,** the paired **renal openings** at the base of the posterior ctenidia and, anterior to the renal pores, the paired **gonopores.**

Can you relate all structures observed in your specimen to the generalized mollusc (Figure 8.1)? What adaptations does the chiton have for clinging tightly to the substrate?

ii. GASTROPODA.   The gastropoda comprise an extremely diverse group, including slugs, snails, whelks, and limpets. In some respects, they depart least from the generalized molluscan plan (Figure 8.1), although this is somewhat obscured by a developmental twisting of the body through 180°, known as **torsion.** The visceral mass turns counterclockwise with respect to the foot, and the mantle cavity and anus occupy an anterior position. The head, with its well-developed sense organs, can be withdrawn into the mantle cavity and protected by the shell. The shell may or may not be coiled, and torsion of the body may be independent of coiling of the shell; that is, the body may be twisted inside a straight, uncoiled shell. In some species, the opening to the shell may be closed by a horny plate, the **operculum.** In some groups, the shell is reduced and there has been a tendency toward detorsion so that the mantle cavity may be reflected laterally or backward. This evolutionary torsion and detorsion not only presents difficulties in studying gastropod anatomy, but poses questions about its adaptive significance.

Depending on the degree of torsion as well as other features, the gastropods are classified into three subclasses. In the Prosobranchia (*proso,* front; *branch,* gill; Gk.), the visceral mass shows a torsion of 180° counterclockwise, with the mantle cavity and openings of the organ systems facing forward. In the Opisthobranchia (*opistho,* rear; *branch,* gill; Gk.), marine forms in which the shell is reduced or absent, the body is flattened and streamlined, and the adults show detorsion so that the mantle cavity is lateral or posterior; the gills are reduced or absent, respiration occurring largely through the body wall. The Pulmonata (*pulmo,* lung; L.) are mostly land or freshwater snails with coiled shells and some torsion. In the pulmonates, the ctenidia have been lost, and the mantle cavity is highly vascularized to function as a lung.

▶   We shall examine a member of each class: the prosobranch marine whelk *Busycon* (Figure 8.4A), the marine opisthobranch (nudibranch) *Aplysia* (Figure 8.2A), and the terrestrial snail *Helix* (Figure 8.5A). Obtain a freshly killed* or preserved specimen of each,† and examine the external features. Note the **bilateral symmetry** and **cephalization.** The anterior **head** bears sensory **tentacles** (how many?) and light-sensitive **eyes.** The **visceral hump** rides as a dorsal protuberance on the ventral, muscular **foot,** and a **shell** is present in all three specimens. Preserved specimens of *Busycon* will have been removed from the shell, and that of *Aplysia* is hidden by folds of the mantle.

▶   *Prosobranchia.*   Observe a living specimen of *Busycon* in an aquarium and notice the large muscular **foot** by means of which the animal can plough through the sand, the vertically held inhalent **siphon,** the extensible **proboscis,** the **tentacles, mouth,** and **operculum.**

▶   Examine the surface features of a preserved specimen of *Busycon,* which will have been removed from its shell. On the foot, locate the thin horny **operculum,** which in life closes the aperture of the shell and protects the animal retracted within it. Near the anterior end of the sole of the foot is the opening of the **pedal gland,** which produces mucus on which the foot glides; by means of mucus and suction the foot can adhere to rocks. The pedal gland also helps

---

*For directions on methods of killing molluscs, see page 271.
†This chapter involves six comparative dissections. Students may work in groups of two or three; each student should dissect at least one gastropod and one or more specimens from the other three classes. In this way, students will perform two or three dissections, depending on the size of the group. The six dissected organisms should be compared as each section of the exercise is completed.

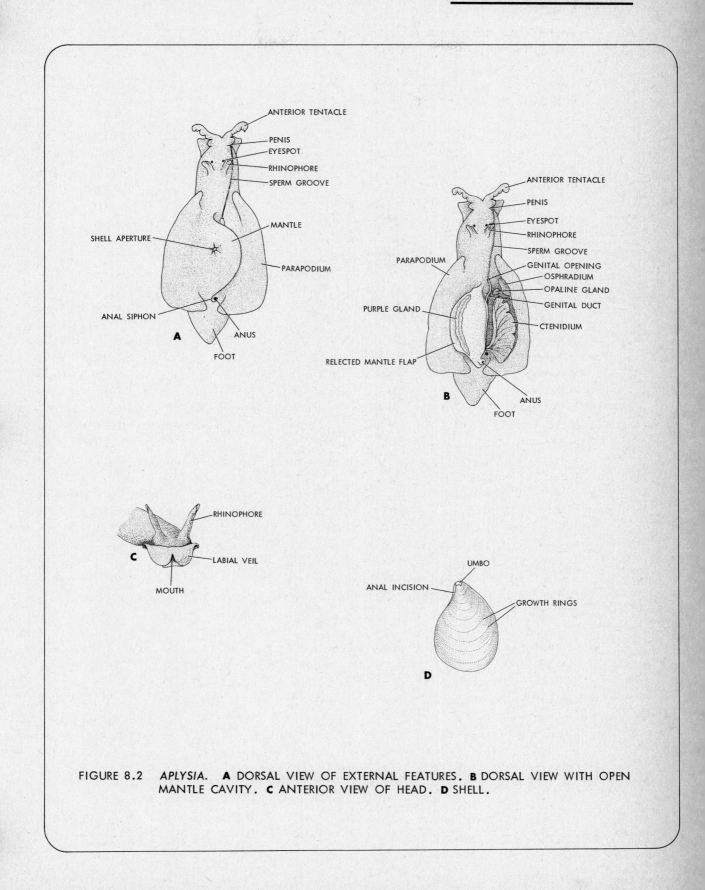

FIGURE 8.2 *APLYSIA.* **A** DORSAL VIEW OF EXTERNAL FEATURES. **B** DORSAL VIEW WITH OPEN MANTLE CAVITY. **C** ANTERIOR VIEW OF HEAD. **D** SHELL.

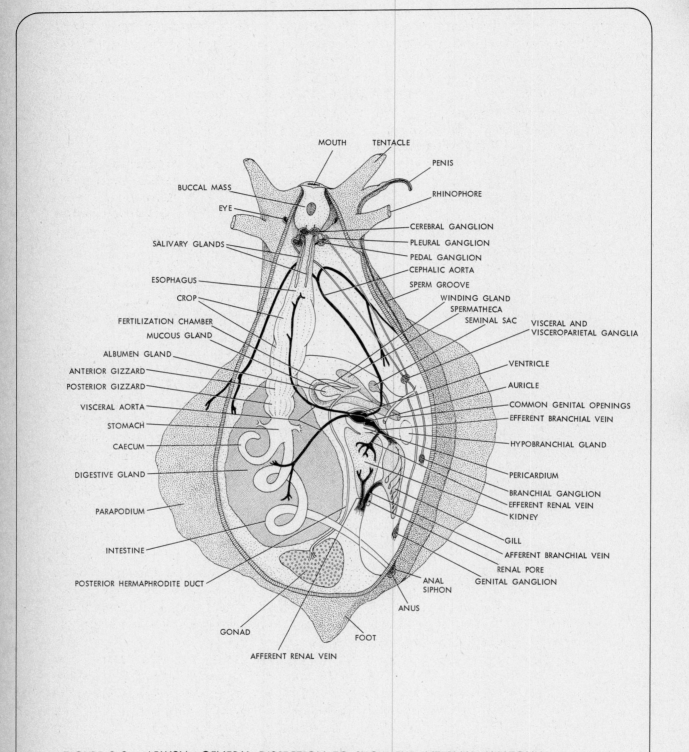

FIGURE 8.3 *APLYSIA.* GENERAL DISSECTION TO SHOW THE INTERNAL ANATOMY.

to mold the egg cases as they are laid. The triangular **mouth** is located at the end of the proboscis protruding from beneath the paired extensible tentacles. Notice the lateral position of the pigmented **eyespots** on the tentacles. In male specimens, a large **penis** can be seen by the right tentacle. The coiled visceral mass is covered by a thin, fairly transparent **mantle,** which thickens to form a **collar** at the base of the visceral mass. The shell is secreted at the edge of the collar. The collar is elongated in one area to form the highly extensible **siphon** through which water is drawn into the mantle cavity. Through the mantle at the apex, locate the right lobe of the brownish **digestive gland** or **liver** on which lies the yellow or orange **gonad** and the straight (female) or coiled (male) **gonoduct.** These two organs fill the first and smallest whorl of the shell. The next whorl is occupied by the left liver lobe, together with the **stomach** and part of the **intestine,** which can be seen on the outer surface of the liver. The large brown **kidney** and the **heart** are located along the dorsal surface to the left of the base of the visceral mass. Anterior to the heart and kidney lies the oblong **ctenidium** and the chemosensory **osphradium** and the outlines of the mucus-secreting **hypobranchial gland.** Where would the heart and kidney be located relative to the ctenidium in the generalized mollusc?

▶ Notice how the visceral mass is shaped to fit the whorls of the shell.

Opisthobranchia. Opisthobranchs probably evolved from a prosobranch type, and, as you will see from your study of *Aplysia,* there has been considerable detorsion and reduction of the shell.

▶ Examine a freshly killed or preserved specimen of *Aplysia** (Figure 8.2**A**), and observe the large muscular grooved **foot.** The grooves enable the animal to climb on rocks and seaweed. On either side are the fleshy **parapodia,** which are reflected laterodorsally to cover the mantle and visceral hump. Turn aside the parapodia and locate the visceral hump in the center of which is a hole, the **shell aperture,** left by incomplete closure of the mantle fold over the shell. The shell is considerably reduced and consists of a thin chitinous plate secreted over the visceral hump by the mantle. The mantle almost entirely encloses the shell; its only contact with the environment is via the shell aperture.

▶ The **mantle cavity** lies on the right and opens to the right. Turn aside the mantle flap covering the mantle cavity (Figure 8.2**B**); inside this flap, locate the **purple gland,** which secretes purple dye that may be squirted out when the animal is molested. (The purple gland in *Aplysia* is homologous to the hypobranchial gland of other gastropods.) The single large fleshy **ctenidium** is located posteriorly on the right. Behind it, find the **anus** in an elongate channel of the mantle floor, the **anal siphon.** The **kidney duct** also opens posterior to the ctenidium. Like all opisthobranchs, *Aplysia* is hermaphroditic, and the **genital duct** opens anteriorly. The **osphradium,** a sense organ for testing the water passing to the ctenidium, is located anterior to the ctenidium. Find the clear grapelike **opaline gland** on the floor of the mantle cavity. In life, this secretes a noxious fluid, an accessory defense mechanism.

▶ Anteriorly, the extensible **head** bears the **mouth** and two pairs of grooved **tentacles.** The pigmented **eyes** are anterolateral to the posterior tentacles, which function as olfactory organs and are called **rhinophores.** To the right of the right anterior tentacle is a small **penis,** from which an open ciliated **spermatic groove** passes back to the genital opening on the visceral hump. Tentacular lip outgrowths form a **labial veil** around the mouth (Figure 8.2**C**); they aid in burrowing.

How does the position of the mantle cavity in *Aplysia* compare with that in *Busycon?* How does the shape of *Aplysia* fit it for its way of life?

*For directions on methods of killing molluscs, see page 271.

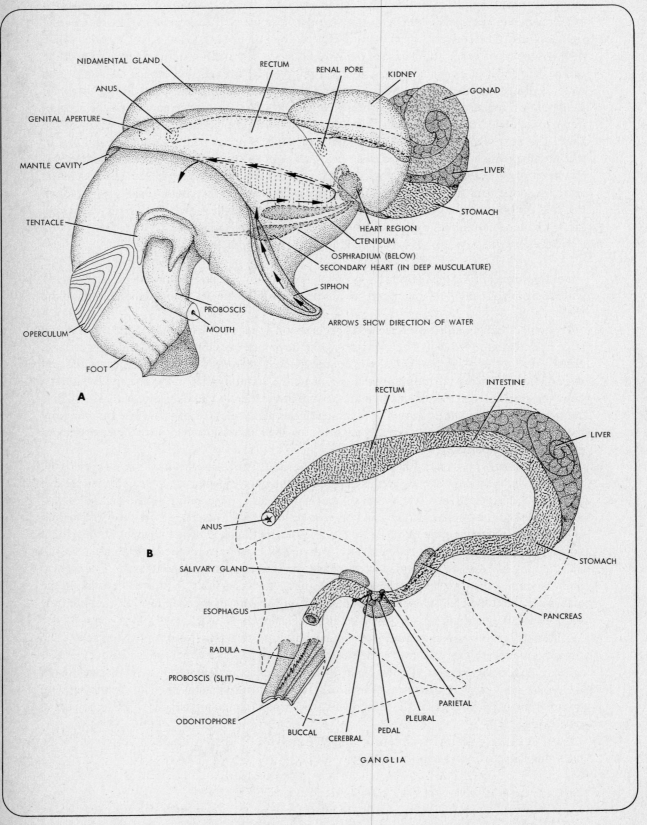

NIDAMENTAL GLAND

RECTUM

RENAL PORE

KIDNEY

GONAD

ANUS

GENITAL APERTURE

LIVER

MANTLE CAVITY

STOMACH

HEART REGION

CTENIDUM

TENTACLE

OSPHRADIUM (BELOW)

SECONDARY HEART (IN DEEP MUSCULATURE)

SIPHON

PROBOSCIS

MOUTH

ARROWS SHOW DIRECTION OF WATER

OPERCULUM

FOOT

A

RECTUM

INTESTINE

LIVER

ANUS

B

STOMACH

SALIVARY GLAND

ESOPHAGUS

PANCREAS

RADULA

PROBOSCIS (SLIT)

ODONTOPHORE

PARIETAL

PLEURAL

BUCCAL

PEDAL

CEREBRAL

GANGLIA

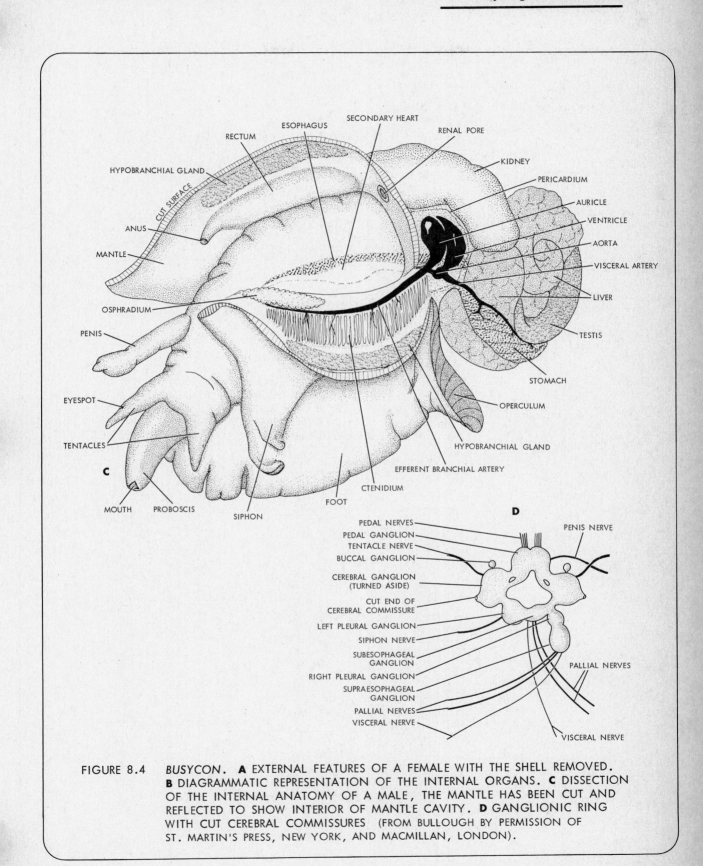

FIGURE 8.4   *BUSYCON.* **A** EXTERNAL FEATURES OF A FEMALE WITH THE SHELL REMOVED.
B DIAGRAMMATIC REPRESENTATION OF THE INTERNAL ORGANS. **C** DISSECTION
OF THE INTERNAL ANATOMY OF A MALE, THE MANTLE HAS BEEN CUT AND
REFLECTED TO SHOW INTERIOR OF MANTLE CAVITY. **D** GANGLIONIC RING
WITH CUT CEREBRAL COMMISSURES (FROM BULLOUGH BY PERMISSION OF
ST. MARTIN'S PRESS, NEW YORK, AND MACMILLAN, LONDON).

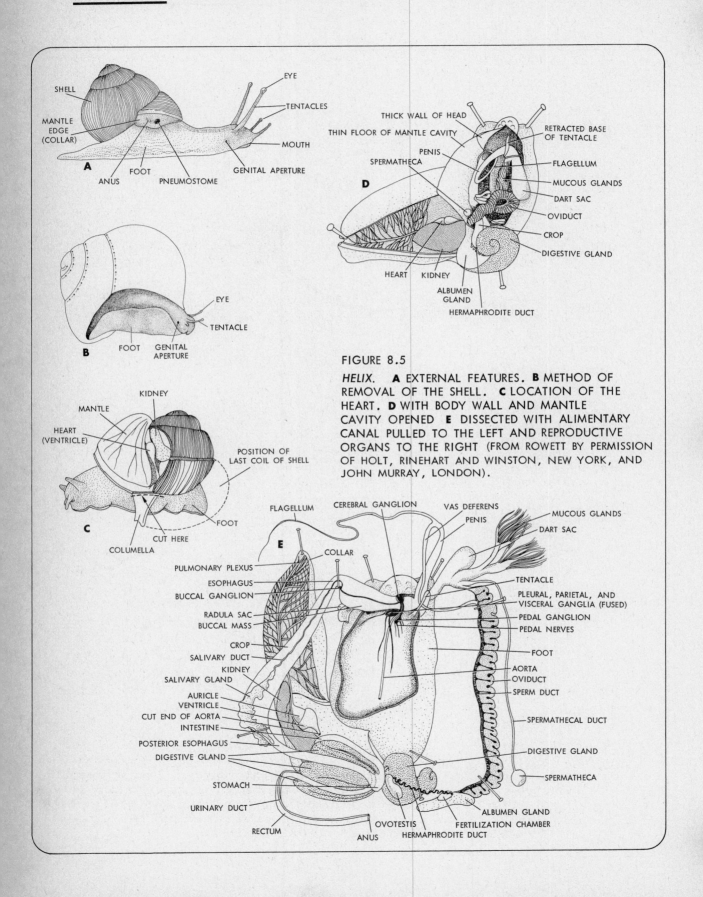

SHELL

EYE

TENTACLES

MANTLE
EDGE
(COLLAR)

MOUTH

ANUS

FOOT

PNEUMOSTOME

GENITAL APERTURE

A

EYE

TENTACLE

FOOT GENITAL
APERTURE

B

KIDNEY

MANTLE

HEART
(VENTRICLE)

POSITION OF
LAST COIL OF SHELL

FOOT

CUT HERE

COLUMELLA

C

THICK WALL OF HEAD

THIN FLOOR OF MANTLE CAVITY

RETRACTED BASE
OF TENTACLE

PENIS

SPERMATHECA

FLAGELLUM

MUCOUS GLANDS

DART SAC

OVIDUCT

CROP

DIGESTIVE GLAND

D

HEART KIDNEY

ALBUMEN
GLAND

HERMAPHRODITE DUCT

FIGURE 8.5

HELIX. **A** EXTERNAL FEATURES. **B** METHOD OF
REMOVAL OF THE SHELL. **C** LOCATION OF THE
HEART. **D** WITH BODY WALL AND MANTLE
CAVITY OPENED **E** DISSECTED WITH ALIMENTARY
CANAL PULLED TO THE LEFT AND REPRODUCTIVE
ORGANS TO THE RIGHT (FROM ROWETT BY PERMISSION
OF HOLT, RINEHART AND WINSTON, NEW YORK, AND
JOHN MURRAY, LONDON).

FLAGELLUM

CEREBRAL GANGLION

VAS DEFERENS

PENIS

MUCOUS GLANDS

DART SAC

E

COLLAR

PULMONARY PLEXUS

ESOPHAGUS

BUCCAL GANGLION

TENTACLE

PLEURAL, PARIETAL, AND
VISCERAL GANGLIA (FUSED)

RADULA SAC

BUCCAL MASS

PEDAL GANGLION

PEDAL NERVES

CROP

SALIVARY DUCT

KIDNEY

SALIVARY GLAND

FOOT

AORTA

OVIDUCT

SPERM DUCT

AURICLE

VENTRICLE

CUT END OF AORTA

INTESTINE

SPERMATHECAL DUCT

POSTERIOR ESOPHAGUS

DIGESTIVE GLAND

DIGESTIVE GLAND

STOMACH

SPERMATHECA

URINARY DUCT

ALBUMEN GLAND

FERTILIZATION CHAMBER

RECTUM

ANUS

OVOTESTIS

HERMAPHRODITE DUCT

▶ *Pulmonata.* Observe a living specimen of the snail *Helix* and note that the head is not clearly separated from the foot (Figure 8.5A). Touch the two pairs of tentacles, and observe how they can be invaginated. Stronger stimulation will cause the animal to retract into the shell.

▶ Hold a freshly killed or preserved specimen of *Helix** in your left hand and, using bone forceps, carefully cut around the spirals of the shell and remove it from the animal (Figure 8.5B). Leave only the central column (**columella**) and, as you cut, be careful not to disturb the soft parts. Observe the **columellar muscle** that attaches the animal to the shell. What is its function aside from anchorage? The **mantle** surrounds the compactly coiled **visceral mass;** the mantle is thickened anteriorly to form a **collar,** which extends around the body and secretes the **shell.** The space enclosed by the mantle is the **mantle cavity.** Extending back from the collar over the first visceral coil, the mantle is highly vascularized. The rest of the mantle is less highly vascularized, and through it can be seen the organs of the visceral hump. With the aid of Figure 8.5C, locate the large **kidney** and, anterior to it, the **heart.** In fresh specimens this will probably pulsate. The dark brown **digestive gland,** with a loop of **intestine** at its surface, the lighter brown **albumen gland,** and the **gonad** fill the remaining whorls to the apex.

▶ Turn the animal so that the ventral side faces you and identify the position of the **mouth** with its two **lateral lips** and one **ventral lip.** Behind and above the right lateral lip, locate the **genital aperture.**

Make sure you can relate all the external anatomical structures you have identified in your gastropod specimens to those of the generalized mollusc.

iii. BIVALVIA. Bivalves are generally sedentary. Foot, visceral mass, and the mantle cavity dominate the body, and the head is completely suppressed (Figure 8.1). Along with the buccal mass, bivalves have lost the radula and the majority are ciliary feeders with large, platelike ctenidia. The extensive mantle encloses the entire body in two symmetrical flaps and secretes the hinged shell.

▶ Obtain a fresh or preserved clam *(Mercenaria)* and note that the body is enclosed by a bivalved **shell** and that head structures are lacking (Figure 8.6). Locate the bulge in the shell **(umbo)** just below the point where the valves are attached to each other by the **hinge ligament.** In *Mercenaria,* the umbones are dorsal. The muscular foot may protrude anteriorly.

▶ The shell valves of a bivalve are lateral; that is, there are right and left valves. Make sure you know which is which. Take a sharp scalpel and carefully insert it between the valves; move the blade along the ventral edge from posterior to anterior, keeping the knife close to the left valve so that you do not damage the specimen, and cut the muscles that effect closure of the shell (Figure 8.6A). In fresh specimens you will need to insert a pair of scissors and twist them with some force to hold the valves apart while you cut the muscles. Spread the valves apart, remove the left valve, and examine the internal anatomy. Underlying the shell and now exposed is the fleshy **mantle** (Figure 8.6B). Notice how it hangs like a sheet, attached dorsally and free ventrally. Through the semitransparent mantle may be seen the dorsally located **pericardium.** Posterior and ventral to the pericardium is the brownish **kidney.** Identify the anterior and posterior **adductor muscles** responsible for shell closure and the anterior and posterior **retractor muscles** for extending and withdrawing the foot. Locate the posterior pigmented **siphons,** formed from the fused edges of the mantle. Water enters through the ventral **inhalent siphon,** circulates through the mantle cavity, and exits via the dorsal **exhalent siphon.** Lift up the mantle and locate the **ctenidia.** Fold back the left ctenidium. The firm ventral protuberance is the **foot,** and the **visceral mass** is dorsal.

*For directions on methods of killing molluscs, see page 271.

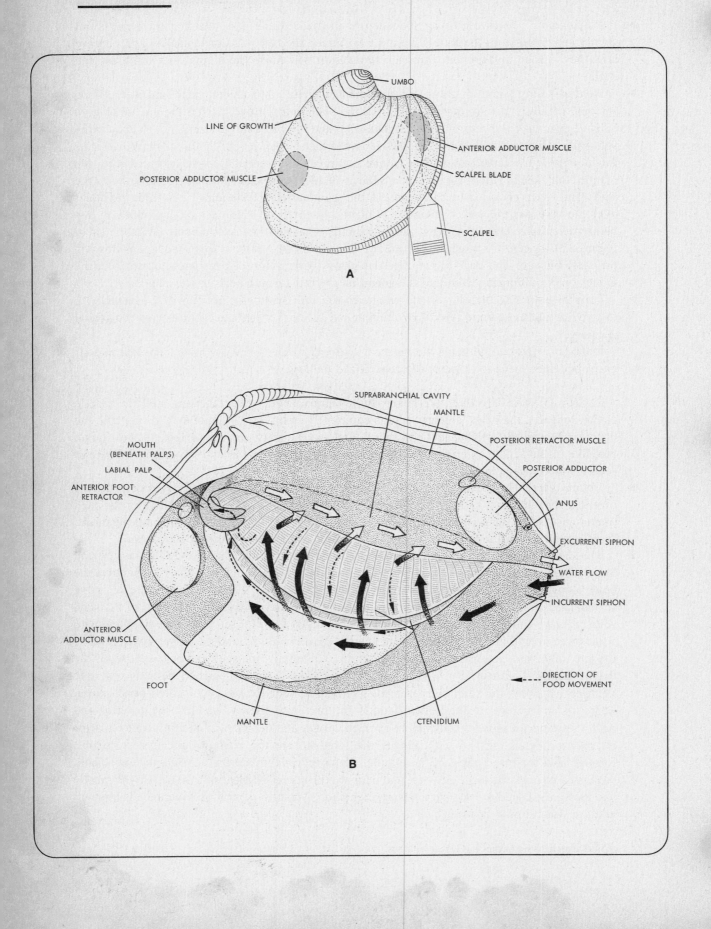

UMBO

LINE OF GROWTH

ANTERIOR ADDUCTOR MUSCLE

SCALPEL BLADE

POSTERIOR ADDUCTOR MUSCLE

SCALPEL

A

SUPRABRANCHIAL CAVITY

MANTLE

MOUTH
(BENEATH PALPS)

POSTERIOR RETRACTOR MUSCLE

LABIAL PALP

POSTERIOR ADDUCTOR

ANTERIOR FOOT
RETRACTOR

ANUS

EXCURRENT SIPHON

WATER FLOW

INCURRENT SIPHON

ANTERIOR
ADDUCTOR MUSCLE

DIRECTION OF
FOOD MOVEMENT

FOOT

MANTLE

CTENIDIUM

B

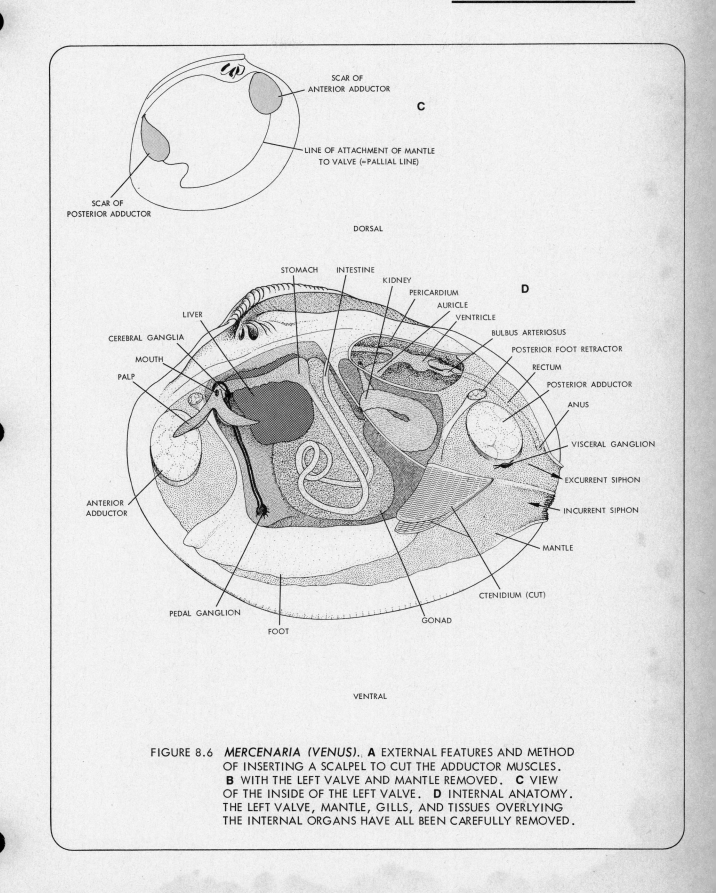

FIGURE 8.6 *MERCENARIA (VENUS).* **A** EXTERNAL FEATURES AND METHOD OF INSERTING A SCALPEL TO CUT THE ADDUCTOR MUSCLES. **B** WITH THE LEFT VALVE AND MANTLE REMOVED. **C** VIEW OF THE INSIDE OF THE LEFT VALVE. **D** INTERNAL ANATOMY. THE LEFT VALVE, MANTLE, GILLS, AND TISSUES OVERLYING THE INTERNAL ORGANS HAVE ALL BEEN CAREFULLY REMOVED.

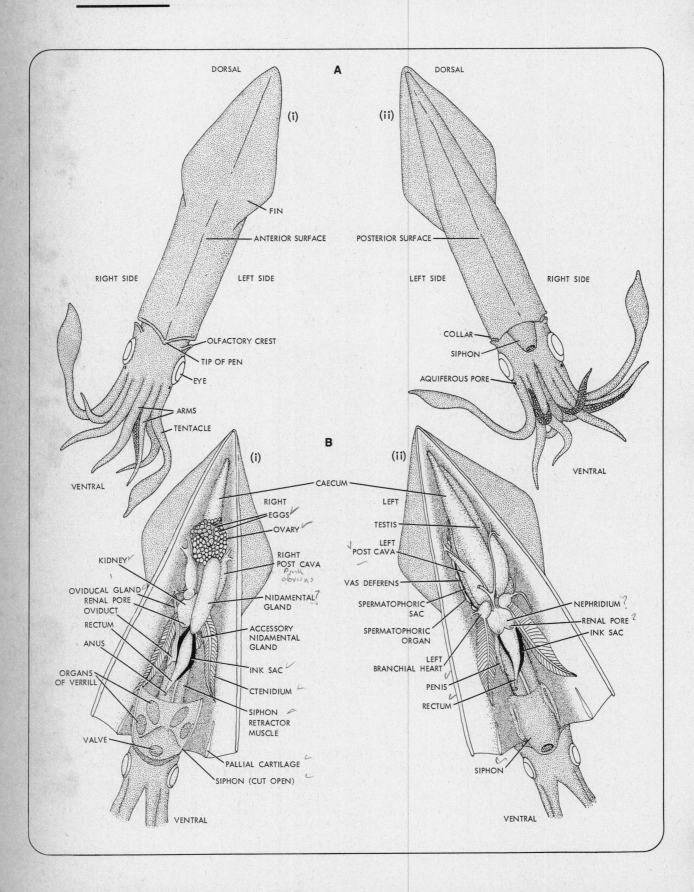

A

(i)

DORSAL

FIN

ANTERIOR SURFACE

RIGHT SIDE

LEFT SIDE

OLFACTORY CREST

TIP OF PEN

EYE

ARMS

TENTACLE

VENTRAL

(ii)

DORSAL

LEFT SIDE

RIGHT SIDE

COLLAR

SIPHON

AQUIFEROUS PORE

VENTRAL

B

(i)

CAECUM

RIGHT

EGGS

OVARY

RIGHT
POST CAVA
Pink
obvious

NIDAMENTAL
GLAND

KIDNEY

OVIDUCAL GLAND
RENAL PORE
OVIDUCT

RECTUM

ANUS

ORGANS
OF VERRILL

VALVE

ACCESSORY
NIDAMENTAL
GLAND

INK SAC

CTENIDIUM

SIPHON
RETRACTOR
MUSCLE

PALLIAL CARTILAGE

SIPHON (CUT OPEN)

VENTRAL

(ii)

LEFT

TESTIS

LEFT
POST CAVA

VAS DEFERENS

SPERMATOPHORIC
SAC

SPERMATOPHORIC
ORGAN

LEFT
BRANCHIAL HEART

PENIS

RECTUM

NEPHRIDIUM

RENAL PORE

INK SAC

SIPHON

VENTRAL

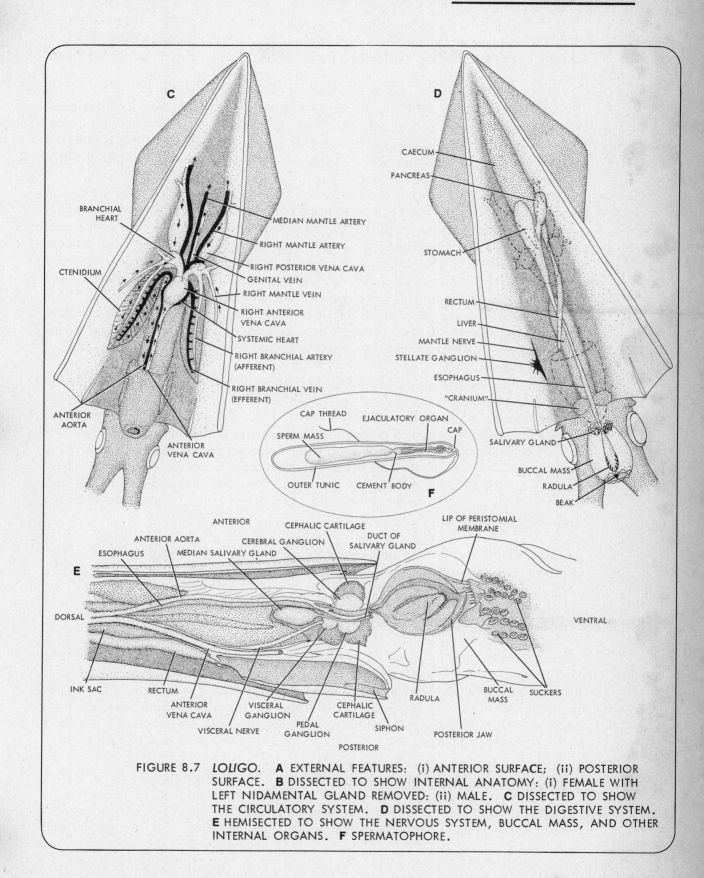

FIGURE 8.7 *LOLIGO.* **A** EXTERNAL FEATURES: (i) ANTERIOR SURFACE; (ii) POSTERIOR SURFACE. **B** DISSECTED TO SHOW INTERNAL ANATOMY: (i) FEMALE WITH LEFT NIDAMENTAL GLAND REMOVED: (ii) MALE. **C** DISSECTED TO SHOW THE CIRCULATORY SYSTEM. **D** DISSECTED TO SHOW THE DIGESTIVE SYSTEM. **E** HEMISECTED TO SHOW THE NERVOUS SYSTEM, BUCCAL MASS, AND OTHER INTERNAL ORGANS. **F** SPERMATOPHORE.

▶ Note the position of the adductor muscles as scars on the shell you have removed (Figure 8.6C). The irregular line also seen in the shell marks the position of attachment of the mantle to the shell. Keep the opened specimen under moist toweling or in a bowl of seawater for future study.

iv. CEPHALOPODA.   As the name indicates (*kephalo,* head; *pod,* foot; Gk.), the head-foot is dominant in cephalopods (Figure 8.1). These are the most highly evolved of all molluscs. They are swift-moving carnivores, and their bodies are quite streamlined. The shell is either not well developed or entirely absent; when present it is internal, except in the most primitive representatives (e.g., *Nautilus*).

▶ Examine a preserved specimen of the squid *Loligo* (Figure 8.7). Note that there is no external shell and that the major part of the body is enclosed by the soft, fleshy **mantle** and is sharply demarcated from the rest of the squid body by the **collar.** Opposite the pointed end (apex) of the animal, you will find the **head-foot** region. The eight **arms** and two longer **tentacles** are derivatives of the foot and surround the mouth. Turn back the arms and tentacles to reveal a muscular membrane running from their bases to the mouth; it is composed of an outer lobed **buccal membrane** with suckers on its inner surface and an inner **peristomial membrane.** Notice the horny beak protruding from the mouth. If your specimen is female, locate the **horseshoe organ,** a glandular area on the buccal membrane. Locate the large, well-developed **eyes** and the small **aquiferous pore** beneath each one leading to the **eye chamber.** Behind the eye is a fold of tissue, the **olfactory crest.**

▶ Orientation at this point may cause problems, so align your specimen in such a way that the following descriptions will not be confusing. Place the animal so that the apex is furthest from you and the arms closest to you. Turn the animal so that the **siphon** (the tubular projection between the collar and the head-foot region) is facing you. It too is a derivative of the foot. The apex, which bears the visceral mass, is dorsal and the arms are ventral. The surface facing you is posterior. The **eyes** are on the left and right sides of the body, and the head structures have moved to a position that is dorsal to the foot; that is, the anteroposterior axis is shortened. What axis of the body has become elongated? Make sure that you can relate all the structures you see to the generalized mollusc (Figure 8.1).

v. SCAPHOPODA.   Tusk or tooth shells are molluscs that burrow. Because of the scarcity of specimens, this class will not be treated in the laboratory.

8–2. MANTLE CAVITY, SHELL, AND RESPIRATION

The mantle is a fleshy cover that incompletely surrounds most of the organs of the body; the space between the fleshy mantle lobes is called the **mantle (pallial) cavity.** The openings of the digestive, excretory, and reproductive systems are all found within the mantle cavity. The mantle also functions in secretion of the shell. However, the most distinctive function of the mantle cavity is to provide a space for housing the delicate ctenidia, thus protecting them from the hazards of the environment and permitting an oriented flow of water across them.

a. Structure of the Mantle and Shell

▶ i. BIVALVIA.   Take the shell that you previously removed from a clam, or remove a shell as described in Section 8–1, and examine its form. On the outside is a thin proteinaceous covering called the **periostracum** (*peri,* around; *ostracum,* a shell; Gk.). This material can be most clearly seen at the hinge. The middle layer, which you can see by breaking off a piece of shell, is the **prismatic** layer; it is composed of calcium carbonate. The innermost layer of the shell, and the one that lies closest to the mantle surface, is the **nacreous** or mother-of-pearl layer. Each of these layers is secreted by the mantle.

▶ Obtain a prepared slide showing a cross section of the mantle lobes, and observe the different cell structure in each of the three lobes (Figure 8.9). The entire outer surface of the mantle secretes the nacreous layer. This same outer lobe, at its tip, secretes the prismatic layer, and the cells at the inner edge of this lobe secrete the periostracum. Concentric lines of growth surround the **umbo,** which is the portion of the shell formed first. Succeeding phases of growth and shell deposition are indicated by the concentric lines that surround the umbo. The middle layer of the mantle is sensory and may have eyes and/or tentacles. Examine a demonstration of the mantle eyes of the bay scallop *Pecten* as well as a cross section through the eye (Figure 8.10). (Details of the eye are discussed in Section 8–7.) The innermost layer of the mantle, which lies closest to the ctenidia, is muscular; when the lobes of the mantle are pressed together, they form the **siphons.**

ii. AMPHINEURA.   The shell of the chiton is divided into eight plates. When detached from its substrate, the animal rolls itself up into a ball, protected by the plates, like a large pill bug. Notice the form of each plate in a representative specimen (Figure 8.8A). What is the role of the **flanges**?

iii. GASTROPODA.   The gastropod shell may be highly ornamented and vividly colored. Observe demonstration specimens.

▶ *Prosobranchia.*   Examine a *Busycon* shell and notice that it is composed of several spirally arranged **whorls.** The number of whorls is constant for the species, although the full number may not be present in a young specimen. Each whorl increases in size from the peak or apex of the shell to the body whorl, which is the largest and most recently formed. The shell is a coiled tube, leading from the body whorl to the apex and wound around a central column, the **columella.** Growth lines may be seen on the external surface. In a sawn-off shell, the three structural layers can clearly be seen, similar to the three layers of the bivalve shell.

▶ Hold your shell with the apex above and the aperture toward you. If the aperture is on your right, the shell is wound dextrally; if the aperture is on your left, the shell is wound sinistrally. The elongated portion of the shell at the aperture is the **siphonal canal,** which in life contains the siphon.

▶ *Opisthobranchia.*   The reduced shell of *Aplysia* is chitinous and almost transparent, with little calcareous material. Insert scissors into the shell aperture of your specimen and carefully cut the mantle edges to remove the shell (Figure 8.2**D**). Growth rings may be seen on the shell, which curves from the pointed **umbo.** A concave area on one side is known as the **anal incision.**

▶ *Pulmonata.*   If you have not already done so, observe a living specimen of *Helix* and notice how the animal retracts into the shell on stimulation. How does the spirally coiled shell compare with that of *Busycon?* Does it appear to be of similar composition? Is there an operculum? Is the shell wound dextrally or sinistrally? Are members of the same species always wound dextrally, always wound sinistrally, or can this vary with the specimen?

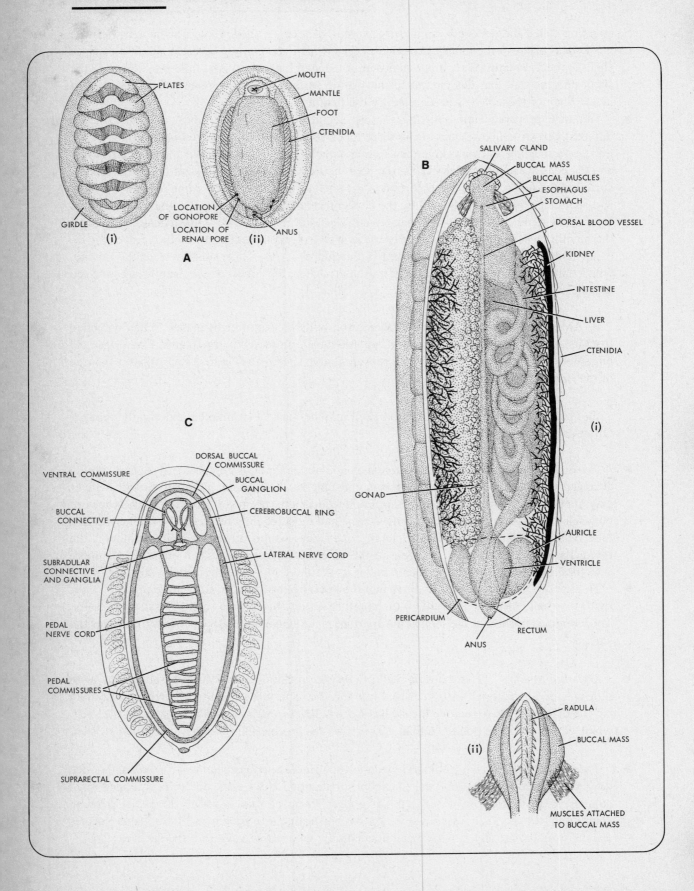

A (i) PLATES, GIRDLE (ii) MOUTH, MANTLE, FOOT, CTENIDIA, LOCATION OF GONOPORE, LOCATION OF RENAL PORE, ANUS

B (i) SALIVARY GLAND, BUCCAL MASS, BUCCAL MUSCLES, ESOPHAGUS, STOMACH, DORSAL BLOOD VESSEL, KIDNEY, INTESTINE, LIVER, CTENIDIA, GONAD, AURICLE, VENTRICLE, PERICARDIUM, RECTUM, ANUS (ii) RADULA, BUCCAL MASS, MUSCLES ATTACHED TO BUCCAL MASS

C VENTRAL COMMISSURE, DORSAL BUCCAL COMMISSURE, BUCCAL GANGLION, BUCCAL CONNECTIVE, CEREBROBUCCAL RING, SUBRADULAR CONNECTIVE AND GANGLIA, LATERAL NERVE CORD, PEDAL NERVE CORD, PEDAL COMMISSURES, SUPRARECTAL COMMISSURE

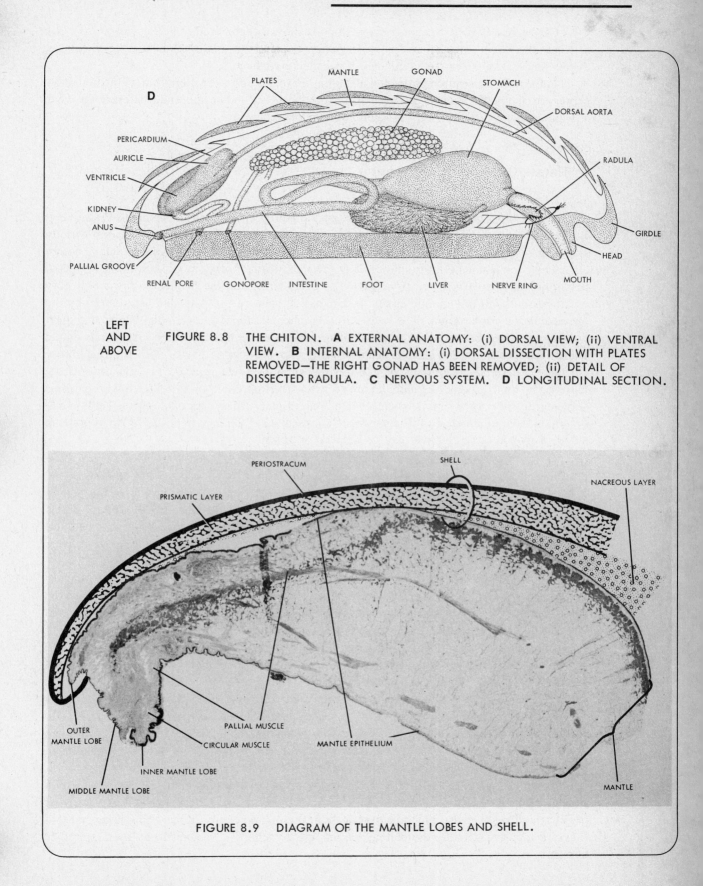

LEFT
AND
ABOVE

FIGURE 8.8 THE CHITON. **A** EXTERNAL ANATOMY: (i) DORSAL VIEW; (ii) VENTRAL VIEW. **B** INTERNAL ANATOMY: (i) DORSAL DISSECTION WITH PLATES REMOVED—THE RIGHT GONAD HAS BEEN REMOVED; (ii) DETAIL OF DISSECTED RADULA. **C** NERVOUS SYSTEM. **D** LONGITUDINAL SECTION.

FIGURE 8.9 DIAGRAM OF THE MANTLE LOBES AND SHELL.

iv. CEPHALOPODA.   Primitive cephalopods show a chambered shell (e.g., *Nautilus*), but in others the shell is internal or absent. In *Loligo* the chitinous endoskeletal element is known as the **pen,** and in *Sepia,* the cuttlefish, it is called **cuttlebone.**

▶ Examine a demonstration of the pen of *Loligo,* or make a cross section through the body of a preserved specimen about 3 to 5 cm from the apex, and locate the endoskeleton. Of what advantage is a reduced or internal shell to cephalopods?

b. Mantle Cavity and Respiration

The most obvious organs of the mantle are the paired ciliated gills or **ctenidia.** Primitively, each is made up of two rows of triangular filaments lying on either side of the main axis. Strictly speaking, a ctenidium is a respiratory structure that includes ciliated filaments (*cten,* comb; Gk.). Once the filamentous structure is lost, through fusion or other modification, the organs of respiration are better referred to as **gills.** The ciliary tracts on the filaments draw respiratory water currents into the mantle cavity on either side. The current is tested as it enters by the chemosensory **osphradium.** The two currents filter through the ctenidial filaments, meet dorsally, and pass backward as an exhalent stream out of the mantle cavity. Gland cells on the filaments secrete mucus, and particles brought into the mantle cavity with the respiratory current are filtered out. Tracts of cilia on the ctenidia direct rejected material toward the midline. In some groups, the ctenidia are so extensive that they serve to filter food organisms sufficient to support the animals, and they are exclusively ciliary-mucoid feeders. Torsion and detorsion have modified the arrangement of the ctenidia in the mantle cavity of gastropods; in some gastropods, the ctenidia may be reduced or absent.

i. AMPHINEURA.   In the primitive molluscs the mantle cavity is posterior, and this is also true of primitive chitons. However, the clinging habit has led to modification of the mantle cavity that, in more highly evolved forms, extends along the sides of the body to the head as narrow grooves between the girdle and the foot. The ctenidia are secondarily multiplied and hang down from the roof of the mantle cavity, extending in a row along the pallial grooves (Figure 8.8A).

▶ Obtain a living specimen of a marine chiton such as *Stenoplax, Mopalia,* or *Chaetopleura* and allow it to attach to a glass plate. Turn the plate over, and, on the ventral surface of the organism, note the large muscular **foot** and the **pallial (mantle) groove.** Place the animal, still attached to the glass surface, in some seawater and add a few drops of a thick suspension of carmine particles or Aquadag. Trace the respiratory flow. The water current enters the pallial groove via temporary passages produced when the animal lifts the girdle at any point along its length. The ctenidia divide the pallial groove into an inhalent groove between the mantle edge and ctenidium and an exhalent groove between ctenidium and foot. After filtering through the ctenidia, the water current passes out posteriorly with the feces and excretory materials. Locate the posterior **anus** on a papilla. Remove one ctenidium and examine it under water in a small watch glass. Add a small amount of carmine suspension. Observe the movement of particles across the ctenidium. What structures produce the respiratory current?

▶ Between the last gill and the anus on each side, locate patches of ciliated sensory epithelium called **osphradia.** They are probably not homologous to the chemosensory osphradia of bivalves and gastropods, which lie anteriorly in the mantle cavity; it has been suggested that they are tactile organs for estimating the amount of sediment brought in by the respiratory currents.

ii. GASTROPODA.   In the gastropods the originally posterior mantle cavity has undergone considerable reorientation and structural modification. Primitively, there is a pair of ciliated ctenidia in the mantle cavity. However, because the gastropods have undergone torsion, the mantle cavity is brought to the front of the body. This may have the advantage that the chemoreceptors of the mantle cavity are now placed in such a way that the water is tested as the animal enters the environment head first, but the twisting of body organs brings about reduction of the pallial organs of the right side of the body, and, in the majority of gastropods, this right ctenidium is lost.

Prosobranchia.   The previously described condition may be easily observed in the limpet *Acmaea*.

▶   Allow a living specimen of *Acmaea* to attach to a glass plate and place the attached animal and plate in a bowl of seawater. Using a Pasteur pipet, add a few drops of carmine suspension or Aquadag in the vicinity of the animal and work out the movement of water through the mantle cavity. Note that, as in the chitons, the pallial groove is around the foot. Water enters anteriorly on the left side, passing over the ctenidium and then along the right pallial groove; another current runs posteriorly on the left side; both exhalent currents converge and exit posterior to the foot.

During the evolution of the gastropods, the ctenidium, which hung free in the mantle cavity, became fused to the mantle wall; this arrangement probably makes for efficient movement of water with less possibility of clogging with silt. The whelk *Busycon* demonstrates this fused condition, with the ctenidium reduced to a single row of filaments on the left side.

▶   Place a living *Busycon* in a bowl of seawater. With a Pasteur pipet, add a small amount of carmine to the water and trace the respiratory current. Notice how the water enters via the siphon, proceeds posteriorly, then turns anteriorly and exits above the tentacles on the head.

▶   Using a preserved specimen of *Busycon,* open the mantle cavity by making a median incision about a centimeter to the right of the middorsal line along the ctenidium until you reach the pericardium. Avoid disturbing the heart, and fold back the mantle to expose the internal organs (Figure 8.4**B**). Observe the single attached **ctenidium** composed of only one row of filaments. Anteriorly, on the inhalent side of the ctenidium, is the brownish chemosensory **osphradium** composed of about 100 triangular filaments covered with epithelium. It has glandular, ciliated, and sensory areas well supplied with nerves. Along the cut edge of the mantle lies the **anus** on a papilla at the end of the rectum, and to its left the mucus-secreting **hypobranchial gland** formed of heavy folds of the mantle. The mucus secreted by this gland helps to consolidate particles rejected by the ctenidia before they leave the mantle cavity.

Opisthobranchia.   In the opisthobranchs, detorsion and relocation of the mantle cavity to the right side has produced modifications in the arrangement of the viscera and structure of the ctenidium. The mantle cavity shows a progressive reduction among members of this group, probably associated with loss of the shell and the transfer of respiratory function to the body wall.

▶   In an opened specimen of *Aplysia* from which the shell has been removed (p. 243), locate the **pericardium** and **kidney** below the transparent mantle (Figure 8.2**B**). Partially remove the mantle to fully reveal the gill and examine it in detail. The ciliated filaments of the prosobranch gill are absent; instead, a double-walled fleshy sheet forms a folded or **plicate gill.** Trace the large **efferent branchial blood vessel** that leaves the gill and shortly enters the **auricle** of the heart (Figure 8.3).

▶   In one group of marine opisthobranchs, the **nudibranchs,** both the shell and ctenidia are lost. The respiratory function is taken over by dorsal projections of the body called **cerata.**

Protective and warning coloration of these cerata make the nudibranchs some of the most attractive of marine creatures. Observe demonstration specimens of *Hermissenda* and *Aeolis*.

Pulmonata. Gastropods show considerable adaptive radiation, and they not only occupy varied marine niches but have ventured forth on land and into fresh water. As one might imagine, the mantle cavity has played a central role in the evolution of air breathing. Air-breathing gastropods, have no ctenidia, and the highly vascularized mantle cavity retains its anterior position.

▶ Examine living specimens of *Helix* or *Limax* and observe the opening and closing of the **pneumostome,** a small aperture on the right side of the body (Figures 8.2 and 8.3). This opens into the expansive mantle cavity, the roof of which is richly supplied with anastomosing blood vessels; the mantle thus serves as a functional lung. Such a respiratory organ can function on land and in water. The common aquarium snails *Lymnaea*, *Physa*, and *Planorbis* are good examples of air-breathing aquatic gastropods.

▶ If you have not already observed the vascularized mantle in a shell-less specimen of *Helix* (p. 237), do so now.

iii. BIVALVIA.   The ctenidium that in the gastropods serves only a respiratory function is so expanded in the bivalves that it lines almost the entire surface of the mantle cavity and serves an additional function in feeding. For this reason members of this group are sometimes referred to as **lamellibranchs** (*lamelli*, plate; *branchia*, gill; L.). Most bivalves possess a single pair of ctenidia, as described in the generalized mollusc. Close observation of your opened specimen, however, will reveal four platelike curtains. This apparent contradiction is explained by the fact that each leaf on one side of the body is half a ctenidium (**demibranch),** the platelike structure being due to elongation of the filaments of the ctenidia, which double back on themselves to form a V. Each individual ctenidium looks like a W in section, with a central axis and a V-shaped reflected gill lamella on each side. Each arm or demibranch is made of many filaments, attached or fused, side by side. The central axis of each ctenidium is fused to the roof of the mantle cavity. The upturned outer filaments of each demibranch are also fused with the mantle roof, and there is some fusion of the inner filaments, separating the lower part of the mantle cavity from the upper, which forms the **suprabranchial chamber.**

▶ The respiratory current and movement of food particles can easily be seen in living specimens of the clam *Mercenaria*. Remove the left shell of a living bivalve (for directions, see Section 8–1). With the mantle reflected back so that the ctenidia are exposed (Figure 8.6**B**), submerge the animal in a bowl containing sufficient seawater that the ctenidia remain wet. By means of a Pasteur pipet, add some carmine or Aquadag suspension to the water in the vicinity of the animal and observe the direction of movement of particles. Water enters the ventral edge of the shell, moves dorsally and anteriorly toward the mouth via the inhalent siphon, passes between the filaments of the ctenidia, enters the suprabranchial chamber, proceeds posteriorly and dorsally, and exits. Particles sieved by the gills are passed by means of cilia to the ciliated **labial palps** close to the mouth (Figures 8.6**B** and **D**). The palps sort the food to some degree. Locate the **food groove** between the labial palps leading to the slitlike **mouth** dorsal and anterior to the foot. Trace the movement of particles in your specimen and sketch the flow of water by a series of arrows.

▶ Current production can also be studied in the isolated ctenidium. Carefully remove a ctenidium from your specimen by cutting along the points of attachment to the body and then transfer it to a dish with a wax bottom. Hold the ctenidium in place by means of two strips of paper with pins at each end, but do not put pins directly through ctenidial tissue. After the ctenidium is covered by a layer of seawater, add some carmine or Aquadag suspension with a Pasteur pipet or a tiny piece of foil. What is the direction of movement? What produces the current? What is the role of the siphon? If time permits, you can measure the time it takes for the aluminum foil to move from one side of the ctenidium to the other under various conditions of, for example, temperature and salinity.

▶ Examine the anatomy of the ctenidium you have removed. Identify the demibranchs. Observe the structure of the filaments under the microscope.

iv. CEPHALOPODA. Cephalopods are active predators, streamlined for quick motion; they therefore require a streamlining of their respiratory current. The walls of the mantle cavity are highly muscularized, and movement of water within the cavity is no longer dependent on ciliary action.

▶ If living specimens are available, observe the orientation of the body and how water drawn into the mantle cavity is used not only as a respiratory current but also in locomotion. The pathway of water may be more easily observed by the addition of small amounts of carmine to the water.

▶ Obtain a preserved specimen of *Loligo,* or return to the specimen previously used (Figure 8.7**A**). With the ventral side of the squid facing toward you, observe the construction of the mantle collar and its relationship to the siphon. The siphon is derived from the foot rather than the mantle, as is the situation in gastropods and bivalves. In life the mantle cavity expands by muscular action, water enters, and the collar locks tightly against the head, leaving the siphon as the only exit pathway from the mantle cavity. The siphon is well equipped with muscles and can be pointed for making directed jet-propulsive movements. Open the mantle cavity by making an incision that runs the entire length of the posterior surface from siphon to apex (Figure 8.7**B**). Dissect with care so as not to disturb the internal organs. Notice that the mantle consists of a thick layer of circular muscles surrounded by integument, with dorsolateral muscularized extensions, the **fins.** Turn the cut mantle edges laterally, breaking the median septum that separates the dorsal part of the mantle cavity into right and left halves, and pin them out to expose the internal organs clearly. Note that the tip of the siphon has a valve that regulates the outflow of water from the mantle cavity. The inner side of the mantle has cartilaginous ridges that keep the inhalent currents separate from the exhalent. There are two **ctenidia** so oriented that the inhalent streams pass over each, then converge and exit as a single exhalent stream. Examine a ctenidium. Is it ciliated?

Save this specimen for future work, keeping the animal moist under wet towels.

8–3. EXCRETION

The paired excretory organs of the molluscs are closely associated with the heart (Figure 8.1). There are two kinds of molluscan excretory organs: brownish **pericardial glands** or **Keber's organs,** derived from a thickening of the pericardial wall (difficult to see in most specimens); and renal **nephridia** or **kidneys,** developed from paired coelomoducts opening from the pericardium via **renopericardial apertures,** and discharging excretory products to the mantle cavity or exterior via **renal pores.** The chitons possess paired symmetrical tubes with dendritic outgrowths, whereas higher gastropods, bivalves, and cephalopods show a more compact kidney. In many molluscs, the digestive glands, like the vertebrate liver, are the primary site of formation of excretory products. These are then extracted from the blood by the pericardial glands and renal organs, or discharged directly into the gut. The blood in the heart has a high hydrostatic pressure, and a clear perfusate passes into the pericardium and then through the renopericardial apertures into the kidneys. In the kidney, nitrogenous wastes (urea, amines, and predominantly ammonia) are extracted from the blood, which is brought to the kidney in the renal portal system, and added to the perfusate from the heart. Inorganic substances may be taken back into the blood, and the composition of the excreta is determined by the extent of renal secretion and reabsorption. Excretory material is passed to the mantle cavity and voided in the excurrent respiratory stream.

The form and arrangement of the kidney and the composition of the excreta vary from species to species, generally in relation to the habitat of the organism. For example, most aquatic molluscs excrete ammonia, but in land snails, where conservation of water is a critical problem, wastes are excreted in solid form as uric acid.

▶   i. AMPHINEURA.   Obtain a freshly killed or preserved specimen of a chiton and remove the shell plates from the dorsal surface, starting with the most anterior. Carefully remove the muscles and skin covering the body organs. Laterally, locate the **nephridia** or **kidneys,** which are paired symmetrical white (preserved specimens) or brownish (fresh material) tubes with fine dendritic outgrowths (Figure 8.8**B**). The tubules lead forward from the renopericardial openings in the anterior wall of the pericardium and then turn back and enlarge to form a small **bladder** that empties posteriorly into the mantle cavity via the renal pore.

    ii. GASTROPODA.   In gastropods the kidney is an expanded thick-walled sac, much folded internally. In higher gastropods, only the left post-torsional kidney is present, the right having become incorporated into the genital ducts. The renopericardial aperture opens into the pericardium anteriorly, but is difficult to see.

▶   *Prosobranchia*.   If you have not already done so (p. 233), identify the kidney in an opened specimen of *Busycon*. It is a large, brown, rectangular organ with a tubular region to the left of a lobular region. This may be seen more clearly after the mantle cavity is opened. The renal pore is a slit in the mantle near the anterior end of the kidney (Figure 8.4**C**).

▶   *Opisthobranchia*.   If you have not yet done so, locate the kidney beneath the transparent mantle near the pericardium to the left of the gill. Observe the blood vessels on the roof of the kidney bringing blood from the mantle and hemocoel. Carefully cut away the kidney anteriorly to show the interior lamellae, being careful not to damage the heart. Locate the efferent renal vein passing from the floor of the kidney to the vessel that leads to the heart (to the auricle). The renal pore opens into the mantle cavity posteriorly and to the right.

The digestive gland of *Aplysia* contains specialized excretory cells that return chlorophyll pigments taken up by the blood to the digestive lumen; urea and uric acid are present in high concentrations in the gland and are discharged directly into the gut.

▶   *Pulmonata*.   If you have not already removed the shell of a specimen of *Helix,* do so now and locate the large kidney in the roof of the mantle cavity close to the heart (Figure 8.5**C**). The kidney does not open directly into the mantle cavity, but is drained by a single duct that loops around the edge of the kidney and then runs alongside the rectum to open outside the mantle cavity. Attempt to trace this duct. Make a slit up from the pneumostome and open it to locate the slitlike opening of the duct to the right of the anus.

    iii. BIVALVIA.   The kidneys of bivalves are paired U-shaped tubes; the lower glandular arm of the U opens from the pericardium and leads to the thin-walled upper arm, which opens into the mantle cavity. Glandular pericardial Keber's organs open into the anterior end of the pericardium in some species.

▶   Return to your opened bivalve from which the left ctenidium has been removed and cut away the free-hanging portion of the left mantle lobe. In the roof of the mantle cavity (left suprabranchial chamber) attempt to locate the **genital pore** on a papilla, just underneath the attachment of the inner demibranch. The **renal pore** is just posterior to it, but small and difficult to see. Carefully remove the rest of the left mantle lobe covering the viscera (Figure 8.6**D**). Trace the thin-walled **kidney duct** posterodorsally from the renal pore. This portion functions as a **bladder;** it turns anteriorly and ventrally as a **glandular region** opening into the pericardium by a minute **renopericardial aperture,** which is difficult to see.

iv. CEPHALOPODA.   The cephalopod kidney is derived from two separate renal organs that have fused: a glandular portion derived from the pericardial epithelium and closely associated with swellings of the blood vessels at the base of the ctenidia, called **branchial hearts,** and surrounding this a nephridial portion, derived from the renopericardial coelomoducts and retaining an opening into the coelom via **renopericardial pores.** The peristaltic contractions of the branchial hearts force fluid through the walls of the blood vessels into the glandular kidney.

▶   Return to your opened specimen of *Loligo* and remove the thin skin covering the organs of the visceral mass. Locate the rectum ending in the anus at the base of the siphon (Figure 8.7**B**). Lateral and dorsal to the rectum are two **papillae,** which are the openings of the **kidney ducts.** Trace the rectum dorsally toward the apex until it disappears medially beneath the paired **kidneys.** If your specimen is a female, the rectum will disappear beneath the large white **nidamental glands.** Remove the left gland to reveal the mottled orange-brown accessory nidamental gland and the left kidney. Notice the swollen branchial heart at the base of the left ctenidium and explore its close attachment to the kidney.

▶   Compare the arrangement of the excretory organs in the specimens you have examined. How is sanitation of the mantle cavity achieved in each case?

8–4 FEEDING AND DIGESTION

Molluscs demonstrate all possible food habits (herbivorous, carnivorous, and omnivorous), and the structure and function of the gut are related to the type of food eaten.

▶   i. AMPHINEURA.   In the living chiton you have observed that the ventral surface bears two openings, one anterior to the foot and the other posterior to it (Section 8–1). If not, do so now. The posterior orifice is the **anus.** The anterior orifice is the **mouth,** which bears a unique molluscan food-gathering organ, the **radula.** Chitons are microphagous herbivores; they rasp the algal surface by running the tooth-bearing radula back and forth. The radula is supported by a cartilaginous rod called the **odontophore,** which is worked by a series of muscles.

▶   Place living specimens of *Mopalia* or *Ischnochiton* on a glass plate covered with a film of algae. Put the glass with the organism attached in an aquarium so that you can observe the chiton's ventral surface and its feeding activity. Algae are rasped and swallowed intermittently, and passed into the elongate **esophagus.**

▶   Make a transverse slit across the mouth opening of a preserved specimen and remove the radula. Using the dissecting microscope, note the orientation of the teeth. The remainder of the digestive tract can be seen in a specimen from which the dorsal shell plates have been removed (see Section 8–3). Observe the middorsal **gonad,** the posterior **heart,** and the **dorsal blood vessel** leading anteriorly from it. From the right-hand side *only,* remove the gonad and its duct, which opens laterally into the pallial groove. Identify the **digestive glands** that lie ventral to the gonad, the anterior **pharynx** with **salivary glands** in front of it, the **buccal mass** with its radiating muscles below the pharynx, and the **esophagus** leading from the pharynx into a large **stomach.** The salivary glands open into the esophagus. Depending on the species, there may also be paired esophageal **sugar glands** on either side of and opening into the esophagus. These secrete amylolytic enzymes, which digest carbohydrates in the food. Pick away the right digestive gland to reveal the stomach. Beneath the stomach lies a diverticulum, the **ventral sac.** The food, wrapped in mucus, is rotated by cilia in the sac, cellulose-splitting enzymes are secreted, and fluid is squeezed into the digestive glands. The long **intestine** leaves the stomach and coils about within the digestive gland, leading to the **rectum,** which stores the feces and terminates at the **anus.** To demonstrate the posterior end of the digestive tract, you will have to sever the blood vessels to the right of the heart and gently lay the heart to one side.

▶   How does the length of the digestive tract compare with the length of the animal? What adaptations to a herbivorous diet are demonstrated by the gut of the chiton?

ii. GASTROPODA

Prosobranchia.   *Busycon* is a carnivorous gastropod; it feeds on any animal of suitable size, particularly bivalve molluscs and decapod crustaceans.

▶   Preserved specimens of *Busycon* are quite satisfactory for dissection of the alimentary tract. Return to your opened specimen, or open one now as described on page 247. If the proboscis is extended, observe the position of the **radula** within the **mouth** at the tip of the proboscis. Specimens in which the proboscis is not extended should be dissected by making a cut between the tentacles to expose the proboscis. Slit the proboscis along its length and reflect the walls laterally. You should now be able to see the esophageal cavity, the radula, and its supportive rod, the **odontophore** (Figure 8.4**C**). Observe the muscles that control the movement of the odontophore and the radula. By cutting the muscle attachments, free the radula and odontophore, and drop the radula into 10% potassium hydroxide for later examination.

Posterior to the proboscis is the elongate **esophagus,** which has a pair of large yellow **salivary glands** on either side. Posterior to these glands is the small **pancreas.** Both glands secrete digestive enzymes, including proteases. Near the salivary glands and forming a ring around the esophagus is a brownish mass of tissue, which consists of the **cerebral, pedal, pleural,** and **parietal ganglia** of the nervous system (Figure 8.4**C**).

Continue to trace and free the remainder of the digestive tract from the visceral hump by cutting it free of the body musculature and identify **stomach,** coiled **intestine,** and **rectum** terminating in the **anus.** The stomach is surrounded by the grayish-brown **liver** or **digestive gland,** which should be carefully picked away. Ducts from the digestive gland lead into the stomach. As you dissect, note the position of the brown **kidney** and the **gonad** (Figure 8.4**C**). Is the gut a straight or twisted tube? What is responsible for its form?

Opisthobranchia.   *Aplysia* browse on macroscopic red, green, and brown algae, which they rasp away with their broad radulas. The digestive tract is quite similar to that of the chitons. What is the adaptive significance of this?

▶   Open the body cavity of your specimen of *Aplysia* by cutting through the dorsal body wall between the genital duct and the opaline gland (Figure 8.2**B**). Keep the blade of your scissors flat and horizontal as nerves and ganglia lie directly underneath. Cut forward in the middorsal line as far as the tentacles and backward along the base of the right parapodium, keeping the opaline gland to the right of the cut. Cut across the body wall behind the anus and continue a short way along the base of the left parapodium. Turn aside the flap of the body wall, carefully freeing the underlying viscera of connective tissue as you do so. Cut off this flap of tissue. Pin the animal out by the parapodia in a wax pan and cover the specimen with seawater.

▶ Trace the course of the gut from the mouth backward (Figure 8.3). Locate the muscular **buccal mass** and elongate **salivary glands** leading posteriorly from it; the narrow **esophagus** is surrounded by a ring of nerve ganglia and leads into a spacious thin-walled **crop.** Behind the crop are the muscular **anterior** and **posterior gizzards,** which disappear into the extensive **digestive gland.** Probe the digestive gland and remove enough of it to expose the small **stomach,** taking care not to damage the gonad at the posterior end of the stomach. A diverticulum from the stomach, the **caecum,** is responsible for fashioning the feces. The stomach leads into a looped **intestine** and **rectum,** terminating posteriorly at the **anus.** Slice away the roof of the stomach to expose the openings of the digestive glands and caecum. Open the two gizzards and observe the chitinous **teeth** within. The first gizzard minces the food, the second strains it. Are the teeth in both gizzards of similar structure? Slice open the buccal mass without disturbing the nerve ring if possible, and observe the paired **jaws** and the broad **radula** on the **odontophore** and the **radular sac.** Remove the radular sac and free the radula. Drop it into 10% potassium hydroxide solution for later examination.

Pulmonata. *Helix* is an active herbivore.
▶ Observe a living specimen of *Helix* as it browses on algae-covered glass.
▶ Return to the specimen of *Helix* from which you have removed the shell, or remove the shell from a specimen as described on page 237. Cut the columellar muscle transversely as indicated by the dotted line in Figure 8.5C. Insert one blade of your scissors through the pneumostome, lift the mantle collar, and cut the mantle away from the body wall under the collar in either direction from the pneumostome. Cut longitudinally along the side of the rectum so that it remains attached to the part of the mantle you have cut away from the body wall. Pin the snail through the head to a wax pan; turn back and pin out the freed mantle lobe, and locate the kidney and heart in the roof of the mantle cavity (the lobe you have pinned out). Starting from the hole in the body wall produced by cutting the columella muscle, cut forward through the floor of the mantle cavity in the middorsal line to the head. Turn aside the flaps and cut them off. Cover the specimen with water. Identify the **reproductive organs (penis, flagellum, dart sac, oviduct, sperm duct,** and **spermathecal gland**), and pull them gently to the right (Figure 8.5D). This should reveal the muscular **buccal mass** leading from the **mouth,** the narrow **esophagus** passing through the nerve ring, together with the **radular retractor muscle** which you should cut. The esophagus leads to a thin-walled **crop** with two elongate **salivary glands** applied to its sides. Trace them forward through the nerve ring to the point at which they open into the buccal mass. Loop the crop around a pin and pull it to your left. Remove the thin mantle left covering the spiral of the visceral hump. Identify the **albumen gland** and **spermatheca** and free them from the connective tissue that holds them to the mass of the **digestive gland.** The esophagus leads from the crop at the base of the visceral hump and coils around to lead into the small, thin-walled **stomach.** The smaller coiled **right digestive gland** and lobed **left digestive gland** open into the stomach. The **intestine** makes an S-shaped turn within the left digestive gland and joins the **rectum,** which terminates in the **anus.** Cut the rectum away from the mantle; free and uncoil the digestive tract and loop the whole thing around pins to the left of your dissection (Figure 8.5E).
▶ Slit open the buccal mass and cut out the radula, which is a broad file with many small unspecialized teeth. Place in 10% potassium hydroxide.
▶ Boil the radulas from the three dissected gastropods in 10% potassium hydroxide for 10 minutes. This will render the teeth more clearly visible. Examine under the microscope and compare the structure of the teeth; relate their structure to the diet of the animal from which the radula came. Note the direction in which the teeth point on each radula. What is the adaptive value of tooth orientation in each case? Where are new teeth formed?

▶   How has detorsion affected the arrangement of the gut and associated glands in *Aplysia?* Compare it with the gut of *Busycon* and *Helix* and note the effects of torsion and coiling in the latter.

iii. BIVALVIA.   Bivalves are principally ciliary-mucus feeders. The ctenidium is the principal organ of food capture in these sedentary animals and also is responsible for some selection of materials.

▶   You may have previously traced the pathway of particles to the mouth in a living bivalve. If not, refer to Section 8–2b and do so now.

Food is trapped in a mucus strand and passed into the short **esophagus.** From the esophagus, food passes to the large **stomach** embedded in the green pulpy **liver** or **digestive gland.** Leading from the stomach is the **style sac** (homologous to the amphineuran ventral sac). In fresh specimens the style sac will contain a gelatinous rod known as the **crystalline style.** In this solid rod are contained enzymes capable of digesting the carbohydrates that form the bulk of the bivalve diet. The enzymes are released as the style rubs against the sac wall and is eroded. The crystalline style is generally not found in preserved or living specimens kept a few days in the laboratory. The thin-walled **intestine** leads from the style sac; embedded in the yellowish **gonad** ventral to the digestive gland, it coils around and passes dorsally to the **rectum** anterior to the pericardium. It passes through the pericardium and is surrounded by the heart. The rectum opens posteriorly at the **anus,** located on a papilla in the region of the dorsal exhalent siphon.

Most of the digestive tract is embedded in the dorsal region of the foot and visceral mass. Carefully pick the tissue away from the left side of your opened specimen, and trace the gut as it loops toward the posterior and then to the anterior end (Figure 8.6**D**).

iv. CEPHALOPODA.   Cephalopods are voracious carnivores and do not depend upon ciliary currents for the capture of food. They do, however, retain some of the microphagous structural components of the gut such as the radula, but this organ is of secondary importance.

▶   Return to your specimen of *Loligo,* or obtain a preserved specimen, and concentrate on the head-foot region. The mouth is surrounded by eight pointed **arms** and two longer **tentacles.** Observe the structure of these arms and tentacles and the arrangement of the **suckers** upon them. Study the organization of the suckers under the dissecting microscope. In life the prey is grabbed by a rapid extension of the two tentacles and brought toward the mouth where it is held firmly in place by the eight arms and killed by an injection of poison. Remove the siphon, and, by a median incision, cut into the head, separating the eyes and exposing the **buccal mass** (Figure 8.7**D**). This is a muscular organ that bears two horny **beaks,** which are used for ripping prey. Pry open the beak and observe the radula and the odontophore. Posterior to the **buccal mass** are a pair of **salivary glands,** which pour their poisonous secretions into the buccal cavity. Trace the thin-walled **esophagus** (surrounded by the **liver**) from the buccal mass to the thick-walled **stomach.** The stomach emerges from the liver tissue to form the **caecum,** which extends to the tip of the visceral mass. The **liver** consists of paired digestive glands fused in the midline; it is a triangular organ with the base located ventrally near the collar. A U-shaped **pancreas** lies anterior to the stomach; its duct unites with that of the liver before passing into the caecum. The **intestine** runs forward from the stomach, shows a diverticulum (**ink sac),** and terminates in the **rectum.**

What is the function of the ink sac? What is the length of the gut relative to the body length? How does this relationship differ in herbivores (e.g., chitons) and carnivores (e.g., squid)? What is the significance of the difference?

▶   How does the location of the anus in each of your dissected specimens assure that the mantle cavity is not fouled with feces?

8–5. BLOOD AND CIRCULATION

Molluscs do not have a spacious coelom, and in all except the cephalopods the body space is composed of large venous sinuses, which act as pooling places for the blood. For this reason the body space is more accurately described as a **hemocoel;** the true coelom is restricted to the pericardium and the cavity of the gonad. Blood that has collected in the sinuses from various parts of the body passes first through the kidney and respiratory organ and then via the efferent branchial artery to the auricles and ventricle of the S-shaped heart. The main propulsive force for distribution of blood in the connective tissue spaces gives the blood an additional function to that of distribution; namely, it acts as a component of the hydrostatic skeleton. The bivalve foot is protruded by the influx of blood and is withdrawn by contraction of longitudinal muscles.

The respiratory pigment of gastropods and cephalopods is the copper-containing protein **hemocyanin.** In rare cases, especially in oxygen-poor environments, the respiratory pigment is **hemoglobin,** e.g., in the mud-dwelling freshwater snail *Planorbis,* the bloody clam *Arca,* and the razor clam *Solen.* Can you offer an explanation for the rarity of respiratory pigments in bivalves?

i. AMPHINEURA.   The simplest arrangement of the open type of molluscan circulatory system is seen in the chitons.

▶   Previously you removed the shell plates from a preserved specimen of *Mopalia* (Section 8–3). If not, do so now. If you have dissected the alimentary canal, you will have observed the tubular **dorsal blood vessel** and the soft median **ventricle** with two more or less transparent **auricles** (Figure 8.8**B**). The former supplies the head and anterior viscera and the latter receive oxygenated blood from the ctenidia via the **efferent branchial arteries.** Cut away the mantle muscle at the right side of your specimen to uncover the ctenidia. Attempt to trace the efferent branchial arteries. Cut across the esophagus and rectum and remove the alimentary canal, heart, gonad, and kidney from your specimen. Beneath the now empty perivisceral cavity can be seen a longitudinal **ventral blood vessel** that collects blood from the head and viscera and two **transverse vessels** that supply blood to the ctenidia on each side.

▶   Diagram the circulatory pathway of the chiton, remembering that the perivisceral cavity functions as a hemocoel.

ii. GASTROPODA

▶   *Prosobranchia.*   Examine a dissected demonstration specimen of *Busycon* in which the circulatory system has been double-injected with colored latex (Figure 8.4). Observe the heart, which lies in the opened **pericardial cavity.** Note that just as there has been a reduction of the ctenidium in this gastropod, so too there is only one **auricle** and one **ventricle.** The auricle receives blood from both the kidney and the ctenidium via the **efferent branchial artery.** Blood then enters the ventricle and is pumped through the body via the aorta. Near the origin of the aorta from the ventricle is the **visceral artery,** which supplies the visceral hump via several smaller arteries. The vessels penetrate deep within the body and are generally difficult to follow. The aorta bends anteriorly and passes below the floor of the mantle cavity parallel with the esophagus as the **cephalic artery.** In the head region the artery enlarges and forms a **secondary heart.** Smaller branches supply the head and foot. Blood supplied to the body via the arterial system pools in venous sinuses and returns to the heart via the ctenidium and also directly from the kidney, so that there is some mixing of venous and arterial blood; thus only partially aerated blood enters the heart. Muscular compression of blood in the venous sinus of the head is partially responsible for protrusion of the proboscis. You may be able to see the large **visceral sinus** along the right side of the visceral hump.

Opisthobranchia. The blood system of *Aplysia* is easier to dissect than that of *Busycon* because of the absence of coiling in the former.

▶ Return to your dissected specimen of *Aplysia,* or open one and remove the shell, and open the mantle cavity as described on page 252. With the help of Figure 8.2**E,** trace the **efferent branchial vessel** along the dorsal edge of the gill to the **heart** lying in the **pericardium.** Open the pericardium, observe the single **auricle** and **ventricle,** and attempt to trace the **efferent renal vein** from its point of entry to the auricle back to the triangular kidney, located posterior and slightly left of the heart. On the roof of the kidney, locate **afferent renal vessels** bringing blood from the hemocoel and mantle. Remove a piece of mesentery from the perivisceral cavity, and notice the holes that allow blood in the hemocoel to pass. At the anterior end of the ventricle, the **aorta** gives rise to a **cephalic branch,** supplying the head, parapodia and anterior viscera, and two **visceral branches,** the anterior one supplying the crop and gizzard and the posterior supplying the digestive gland and intestine. These branches were probably broken during the dissection of the gut, but trace them from the ventricle as far as you are able.

▶ Return to the gill and note that blood spaces pass from afferent to efferent sides. Trace the afferent branchial vein from its entry to the gill back to where it merges into a venous sinus.

▶ Small nudibranchs such as *Aeolis* or *Hermissenda* are convenient for studies of the heart *in situ.* Recall that the mantle cavity and gill have been lost in these organisms, and blood is sent to the dorsal **cerata** and skin for aeration. On its return to the heart, the blood is mixed with that from the viscera; compare this situation with that in *Busycon* and *Aplysia.* Observe specimens under a dissecting microscope and attempt to see the pulsations of the heart, which beats between 50 and 100 times per minute. Do auricle and ventricle contract simultaneously? How is backflow of blood from the blood vessels prevented? By what mechanism does the heart fill? To where does blood flow from the ventricle?

▶ *Pulmonata.* The activity of the gastropod heart can also be seen in a living specimen of *Helix.* Remove the shell of a live specimen as directed in Section 8–1 (p. 237). Identify the heart, which is usually located under the next to the largest coil of the shell (Figure 8.5**B**). The beating heart within the pericardium, just above the whitish kidney, can be seen through the thin, transparent mantle.

▶ If you have previously dissected a specimen of *Helix,* return to that dissection and locate the heart to the right of the pinned out flap of highly vascularized mantle (Figure 8.5**D**). Open the pericardium and observe the **aorta** leading from the posterior end of the ventricle. Trace its **cephalic** and **visceral branches** as far as you are able. Blood from these vessels supplies a system of venous sinuses and returns to the heart via the highly vascularized mantle. Try to locate the **efferent branchial vessel** leading from the mantle to the thin-walled auricle. Just before it enters the auricle, it is joined by blood from the kidney in the **efferent renal vessel,** producing a mixing of arterial and venous blood.

iii. BIVALVIA. The circulatory system of bivalves is not easily discernible in fresh material except for the beating ventricle.

▶ Carefully remove the left valve from a living clam *(Mercenaria)* (for directions see Section 8–1). Expose the pericardial region, which is dorsal and anterior to the posterior adductor muscle, and identify the parts of the heart (Figures 8.6**B** and **D**). The long pyramidal **ventricle** loops over the intestine, and two triangular thin-walled **auricles** open from it laterally. Blood enters the auricles from thin-walled sinuses and leaves the ventricle via two **aortae:** the anterior one runs dorsally to the intestine, the posterior one runs ventrally to the intestine where it forms a conspicuous swelling, the **bulbus arteriosus.** If time permits, try to affect the rate of beat of the ventricle by increasing the water temperature or dropping 0.5 *M* KCl on the heart.

iv. CEPHALOPODA.   The active swimming and carnivorous habits of the cephalopods require a more efficient circulatory system—one that is closed and contains capillaries. There is no hemocoel and the blood does not play a role in locomotion. The body cavity, which is more spacious in most cephalopods than in any other molluscan class, is a true coelom. Otherwise, the circulatory system of cephalopods is similar to that of other molluscs except for some additional components. Interpolated between the body and the ctenidia are a pair of accessory pumping devices known as **branchial hearts.**

▶ Examine a dissected demonstration specimen of the squid in which the circulatory system has been double-injected with colored latex. Trace the pathway of circulation (Figure 8.7C). The blood flows from the **systemic heart** to the body via the **anterior** and **posterior aorta,** which give rise to various arteries supplying the head, mantle, and visceral organs. Blood drains from the body regions in the **veins** and pools in large **anterior** and **posterior vena cavae,** passes to the ctenidia via the branchial hearts, and then returns to the systemic heart.

8–6. NEUROMUSCULAR SYSTEM

a. Nervous System

The structure of the nervous system in molluscs ranges from the simple "ladder" type to a system in which there is extreme fusion of ganglia, forming a true brain. Basically common to all systems are a series of paired ganglia comprising a **cerebral** or **circumoral ring** (innervating and controlling the head region), a pair of **pallial** or **palliovisceral** ganglia (centers for nervous function in the mantle and visceral mass), and paired **pedal ganglia** (innervating the foot).

i. AMPHINEURA.   The amphineuran nervous system is of the "ladder" type, with few localized ganglia.

▶ Return to the chiton from which you removed the shell plates to reveal the digestive tract (section 8–4), or take another specimen and proceed as indicated in the next paragraph. Just anterior to the buccal capsule you may be able to locate the **circumoral nerve ring,** which consists of a dorsal **cerebral band** and a ventral **labial (pleural) commissure** (Figure 8.8C). The ganglia of the circumoral ring innervate and control the odontophore. Running posteriorly from the circumoral nerve ring are two pairs of longitudinal cords with cross-connectives that give the ladder-like appearance. One pair runs to the foot **(pedal system)** and the other to the mantle edge **(palliovisceral system).**

▶ If your specimen is not favorable for dissection of the amphineuran nervous system, take a fresh specimen and dissect from the ventral aspect, carefully removing the foot from its attachments to reveal the visceral mass. Then proceed as described in the preceding paragraph.

ii. GASTROPODA.   The gastropod subclasses show increasing cephalization, with fusion of some of the ganglia.

The typical gastropod nerve ring around the anterior end of the gut (usually the esophagus) consists of paired dorsal **cerebral ganglia,** two **pedal** and two **pleural ganglia** laterally and ventrally. The cerebral and pedal pairs are linked in the midline by **cerebral** and **pedal commissures,** and the cerebrals are linked with the pleurals and pedals and the pleurals with the pedals by short **connectives.** The pleural ganglia give rise to a **visceral loop** connecting them to the paired **parietal** and **visceral ganglia** from which nerves pass to the mantle, its organs (heart, kidney, ctenidia, etc.), and the viscera. During torsion, this visceral loop is twisted to resemble a figure 8. The right parietal ganglion crosses above the gut to lie on the left below the esophagus and is known as the **supraintestinal ganglion.** The left parietal passes below the gut to the right and is known as the **subintestinal ganglion.** In the higher gastropods, cephalization brings the parietal ganglia into the nerve ring and the visceral figure 8 is progressively obscured. The Opisthobranchia show detorsion of the nervous system; with the movement of the mantle cavity backward along the right side, the visceral loop untwists.

▶ *Prosobranchia.* Recall that during your dissection of the digestive tract of *Busycon* (Section 8–4), you saw that all the ganglia except the visceral pair were arranged in the form of a ring surrounding the esophagus. (If you did not perform this dissection, do so now.) Remove the salivary glands, cut through the esophagus anterior to the nerve ring together with the cephalic artery that crosses it, and remove the proboscis and its sheath. From the dorsal **cerebral ganglia,** attempt to trace the large nerves supplying the tentacles. Cut the cerebral commissure between the cerebral ganglia, reflect them laterally, and carefully remove the esophagus to reveal the other ganglia of the nerve ring (Figure 8.4**D**). Gelatinous connective tissue that may obscure the ganglia should be carefully cleared. Examine the nerve ring with a hand lens, and observe the small paired **buccal ganglia** located on the ventral side of the esophagus and connected to the cerebral ganglia. The anterior **pedal ganglia** give rise to nerves supplying the foot and, in male specimens, the right pedal ganglion gives rise to a large **penis nerve.** The **parietal ganglia** are partially fused with the **pleural ganglia,** imparting an asymmetrical arrangement to the ventral aspect of the nerve ring. The left pleural is fused with the **subesophageal** (parietal) ganglion, and the right pleural is fused with the **supraesophageal** (parietal) ganglion, which is free and directed backward. From the subesophageal ganglion arise two **pallial nerves** and the **visceral nerve,** which cross to the **right;** the supraesophageal ganglion bears two **pallial nerves** and the **visceral nerve,** which cross to the **left.** Attempt to trace these nerves and locate the point at which they cross. The **visceral ganglion** itself, probably a fused pair, lies below the external opening of the kidney.

▶ Compare the nervous system of *Busycon* with that of the chiton, and note the effects of increased cephalization and torsion upon it.

Opisthobranchia. The nervous system of *Aplysia* demonstrates the untwisted arrangement of the visceral loop produced by detorsion (Figure 8.3).

▶ Return to your opened specimen of *Aplysia,* or open one as directed on page 252. Locate the nerve ring anteriorly around the esophagus. Sever the esophagus behind the nerve ring and lift it out from the ring. Five pairs of sensory nerves pass anteriorly from the **cerebral ganglia,** and connectives run forward to paired **buccal ganglia** located ventral to the buccal mass. Locate the **pedal ganglia,** connected to the cerebrals and to each other by slender connectives, and the **pleural ganglia** attached by connectives to the cerebrals and pedals. The pedal and pleural ganglia supply seven main pairs of nerves to parapodia, mantle, and viscera. The pleural ganglia give rise to the **visceral nerves,** which pass posteriorly and to the right to paired ganglia beneath the mantle floor. The **right** ganglion is the **supraintestinal** (parietal) ganglion and the **left subintestinal** ganglion is fused with the **visceral** ganglion, forming the slightly larger left **visceroparietal** member of the pair. Attempt to trace the visceral nerves to the posterior ganglia.

From the visceroparietal ganglion, a nerve runs to the **genital ganglion;** from the right parietal, a nerve runs to the **branchial ganglion,** which supplies the gill.

▶ Compare the position of the subintestinal and supraintestinal ganglia in *Aplysia* to that in *Busycon* and notice the reversal of right to left and vice versa.

Pulmonata.   *Helix* shows further cephalization reflected by fusion of the pleural, parietal, and visceral ganglia to form paired subesophageal ganglia, and the cerebral ganglia are themselves fused to form a single supraesophageal ganglion. The figure 8 of the visceral loop is completely obscured.

▶ If you have dissected a specimen of *Helix*, return to it now or open a specimen as described on page 253. The digestive system has been turned to your left so that one side of the nerve ring should be visible (Figure 8.5E). The ganglia and nerves forming the ring are enclosed by connective tissue, which should be dissected away. Observe the ganglia of the ring and its connectives. Sensory nerves pass anteriorly from the **cerebral ganglia.** Cerebrobuccal connectives lead to **buccal ganglia** at the base of the salivary ducts. From the fused **subesophageal ganglia** arise a pair of lateral **mantle nerves,** and, posteriorly in a common sheath of connective tissue, a **genital nerve** on the left and a median **mantle nerve** on the right. Beneath them, numerous **pedal nerves** pass back from the **pedal ganglia,** located lateral to the subesophageal ganglion. Dissect out as many ganglia as possible, and attempt to trace the nerves that lead from them.

iii. BIVALVIA.   Bivalves possess paired ganglia organized into three centers: **cerebro-pleural, pedal,** and **visceral** (Figure 8.6D).

▶ Return to one of your opened specimens of a bivalve, or open a specimen now (see Section 8–1 for directions). The visceral mass should be carefully dissected away from the esophagus. Carefully strip away the anterior foot retractor muscle. It may be difficult to locate the ganglia, but in favorable material the **cerebral ganglion,** a yellowish pinhead-sized organ, will be found just posterior and slightly dorsal to the anterior adductor muscle. It is connected to the cerebral ganglion of the other side by a commissure, which passes anterior to the esophagus. Two nerves pass from each ganglion; one of these travels posteriorly to connect with the **visceral ganglion** of the same side, and the other passes into the foot to unite with the **pedal ganglion.** Anteriorly, the cerebral ganglia give rise to **pallial nerves.** Attempt to trace the connectives to the visceral ganglion, which is located ventral to the posterior adductor muscle. (You may have to cut the lamellae of the ctenidia to see this area.) The pedal ganglion is embedded in the foot; a median sagittal section of the foot running into the visceral mass usually exposes this ganglion, which is situated just dorsal to the muscular portion of the foot itself. In the actively swimming bivalve *Pecten* (the bay scallop) some of the ganglia are fused (e.g., cerebral and visceral) so that a "brain" is formed.

What factors seem to be involved in decentralization or centralization of nervous centers in bivalves?

iv. CEPHALOPODA.   The ultimate in invertebrate cephalization is seen in the cephalopods (Figure 8.7E). Evolutionary fusion of the ganglia followed by extensive differentiation makes it difficult to homologize the ganglia with those of other molluscs, however.

▶ Hemisect a preserved specimen of the squid *Loligo* by cutting from the apex to between the eyes (Figure 8.7E). Most of the paired ganglia of the central nervous system are located in the head region, and there is considerable fusion. Four pairs of fused ganglia form a ring around the esophagus. Above the esophagus lies the **supraesophageal ganglion,** which represents the fused cerebral ganglia; below lie the paired and fused **visceral ganglia.** The paired **pleural ganglia,** which form the sides of the esophageal ring, cannot be seen in your section. Note how these ganglia, which form the brain, are encased in a cartilaginous **cranium.**

▶ The ganglia give rise to nerves, which innervate the various parts of the body. Turn one half of your specimen over and carefully remove the eye by cutting the eye muscles and optic nerve; place the eye in a dish of water and save for examination at another time. Notice the large **optic ganglion** behind the eye; from this ganglion optic nerves run to the pleural ganglia. A pair of large **mantle nerves** run from the visceral ganglion to the large **stellate ganglia,** located on the inner dorsal surface of the mantle at the level of the tip of the ctenidia (Figure 8.7**D**). The stellate ganglia are the motor centers of the mantle and give rise to its **giant fiber system,** which is favorite material for neurophysiologists. Attempt to trace the mantle nerve from its origin to the stellate ganglion in one half of your specimen.* In the midline the visceral ganglia give rise to a pair of **visceral nerves,** which are fused for part of their length.

▶ Diagram the layout of the nervous system in the specimens you have examined, and compare and contrast the arrangement of nervous elements in the representatives of each class.

b. Locomotion

▶ i. AMPHINEURA AND GASTROPODA.   Movement of chitons and gastropods involves peristaltic waves of the foot and is not unlike the crawling motion of flatworms. The waves can most easily be seen by allowing a living animal to attach to a glass plate and viewing the animal from the ventral side. Note how the expanded part of the foot remains in contact with the glass plate by means of mucus secreted by the pedal glands, and how the contracted region advances forward,

▶ Locomotor activity in gastropods may not be restricted to the sole of the foot. The role of the **columellar muscle** in retracting the animal into its shell is one example; this muscle serves other functions as well. It is particularly important in the righting reaction. Take a snail such as *Helix* or *Nassarius* and turn it over on its back; note how the animal rights itself. By contraction of the circular elements in the columellar muscle, the sole is pushed outward, and when it attaches to the substrate, the sharp contraction of the longitudinal muscles turns the animal upright.

▶ Swimming in some gastropods, such as the nudibranchs, is accomplished by the use of lateral projections of the foot and mantle **(parapodia).** Drop some *Aplysia* or *Hermissenda* into a bowl of seawater and observe the graceful swimming motion, if you have not done so previously. Is flexion of the body involved? One species of *Aplysia (saltator)* forms a tunnel by keeping the parapodia closed and shoots a jetstream of water backward that propels the animal forward.

ii. BIVALVIA.   Bivalves employ the foot in burrowing; protrusion is effected by the contraction of transverse muscles, which act on the blood in the sinuses of the foot.

▶ Place some living clams (e.g., *Ensis, Solen*) on a bed of soft sand, and watch how the foot is protruded, spread out, and anchored, making the foot hatchet-shaped (thus the alternative class name, **Pelecypoda;** see Classification of Mollusca, p. 270), and then how the rest of the body is moved by contraction of the longitudinal muscles in the foot. Note the arhythmic

*If you have a female and have not removed the nidamental gland from one half, do so now (see Section 8–3) and use this half for this portion of the exercise.

character of such movements compared with similar movements in gastropods. The foot of *Mytilus* is not used in burrowing, but for attachment. A gland in the foot secretes threads (called **byssus threads**) used for anchoring the mussel. Study a demonstration specimen of *Mytilus* attached to a rock.

▶ The muscles of bivalves, particularly the adductor muscles, may perform other functions as they close the valves. The scallop *Pecten* is able to swim by using the large adductor muscle to clamp shut the valves, then squirting out water and jetting away. Observe a demonstration of this.* It is this adductor muscle that one eats under the name **scallop.**

▶ A more unusual function of the adductor muscle is closure of the shell for protection. The adductor is a muscle that shortens rapidly and can maintain its tonus over long periods of time. Force a closed pair of scissors between the valves of a live clam or mussel, and turn them so that they hold the valves open. After a while remove the scissors. Do the valves snap shut? File away the edge of the valves so that a pair of hooks can be inserted between them. Attach one of the valves to a clamp, and on the other hook add some weights. Compare the nature and amount of force necessary to open the valves: quick pull versus long, sustained pull. What are the antagonists of the adductors?

▶ iii. CEPHALOPODA. From the complex structure of the cephalopod nervous system, one would expect these organisms to be capable of a variety of neuromuscular activities. They are particularly well demonstrated by the feeding and swimming movements of the squid, which you should observe if you have not already done so (Section 8–2). In *Loligo,* which is a rapid swimmer, the giant fiber system of the mantle is effective in producing quick-firing, rapid muscular contractions rather than slow, prolonged tonus. Students interested in this aspect of cephalopod biology should consult the reference by Bullock and Horridge (1965).

8–7. SENSE ORGANS AND BEHAVIOR

The organization of the nervous system is indicative of the habits of an organism. Those organisms that move actively show a greater degree of cephalization and better-developed sense organs than those that are sluggish; the latter have less-centralized nervous systems, without concentrated ganglia and complex organs such as a brain.

a. Sluggish and Sedentary Forms

Amphineurans and most bivalves have relatively poorly developed sense organs. Amphineura spend most of their lives closely applied to the substrate by means of the girdle and foot, and head structures are reduced. There are no eyes or tentacles, but the animal is sensitive to light and shadow. The dominant sensory area is found on the valves, where there are pits containing simple epidermal projections called **aesthetes,** which function as tactoreceptors and photoreceptors. They are most numerous near the growing points of the shell plates, where they have not yet been worn away.

▶ Attempt to identify aesthetes in specimens of chitons available in the laboratory.

Because the head is absent in bivalves, the main sensory apparatus is restricted to the mantle edge, that portion of the soft part of the animal that is most immediately in contact with the

Lima is a good substitute.

environment. The siphons of the clam *Mya* respond to sudden illumination by withdrawal. The **photoreceptors** are epidermal, unicellular, and not unlike those found in the earthworm skin.

▶   If living clams such as *Mya* or others with long siphons are available, allow an animal to protrude its siphons in subdued light, and then note the reaction to a beam of light focused to a point on various body regions. Where is the most light-sensitive area? Are photoreceptors situated on areas of the body other than the siphons?

Other sense organs are the **palps** and the **statocyst** (embedded in the foot), but these are not easily studied.

b. Active Forms

▶   In the gastropods torsion has brought the chemoreceptor of the mantle cavity, the **osphra-dium,** to a forward position. This supplements the other organs of sensation associated with the head, such as eyes and tentacles. Examine a specimen of *Busycon* and identify the osphra-dium, eyes, and tentacles (Figures 8.2 and 8.4). **Chemoreception** is easily demonstrated by offering some clam juice to live *Busycon;* note their accelerated crawling behavior.

Photoreception can be studied in gastropods such as *Helix, Limax,* or *Littorina.* Many of these animals use light as a cue, directing themselves with reference to the light beam as if it were a compass. The cuplike eyes are normally borne on stalks, contain a spherical lens, and are covered by a double-layered cornea. Only light falling on the eye within 35° and 130° of the body can reach the retina, so that orientation angles outside these limits cannot be used.

▶   Determine whether animals orient toward light or dark by placing some specimens in a container that is half in darkness and half in light. Where do the majority of animals settle? Why?

▶   Allow the animals to crawl in diffuse subdued light, and then illuminate with a horizontally directed beam. How do the animals orient themselves (i.e., what is the angle of movement relative to the light source)? Repeat, using a vertical light source. What sort of movement results if two light sources are used? Why? Remove one eye and determine the orientation of an animal when light falls on the blinded side from behind and when the functional eye is illuminated. Are both eyes necessary for the animal to orient itself with respect to a directed light beam? What can you conclude about the role of light intensity versus angle in oriented motion from the results of these experiments? See the paper by Newell (1958) for further details of this **compass reaction.**

▶   Gastropods such as *Helix* have a strong tendency toward **geotaxis,** related to the presence of **statocysts,** which are usually located in the foot. Allow a snail to crawl along on a glass plate and note the change in its direction of movement as one edge of the plate is lifted, for example, the edge in front of the snail. Repeat the experiment, but this time reverse the edge that is lifted. How long is it before the snail responds? What is the response? Attach a string to the shell of the animal by means of wax or glue. As the animal crawls on a vertical or horizontal surface, displace the shell by means of a gentle upward tug on the cord. What is the response? What do these experiments tell you about gravitational responses? Do the same responses occur when these experiments are performed with the snail submerged?

▶   By most standards gastropods are slow moving, but some molluscs qualify as swift swim-mers. The bay scallop *Pecten* is one of these. You may already have observed its method of swimming (Section 8–7b); if not, do so now. The sensory apparatus of *Pecten* is restricted to the mantle edge, as in other bivalves, but the structures are elaborate. Observe particularly the structure of the eye, and relate its organization as seen in cross section (Figure 8.10) to that of the human eye. What is the function of such an eye? The mantle edge near the mouth (the *velum*) is chemosensitive. Place a small amount of ground starfish gonad close to a scallop in a bucket of seawater, and observe the reaction.

▶ As anticipated, the sensory apparatus of cephalopods is the best developed among all molluscs, and perhaps among all invertebrates. Carefully remove an eye from a preserved specimen of *Loligo* by cutting the eye muscles and optic nerve, or use the eye that you removed in Section 8–6a. The outermost covering is the **false cornea,** which is underlain by the **true cornea.** Remove the corneas and identify the spherical **lens,** the colored **iris** and **choroid coat,** and the innermost lining, the **retina.** To observe the organization *in situ*, remove the other eye and hemisect it by cutting from lens to optic nerve. How does the organization of the eye of the squid (Figure 8.10) compare with that of the human?

▶ Immediately posterior to the eye, locate the **olfactory crest,** behind which is a concavity called the **olfactory groove** containing a sensory device that may function in testing the water.

▶ **Statocysts** may be found embedded in the cephalic cartilage just beneath the visceral ganglion. How do these function?

▶ If live squid are available, observe their swimming movements and responses to light and touch. Note the color changes caused by the presence of pigment spots (**melanophores**).

The fascinating study of learning in Cephalopoda is beyond the scope of the present course. Students interested in this area should consult the reference by Wells (1962).

8–8. REPRODUCTION AND DEVELOPMENT

The primitive mollusc probably had a pair of gonads that opened directly into the pericardium; from here, gametes were discharged into the exhalent stream. It is uncertain whether these early forms were dioecious or hermaphroditic. Fertilization was external. The larval form was a **trochophore** much like that of the Annelida, which later took on the **veliger** form more typical of present-day representatives (Figure 8.11).

a. Reproductive Anatomy

i. AMPHINEURA.   In chitons the sexes are separate, although it is difficult to determine the sex of a chiton from external characteristics. The gonad lies in a median dorsal position, and lateral gonoducts lead directly to the mantle cavity (a departure from the ancestral condition).

▶ If you have not located the gonad in a previously dissected specimen (Section 8–3, Figure 8.8B), make a longitudinal section through a preserved specimen and identify the gonad (Figure 8.8D).

Fertilization in chitons is external; free-swimming larvae hatch from the yolky green, yellow, or orange eggs in a week to 10 days, eventually developing into the adult form. Some species may brood their young in the pallial grooves.

ii. GASTROPODA.   Sexual dimorphism is not always apparent in molluscs, but some gastropods do show this condition. Others are hermaphroditic with complex arrangements to prevent self-fertilization, including a tendency toward protandry among the prosobranchs. All gastropods have a single gonad.

Prosobranchia.   In more primitive gastropods, the gonad may open directly into the right kidney or its duct. In higher prosobranchs, the renal function of the right kidney is lost and it

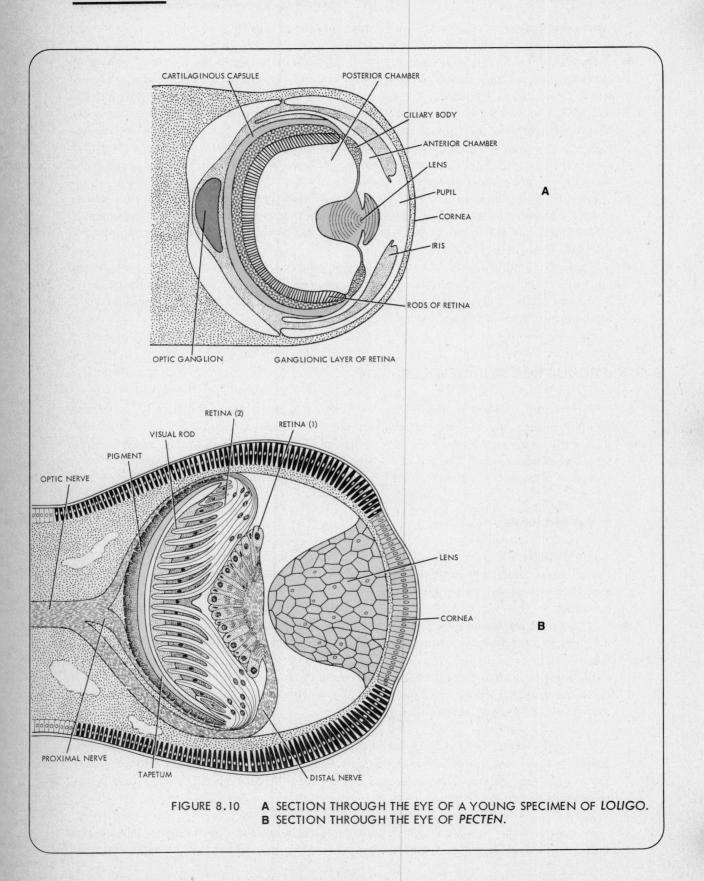

CARTILAGINOUS CAPSULE

POSTERIOR CHAMBER

CILIARY BODY

ANTERIOR CHAMBER

LENS

PUPIL

CORNEA

IRIS

RODS OF RETINA

A

OPTIC GANGLION

GANGLIONIC LAYER OF RETINA

RETINA (2)

VISUAL ROD

RETINA (1)

PIGMENT

OPTIC NERVE

LENS

CORNEA

B

PROXIMAL NERVE

TAPETUM

DISTAL NERVE

FIGURE 8.10 **A** SECTION THROUGH THE EYE OF A YOUNG SPECIMEN OF *LOLIGO*. **B** SECTION THROUGH THE EYE OF *PECTEN*.

survives as a short section of the genital duct, itself formed from a section of the right side of the mantle. The kidney is thus unpaired. This condition obtains in *Busycon*.

▶   If you have not previously identified the gonads and their ducts, return to your dissected specimen of *Busycon* or open a preserved specimen (see Section 8–3, p. 247, for directions). If it is a male, attempt to identify the **penis** and the reddish **testis,** which is situated in the visceral mass over the digestive gland. A convoluted **vas deferens** leads from the testis to the penis (Figure 8.4**B**). In a female, identify the large, yellow **nidamental gland** that overlies the brownish digestive gland and may have been disturbed during your dissection of the gut (Figure 8.4**A**). You may be able to trace the **oviduct** leading away from it to the **genital aperture** on a papilla to the right of the anus. Cut open the nidamental gland in a line parallel to the oviduct within and trace the oviduct to the ovary.

Fertilization is internal; the large penis of the male is inserted into the end of the female genital duct, which thus forms a functional **vagina.** The nidamental gland secretes disc-shaped egg cases containing 20 or more eggs and a cord to which the egg cases are attached. The cord is fastened to a stone or a buried object. The eggs are supplied with large quantities of yolk, and the young embryos feed on each other within the egg case, the survivors emerging as miniature adults.

Some prosobranchs, previously thought to be dioecious, are in fact protandrous hermaphrodites. They begin life as males, pass through a hermaphroditic stage, and finish life as females. The slipper limpet, *Crepidula fornicata,* is probably the best known example, forming a chain of individuals piled one on the other; the largest, basal member is a female, and the smallest, youngest individual is a male at the top of the chain. Members of such chains are oriented with their right anterior margins in contact, facilitating insertion of the penis into the female gonopore.

▶   If available, observe a mating chain of *Crepidula* and attempt to locate the gonopores.

Opisthrobranchia.   Opisthobranchs are hermaphroditic and, as is frequently the case in hermaphroditic animals, have rather complex genitalia. The single gonad is subdivided into male and female parts. The prosobranch nidamental gland is represented by a large mucous gland that secretes a jelly-like egg mass. This gland, together with the albumen gland, sperm receptacles, and seminal bursa, lies deep within the body cavity, and each gland may be subdivided or even cut off from the main genital passage.

▶   Pull the digestive tract of an opened specimen of *Aplysia* (Section 8–4, p. 252) to your left and hold it in position by looping it around a pin. Locate the large **ovotestis,** which is the most posterior organ in the visceral mass, closely applied to the digestive gland. Trace the coiled **hermaphrodite duct,** which runs from the anterior surface to the genital mass anterior to the heart (Figure 8.3). It loops over the small **spermatheca** and divides into two branches. One branch swells to form a **fertilization chamber,** out of which lead the other glands and ducts of the female system. Unravel the organs of the genital complex carefully and locate the compact, globular **albumen gland** and the thick, banana-shaped **mucous gland** with a highly convoluted portion, sometimes known as the **winding gland,** leading out of the fertilization chamber. From the anterior end of the fertilization chamber arises the straight **vagina,** with the spermatheca as a lateral diverticulum. The unconvoluted end of the mucous gland leads to the sacculated **oviduct** running alongside the vagina. The second branch of the hermaphrodite duct leads directly to the almost invisible **sperm duct,** which also runs along the oviduct. This triple duct is really one **hermaphrodite duct** with grooves dividing it into three compartments. Slice across it to demonstrate the relationship of the three separate parts. The anterior end of the hermaphrodite duct swells into the **bursa seminalis** out of which leads a small **seminal sac.** The seminal sac apparently absorbs seminal fluid and debris introduced at copulation. The tubular common

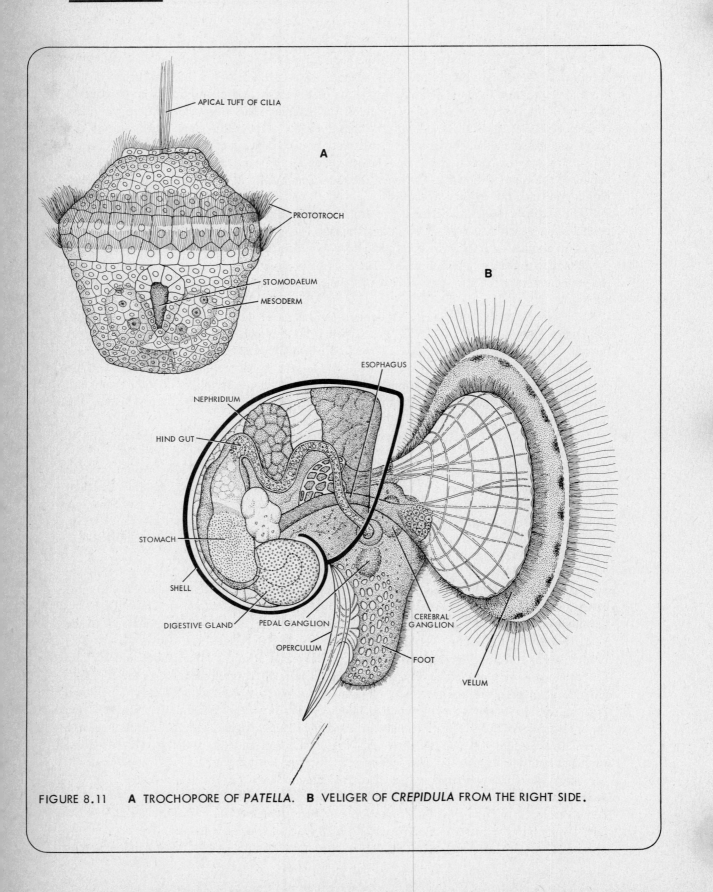

FIGURE 8.11   **A** TROCHOPORE OF *PATELLA*. **B** VELIGER OF *CREPIDULA* FROM THE RIGHT SIDE.

hermaphrodite duct leads from the bursa seminalis to the **common genital opening** or **gonopore,** the sperm duct continues anteriorly from here as an open ciliated sperm groove to the **penis.**

During mating, *Aplysia* forms a copulatory chain, each animal acting as a male to the one in front and a female to the one behind. Sperm passes from the gonad along the sperm duct to the penis, which is introduced into the gonopore of the animal in front. Eggs are fertilized in the fertilization chamber, albumen and mucus coating are applied, and the eggs pass down the oviduct to be laid in gelatinous strings from the gonopore. The eggs hatch as swimming veliger larvae.

Pulmonata. Adaptation to life on land and in fresh water required fertilization to be internal, the larval life to be shortened, and the eggs to be protected from the environment after laying, or even retained. The pulmonates exhibit special genital ducts for the passage of sperm at fertilization and glands for providing nutritive and protective layers for the eggs. Like opisthobranchs, pulmonates are generally hermaphroditic, and their reproductive organs are similarly complicated.

▶   Return to your opened specimen of *Helix* or open one as described on page 253. During your dissection of the gut, the male and female organs were freed and laid out to the right (Figure 8.5**E**). If you did not dissect and identify these organs at that time, do so now. The whitish gonad, or **ovotestis,** is located at the inner surface of the coiled right digestive gland. The convoluted **hermaphrodite duct** leads to the **albumen gland** in which is embedded the **fertilization chamber.** From here, the common hermaphrodite duct runs anteriorly. It consists of a sacculated outer **oviduct** and a smooth-walled inner **sperm duct.** Cut across the hermaphrodite duct and notice that the lumens of the two channels are incompletely separated by an internal septum. Anteriorly, the sperm duct separates from the oviduct, which is joined by the **spermathecal duct.** Trace this duct back to the spherical **spermatheca,** located near the albumen gland. There may be a narrow diverticulum from the spermathecal duct, the **caecum.** This is present in *Helix aspersa,* but not in the larger *Helix pomatia.* The oviduct opens into a short **vagina,** which joins the **penis** at its anterior end to open at the common **genital aperture.** Into the oviduct opens a tufted pair of **mucous glands** and a muscular **dart sac,** which secretes a calcareous **spicule.** At copulation, this is jabbed into the side of another snail as part of the mating ritual. The sperm duct runs from the common duct toward the penis as the **vas deferens.** Just before it joins the penis, the vas deferens gives rise to a **flagellum,** a narrow duct in which the sperm are packaged into **spermatophores.** The penis leads to the common genital aperture through which it may be protruded and retracted again by a posterior **muscle.**

Helix shows a more or less synchronous ripening of eggs and sperm in the same gonad, although some authors report a tendency to protandry. During copulation, animals pair and indulge in a mating dance, if it is possible for snails to dance. After much maneuvering, which may take 2 to 3 hours, the gonopores are juxtaposed and an exchange of spermatophores occurs. Fertilization occurs in the fertilization chamber and the large eggs are supplied with albumen and a calcareous shell from the sacculated portion of the oviduct. The eggs are laid in the ground and hatch as tiny snails. Some species of *Helix* are ovoviviparous; the eggs are held until hatching in the widened end of the oviduct, which forms a functional uterus.

How does the reproductive system of *Helix* differ from that of *Aplysia?* What adaptations to terrestrial life are demonstrated by *Helix?*

iii. BIVALVIA.   Most bivalves are dioecious, although it is quite difficult to tell the sexes apart. Other species may be hermaphroditic, either protandrous or synchronous. In most species the gonads are paired and fused in the midline.

The gonads of *Mercenaria* are embedded in the visceral mass (Figure 8.6**D**). Identify the gonad in an opened specimen if you have not already done so (see Section 8–3 for directions). Gametes are discharged via gonoducts into the exhalent stream. The free-swimming larva settles and develops into a young adult within a few weeks.

iv. CEPHALOPODA.　In cephalopods the sexes are always separate; the gonad is at the apex of the body and its ducts open directly into the coelom. Fertilization is internal, and there may be complicated courtship behavior and parental care of the young.

▶　Return to the specimen of *Loligo* that you dissected previously, or open a specimen now (see Section 8–2b for directions). If your specimen is a female, identify the single large **ovary** at the apex of the visceral mass. You may already have removed the large, white **nidamental glands** from one side (Section 8–3). If not, identify them now, together with the orange-speckled **accessory nidamental glands** beneath their ventral ends. The **oviduct** is a transparent tube, packed with eggs, leading by a small ciliated funnel from the vicinity of the ovary. The oviduct loops ventrally, dorsally, and ventrally again as the glandular, thicker walled **oviducal gland.** It terminates in a flared opening. The large, yolky eggs are shed from the ovary into the coelomic cavity and are picked up by the ciliated funnel of the oviduct. As the eggs pass along the oviduct, they receive an elastic membrane from the nidamental glands and a gelatinous coat from the oviducal glands.

▶　If your specimen is a male, remove the left gill and branchial heart. The single large, white **testis** is located at the apex of the visceral mass. It opens directly to the coelom by a slit at its anterior end. Near the opening lies the ciliated opening to the **vas deferens,** an opaque, white, coiled tube leading ventrally between the **spermatophoric sac** on the right and the thick-walled **spermatophoric organ** on the left, and opening to the exterior by the **penis** to the left of the rectum. (The penis is not a muscular intromittent organ and is no more than the end of the vas deferens).

▶　Remove some spermatophores from the sac and place in a dish of seawater. Examine with a lens or dissecting microscope, and, with the aid of Figure 8.7**F,** locate the various parts. When the spermatophores are released, the cap breaks open, the ejaculatory organ turns inside out and pulls the sperm mass and cement gland with it; the cement gland fastens the sperm securely wherever it happens to land. This is usually the **horseshoe organ,** a glandular area located on the buccal membrane of the female. During copulation, the male squid exhibits courtship behavior that culminates in the deposition of spermatophores on the female. In male squid, the fourth arm to the left is specially modified to pick packets of spermatophores from the opening of the vas deferens (penis) and transfer them to the female. Known as the **hectocotylus,** the distal suckers of this arm are modified as **papillae.**

▶　Locate the horseshoe organ in a female and the hectocotylus in a male.

Following courtship and fertilization, the female holds the gelatinous egg mass in her arms and attaches it to a suitable spot, usually rocks below the low tide mark. Young squid hatch within 2 to 3 weeks.

b. Development

Cleavage in molluscan eggs is spiral and determinate, similar to cleavage in annelid eggs. The free-swimming larva is at first a **trochophore,** which later develops a velum-like fold and becomes the characteristic molluscan **veliger** larva. During the veliger stage the larva undergoes **torsion;** Garstang has suggested that this is a larval adaptation that enables the headparts to be withdrawn into the shell.

▶　To gain some appreciation of the developmental events and the larval forms, examine a series of prepared slides of the development of the slipper-shell *Crepidula* showing cleavage, blastula, gastrula, trochophore, and veliger stages. What is meant by the term **mosaic development?** How do larval characteristics unite the annelids and the molluscs? What adult characters make for a separation of the annelids from the molluscs?

The restricted time available to you prevents detailed investigation of the development of molluscs, but interested students should consult the references by Costello (1957) and Raven (1958) for further information.

References

Barnes, R. D. *Invertebrate Zoology,* 3rd ed. Philadelphia: W. B. Saunders, 1974.

Barrington, E. J. W. *Invertebrate Structure and Function.* Boston: Houghton Mifflin Co., 1967.

Borradaile, L. A., and F. A. Potts. *The Invertebrata.* London: Cambridge University Press, 1961.

Brown, F. A., Jr. (ed.). *Selected Invertebrate Types.* New York: John Wiley & Sons, 1950.

Bullock, T. H., and G. A. Horridge. *Structure and Function in the Nervous System of Invertebrates.* San Francisco: W. H. Freeman and Co., 1965.

Bullough, W. S. *Practical Invertebrate Anatomy.* New York: St. Martin's Press, 1962.

Costello, D. P., M. E. Davidson, A. Eggers, M. H. Fox, and C. Henley. *Methods for Obtaining and Handling Marine Eggs and Embryos.* Woods Hole, Mass.: The Marine Biological Laboratory, 1957.

Hyman, L. H. *The Invertebrates.* Vol. VI. *Mollusca I.* New York: McGraw-Hill Book Co., 1968.

Kandel, Eric R. Nerve cells and behavior, *Sci. Amer.* **223**(1):57–70, 1970.

Kennedy, D. Small systems of nerve cells, *Sci. Amer.* **216**(5):44–52, 1967.

Meglitsch, P. *Invertebrate Zoology.* New York: Oxford University Press, 1967.

Morton, J. E. *Molluscs. An Introduction to Their Form and Function.* London: Hutchinson and Co., 1958.

Newell, G. E. An experimental analysis of the behavior of *Littorina littorea* under natural conditions and in the laboratory, *J. Mar. Biol. Ass. U.K.* **37**:241–266, 1958.

Nicol, J. A. C. *The Biology of Marine Animals.* New York: Interscience Publishers, 1960.

Parker, T. J., and W. A. Haswell. *Text-book of Zoology.* New York: Macmillan Publishing Co., 1951.

Prosser, C. L. *Comparative Animal Physiology,* 3rd ed. Philadelphia: W. B. Saunders Co., 1973.

Potts, W. T. W. Excretion in the molluscs. *Biol. Rev.* **42**(1):1–41, 1967.

Purchon, R. D. *The Biology of the Mollusca.* Oxford and New York: Pergamon Press, 1968.

Ramsay, J. A. *Physiological Approach to the Lower Animals.* London: Cambridge University Press, 1952.

Raven, C. *Morphogenesis: An Analysis of Molluscan Development.* Oxford: Pergamon Press, 1958.

Rowett, H. G. Q. *Dissection Guides,* Vol. V. *Invertebrates.* New York: Holt, Rinehart and Winston, 1957.

Russell-Hunter, W. D. *A Biology of Lower Invertebrates.* New York: Macmillan Publishing Co., 1968.

Stasek, C. R. Feeding and particle-sorting in *Yoldia ensifera* (Bivalvia; Protobranchia) with notes on other nucuannelids, *Malacologia* **2**:349–66, 1965.

Tompsett, D. H. *Sepia. L.M.B.C. Memoirs XXXII.* Liverpool: University of Liverpool Press, 1939.

Wells, M. J. *Brain and Behavior in Cephalopods.* Stanford, Calif.:Stanford University Press, 1962.

Welsh, J. H., R. I. Smith, and A. E. Kammer. *Laboratory Exercises in Invertebrate Physiology.* Minneapolis: Burgess Publishing Co., 1968.

Wilbur, K. M., and C. M. Yonge. *Physiology of Mollusca.* Vols. I and II. New York: Academic Press, 1964 and 1966.

Yonge, C. M. The pallial organs in the aspidobranch gastropoda and their evolution throughout the Mollusca, *Phil. Trans. Roy. Soc.,* Ser. B, **232**:442, 1947.

An Abbreviated Classification of the Phylum Mollusca (After Morton and Yonge, 1964)

Phylum Mollusca

Class I Monoplacophora

Almost bilaterally symmetrical molluscs with ventral foot and one-piece shell.

Class II Amphineura

Elongate, bilaterally symmetrical, mouth and anus terminal, mantle extensive and dorsolateral.

Subclass 1 Polyacophora

Flattened, broad ventral foot, mantle bearing eight dorsal plates. Ctenidia numerous, e.g., *Chaetopleura, Mopalia.*

Subclass 2 Aplacophora

Aberrant, wormlike, deep-water; mantle covers entire body except for ventral groove.

Class III Gastropoda

Asymmetrical well-developed head, flat foot, one-piece shell. Visceropallium undergoes torsion of 180°.

Subclass 1 Prosobranchia

Pronounced torsion, shell in one piece and in a helical spiral. Two ctenidia or reduction in number, e.g., *Acmaea, Busycon.*

Subclass 2 Opisthobranchia

Shell reduced; tendency toward detorsion; mantle cavity moves to posterior along right side; e.g., *Aplysia, Aeolis, Hermissenda.*

Subclass 3 Pulmonata

No ctenidium, mantle cavity forms a lung, detorsion with cephalized nervous system, e.g., *Helix, Limax, Lymnaea, Physa, Planorbis.*

Class IV Bivalvia (Pelecypoda)

Bilaterally symmetrical, rudimentary head, no radula, enlarged ctenidia, laterally compressed, bivalved shell, e.g., *Mercenaria, Mytilus.*

Class V Cephalopoda

Bilaterally symmetrical, 8 or 10 arms or tentacles surround head, nervous system highly developed, e.g., *Loligo, Sepia, Octopus, Nautilus*.

Class VI Scaphopoda

Bilaterally symmetrical, mantle and shell elongate forming a tapered tube open at both ends. No ctenidium.

Methods of Killing Mollusca

1. *Helix* is best killed in stoppered bottles full of water over a 24-hour period; the animals will then die of asphyxiation in an extended state. They may be killed by plunging them into boiling water, but will then produce large amounts of mucus and die in a retracted state.

2. With the exception of *Loligo,* which dies on removal from the water, marine molluscs should be narcotized by the addition of magnesium sulfate crystals (Epsom salts) to the dish of water in which the animals have been placed. After several hours, they will be relaxed and ready for dissection. *Aplysia* may secrete copious quantities of purple mucus, which should be rinsed off thoroughly before dissection is attempted.

Equipment and Materials

Equipment

Microscope and lamp
Microscope slides and cover slips
Tinfoil or aluminum foil
Dissecting pan
Dissecting instruments
Pins
Hooks and weights
Clamp
Glass plate
File

Solutions and Chemicals

0.5 *M* KCl
Carmine suspension
Aquadag
Ground starfish gonad
Magnesium sulfate (Epsom salts)

Preserved Material

Helix
Busycon, normal and injected
Busycon, shell
Mercenaria (Venus)
Nautilus
Loligo*
Loligo (double injected)*
Chaetopleura
Mopalia or *Ischnochiton*
Aplysia

Living Material†

Helix (2,6,11)
Busycon (7,12)
Mytilus (9,12)
Mercenaria (Venus) (9,12)
Loligo (7,9,12)
Mopalia (9)
Acmaea (9,12)
Aplysia (3,7,19)
Hermissenda (7,9)
Aeolis (7,12)
Lymnaea or *Planorbis* (3,11)
Pecten or *Lima* (7,12)
Anodonta (3,5,11)
Mya (12)

*Monterey squid, available in frozen 5 lb blocks from many supermarkets or fish stores, are reported to provide relatively inexpensive material for anatomical studies. They may be superior to preserved or injected material, but are very perishable after thawing.
†Numbers in parentheses identify sources listed in Appendix B.

Chaetopleura (7,12)
Stenoplax (9)
Littorina (7,9,12)
Ensis (12)
Crepidula (12)

Prepared Slides

Crepidula, cleavage
Crepidula, trochophore and veliger
Pecten, eye
Clam mantle

ECHINODERMATA

Contents

*This portion of the exercise should be started at the beginning of the laboratory period to allow time for completion.

Echinoderms (*echino,* spiny; *derma,* skin; Gk.) are familiar seashore animals known commonly as starfish, sea urchins, sand dollars, sea cucumbers, and sea lilies. The group gets its name from the presence of **calcareous ossicles** (plates), which form a **dermal endoskeleton.** They are exclusively marine, and more than any other Metazoa they depart radically from the main evolutionary stream of coelomate animals.

As adults the **pentaradiate symmetry** of echinoderms is in striking contrast to that of other Metazoa. Correlated with this are sluggish habits, lack of cephalization, and absence of segmentation. This radial condition is a secondary acquisition, however, for all the larvae are bilaterally symmetrical. In addition, during embryonic development the coelom characteristically forms as an outpocketing of the gut (i.e., coelom formation is **enterocoelic**); this is also characteristic of the chordates but not of most other invertebrates. A unique feature of the group is a derivative of the coelom known as the **water vascular system,** which is visible externally as numerous projections or **tube feet (podia).** There is no true circulatory system, and the coelomic fluid acts as the principal medium for transport of food, respiratory gases, etc. The gut is complete and muscularized. There are no excretory organs; therefore echinoderms have little capacity for ionic exchange. The reduced ability to osmoregulate explains why the group has never invaded fresh water successfully.

Echinoderms are presumed to be more closely related to the chordates than any other invertebrate group (except the protochordates) because of their mesodermally derived endoskeleton, method of coelom formation, and larval development, as well as several other features.

9–1. BODY ORGANIZATION

Before proceeding any further with this exercise, read General Directions for Laboratory Work on Echinoderms at the end of this exercise (p. 304).

The reason that echinoderm radial symmetry is based upon five radii remains enigmatic. Because of their basic pentaradial symmetry, the surfaces of echinoderms cannot be dorsal and ventral, but are referred to as **oral** and **aboral.** In general, the surface bearing the mouth is oral and that bearing the anus, or opposite the mouth, is aboral.

i. ASTEROIDEA.   The simplest and perhaps the most familiar of all echinoderm anatomical plans is that found in the starfish (Figure 9.1A).

▶ Obtain a specimen of the starfish *Asterias* or *Pisaster,* and note that the organism is nearly flat and that one surface bears the **mouth.** In life the organism orients with this surface down, toward the substrate. Turn the specimen so that the oral surface faces you. The mouth is located in the central **disc,** and radiating outward from the disc are five **arms** or **rays.** Running down the center of each ray is the open **ambulacral groove** with its many **tube feet** arranged in four rows. A single modified tube foot or **tentacle** occurs at the top of each ray. It lacks a sucker and has a light-sensitive pigmented **eyespot** at its base. If you separate the tube feet, you may be able to see the thick, white **radial nerve cord** that runs down the center of each ray. Along the sides of the ambulacral groove are a series of movable **spines** that can be held across the groove and function in protecting the structures that run along it. The spines surrounding the mouth are larger than the others and help to push food into the mouth.

▶ Turn the animal over so that the aboral surface faces you. Again there is a central disc with five rays. Conspicuous at the edge of the central disc is a round, stonelike structure, the **madreporite.** The microscopic **anus** is located in the next interradius clockwise from the madreporite. Situated between two rays, the anus and madreporite impose a superficial bilateral symmetry upon the organism. The rays on either side of the madreporite are together referred to as the **bivium** and the remaining three rays as the **trivium.** The aboral surface is covered with calcareous **tubercles, warts** and **spines,** pincer-like **pedicellariae,** which are modified spines, and small fleshy **papulae (dermal branchiae),** which are protrusions of the coelom and emerge between the plates of the endoskeleton. Place the starfish so that at least part of it lies in a small bowl of seawater, and observe each of these more closely under the dissecting microscope on both the oral and aboral surfaces. Are these structures arranged in any distinct pattern on the surface? Observe the **suctorial discs** at the tips of the tube feet. Are the arms of the starfish flexible? In what plane is the animal flattened? Notice that the entire external surface of the starfish is ciliated.

ii. OPHIUROIDEA.   The brittle stars, or ophiuroids, are superficially similar to the asteroid pattern in their general anatomical arrangement, but the long and sinuous arms are more distinctly set off from the central disc (Figure 9.2A).

▶   Obtain a living specimen of *Ophioderma* or *Ophioplocus* and orient the animal so that the oral surface faces you. The mouth is located in the center of the disc and is surrounded by five triangular **jaws.** The arms bear reduced tube feet called **tentacles,** which project between the skeletal plates or **shields** of the oral surface of the arms (Figure 9.2B). Note that these areas are not true ambulacra; the skeleton of the arms is highly specialized and forms a kind of vertebral column. Move the arms and observe how flexible they are. Compare and contrast the movements of this organism with those of *Asterias (Pisaster)*. Distal to the oral shields and at the edge of the disc next to the arms are ten pairs of **genital bursae** *(Ophioderma)*. (The five outer pairs are not present in *Ophioplocus*.) The gametes may be discharged into these bursae. In some viviparous species, the bursae act as brood pouches, but not in *Ophioderma* or *Ophioplocus*.

▶   Examine your specimen under a dissecting microscope, noting particularly the roughened nature of the spines, which are used for gripping the substrate, and the absence of suctorial discs on the tube feet. Now turn the animal so that you can see the aboral surface. Is there a madreporite? What is its location? Can you find pedicellariae and/or dermal papulae?

Some ophiuroids have branched arms, e.g., the basket star *Gorgonocephalus*. Examine a demonstration specimen.

iii. ECHINOIDEA.   The echinoids, or sea urchins and sand dollars, are globose, oval, or flattened disc-shaped organisms with no arms. In life the animal is oriented with the oral surface downward and the body is covered with spines that are borne on a solid shell or **test.**

▶   Obtain a living specimen of the sea urchin *Arbacia* or *Strongylocentrotus* (Figure 9.3). Note that the oral-aboral axis is longer than in the echinoderms you have examined previously. Turn the animal so that the oral side faces you (Figure 9.3A). In the center of the oral surface is the **mouth,** which bears five protrusible **teeth;** these are the principal chewing organs of the jaw apparatus known as **Aristotle's lantern.** A soft, membranous area called the **peristome** surrounds the mouth; it has specialized **buccal tube feet** associated with it that are probably chemoreceptive in function. At the edge of the peristome are five pairs of **gills** that are evaginations of the body wall, communicate with the body cavity, and are presumed to have an accessory respiratory function. Turn the animal to the aboral side (Figure 9.3B) so that you can see the centrally located **periproct,** the excentric opening of the **anus,** and the **genital plates** with **pores** for exit of the genital products. Where is the **madreporite?** Between each genital plate and at the aboral extremity of each ambulacrum is an **ocular plate** pierced by a single pore through which, in life, passes a **light-sensitive tube foot.**

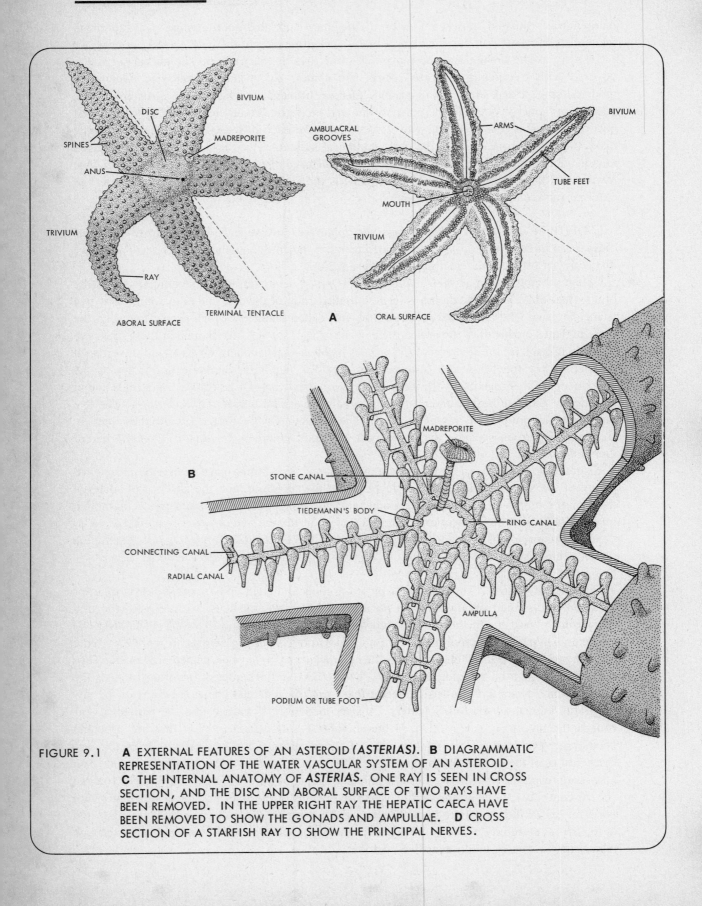

FIGURE 9.1 **A** EXTERNAL FEATURES OF AN ASTEROID (*ASTERIAS*). **B** DIAGRAMMATIC REPRESENTATION OF THE WATER VASCULAR SYSTEM OF AN ASTEROID. **C** THE INTERNAL ANATOMY OF *ASTERIAS*. ONE RAY IS SEEN IN CROSS SECTION, AND THE DISC AND ABORAL SURFACE OF TWO RAYS HAVE BEEN REMOVED. IN THE UPPER RIGHT RAY THE HEPATIC CAECA HAVE BEEN REMOVED TO SHOW THE GONADS AND AMPULLAE. **D** CROSS SECTION OF A STARFISH RAY TO SHOW THE PRINCIPAL NERVES.

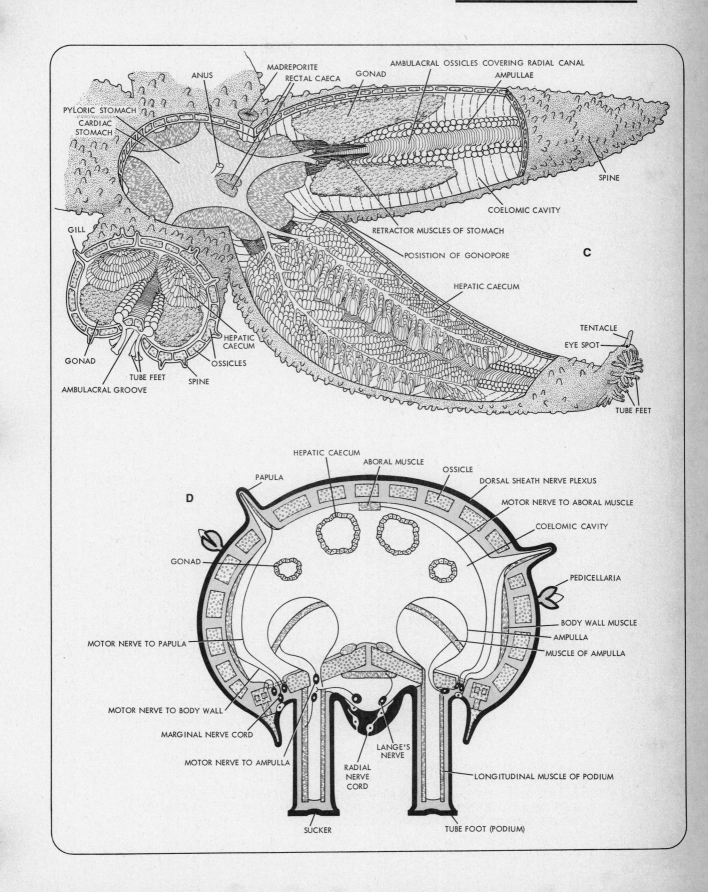

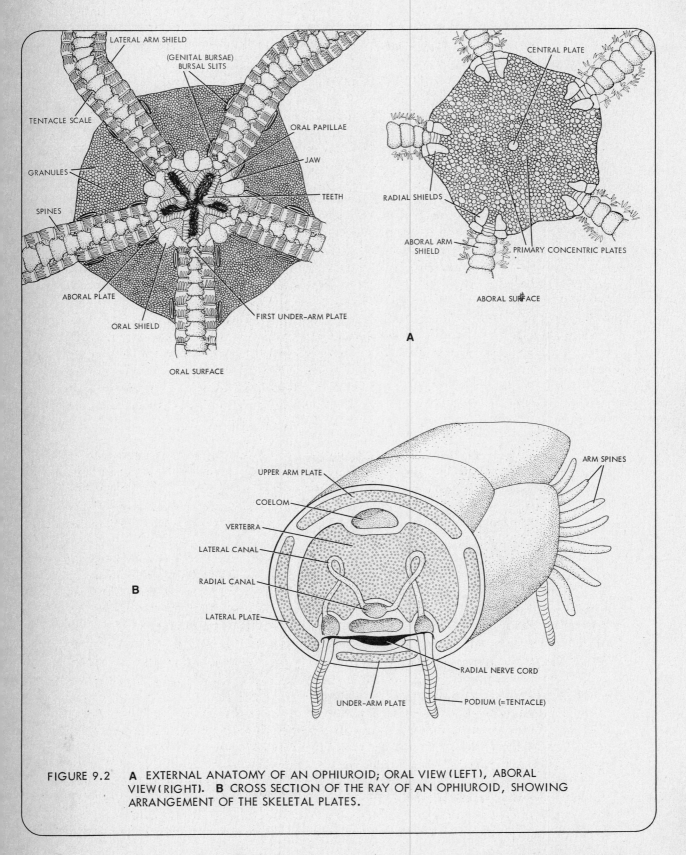

LATERAL ARM SHIELD

(GENITAL BURSAE) BURSAL SLITS

TENTACLE SCALE

ORAL PAPILLAE

GRANULES

JAW

TEETH

SPINES

ABORAL PLATE

ORAL SHIELD

FIRST UNDER-ARM PLATE

ORAL SURFACE

CENTRAL PLATE

RADIAL SHIELDS

ABORAL ARM SHIELD

PRIMARY CONCENTRIC PLATES

ABORAL SURFACE

A

UPPER ARM PLATE

COELOM

VERTEBRA

LATERAL CANAL

RADIAL CANAL

LATERAL PLATE

ARM SPINES

RADIAL NERVE CORD

UNDER-ARM PLATE

PODIUM (=TENTACLE)

B

FIGURE 9.2 **A** EXTERNAL ANATOMY OF AN OPHIUROID; ORAL VIEW (LEFT), ABORAL VIEW (RIGHT). **B** CROSS SECTION OF THE RAY OF AN OPHIUROID, SHOWING ARRANGEMENT OF THE SKELETAL PLATES.

▶ Study the surface of a living sea urchin in a bowl of seawater under the dissecting microscope. The entire external surface is ciliated. Examine the jointed, movable **spines** anchored to the plates of the test by a **ball and socket joint.** Locate the tube feet and notice that they are restricted to five regions of the test, the **ambulacra.** These regions are closed (i.e., they are covered over by plates of the endoskeleton) and the tube feet emerge through paired perforations in the plates. The ambulacra are separated from each other by five **interambulacral** regions with no tube feet. Can you locate suctorial discs on the tube feet? Are there dermal branchiae and/or pedicellariae?

The globose form of the echinoid is modified in some groups, where the organisms are flattened in the oral-aboral plane, the anus has shifted in position, and the ambulacra of the aboral surface are typically arranged into **petaloids.**

▶ Examine the aboral surface of a specimen of the sand dollar *Echinarachnius* (Figure 9.4). Observe the arrangement of the ambulacral regions on the surface into petaloid shapes. Each ambulacrum contains a double row of tube feet running from the petaloid aboral end to the edge of the disc. At the aboral end of each ambulacrum is an **ocular plate** through which passes a single **light-sensitive podium.** Between the ocular plates are four larger **genital plates,** each bearing a **genital pore** and a larger **madreporite.** Are the tube feet of this surface locomotory in function? Are the spines of the aboral surface uniform in size and distribution?

▶ Turn the animal so that the oral surface faces you. Identify the centrally located **mouth,** the teeth of **Aristotle's lantern,** and the five **ambulacral grooves** that radiate from the mouth and branch irregularly until each touches its neighbor. This branching pattern results in a complete ring of tube feet around the edge of the disc. Are all the spines the same size? What are the functions of the spines and tube feet on the oral surface? The **anus** is no longer aboral in position as in the sea urchin; it is found on the edge of the disc. Locate the anus in your specimen. How is the form of the sand dollar adapted to its habitat?

iv. CRINOIDEA.   Feather stars and sea lilies or crinoids are the most primitive class of living echinoderms. Their anatomy is based on the same pentamerous plan as that of other echinoderm groups, but many of them have aboral **stalks,** and in life are oriented with the oral surface upward. They have long, filamentous arms with branching **pinnules** that bear the tube feet and run from a central rachis. Some of the more common modern representatives have lost their stalks and swim freely, e.g., *Antedon.*

▶ Examine a plastic mount of *Antedon* or *Heliometra* (Figure 9.5A). The disclike spherical body is covered by a leathery skin, the **tegmen,** in which are embedded minute calcareous plates. Distinguish the oral surface (Figure 9.5B) with the small central **mouth** and the interradial **anus** on a muscular **papilla,** which directs the feces away from the mouth. Can you think of an evolutionary explanation for the location of the anus on the oral surface? Five ciliated **food grooves** radiate from the mouth and divide to form ten grooves, which pass onto the ten filamentous **arms.** With the aid of a lens, observe that along the sides of the ciliated grooves are rows of yellowish **sacculi,** dermal vesicles of unknown function. Large numbers of **podia** covered with **mucous glands** project inward over the grooves. The tegmen is perforated by a number of **pores** that open into the coelom. There is no madreporite. The aboral surface (Figure 9.5C) has a central skeletal ossicle, the **centrodorsal ossicle,** from the edges of which arise 17–40 hooked **cirri,** each composed of a column of jointed ossicles; the last ossicles at the tip of each cirrus are pointed and hooked. The cirri are used for gripping rocks. The arms arise at the aboral surface from whorls of **brachial ossicles,** arranged in a pentaradiate pattern. These ossicles are together called the **calyx** and may be hidden by the cirri. The arms are supported by a series of calcareous ossicles and each arm branches once near the base. When the animal swims, five arms beat down while the alternate five arms beat upward. From each arm ossicle

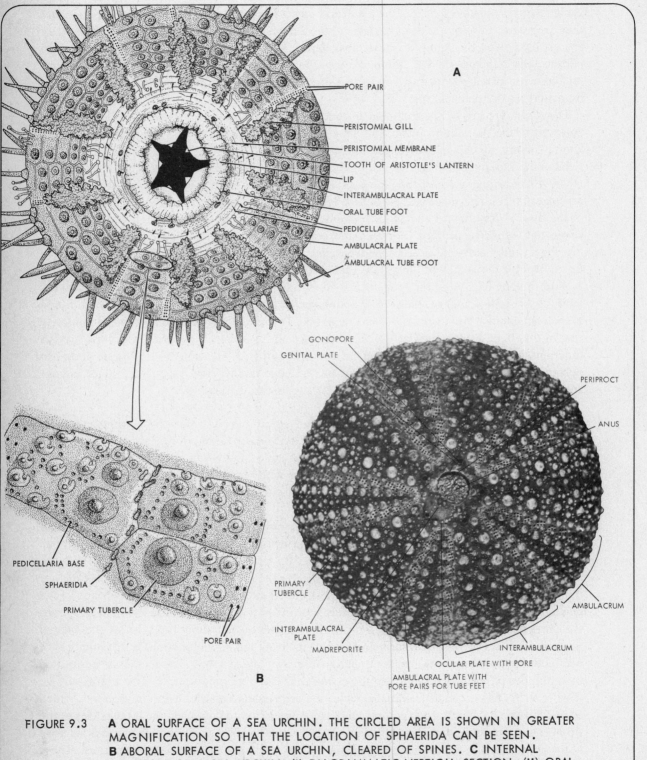

A

PORE PAIR

PERISTOMIAL GILL

PERISTOMIAL MEMBRANE

TOOTH OF ARISTOTLE'S LANTERN

LIP

INTERAMBULACRAL PLATE

ORAL TUBE FOOT

PEDICELLARIAE

AMBULACRAL PLATE

AMBULACRAL TUBE FOOT

GONOPORE

GENITAL PLATE

PERIPROCT

ANUS

PEDICELLARIA BASE

SPHAERIDIA

PRIMARY TUBERCLE

PORE PAIR

PRIMARY TUBERCLE

INTERAMBULACRAL PLATE

MADREPORITE

OCULAR PLATE WITH PORE

AMBULACRAL PLATE WITH PORE PAIRS FOR TUBE FEET

AMBULACRUM

INTERAMBULACRUM

B

FIGURE 9.3 **A** ORAL SURFACE OF A SEA URCHIN. THE CIRCLED AREA IS SHOWN IN GREATER MAGNIFICATION SO THAT THE LOCATION OF SPHAERIDA CAN BE SEEN. **B** ABORAL SURFACE OF A SEA URCHIN, CLEARED OF SPINES. **C** INTERNAL ANATOMY OF A SEA URCHIN: (i) DIAGRAMMATIC VERTICAL SECTION; (ii) ORAL AND ABORAL HALVES HAVE BEEN SEPARATED TO SHOW THE ARRANGEMENT OF THE ORGANS WITHIN THE TEST.

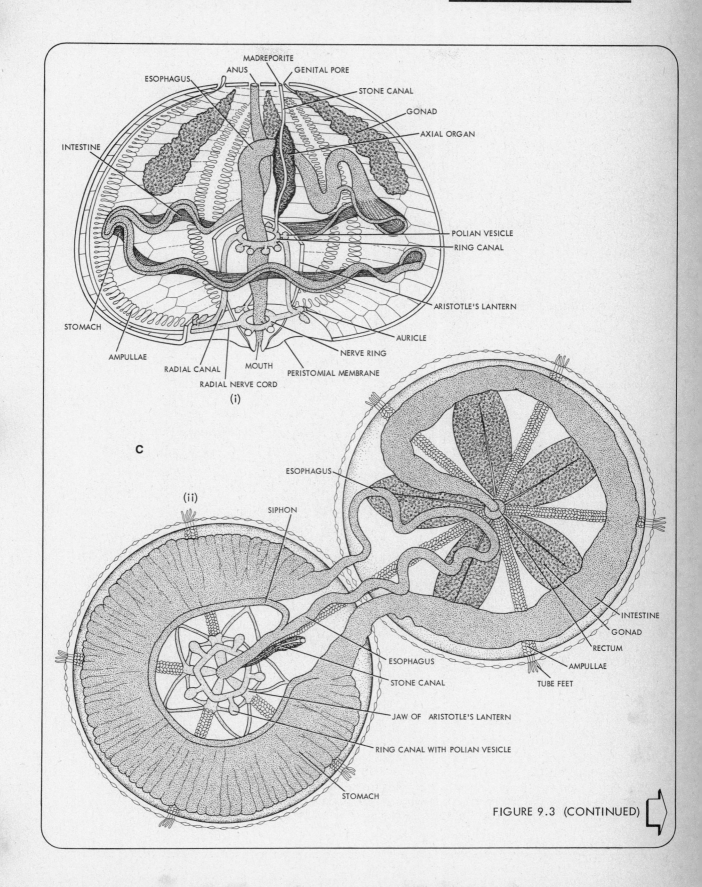

FIGURE 9.3 (CONTINUED)

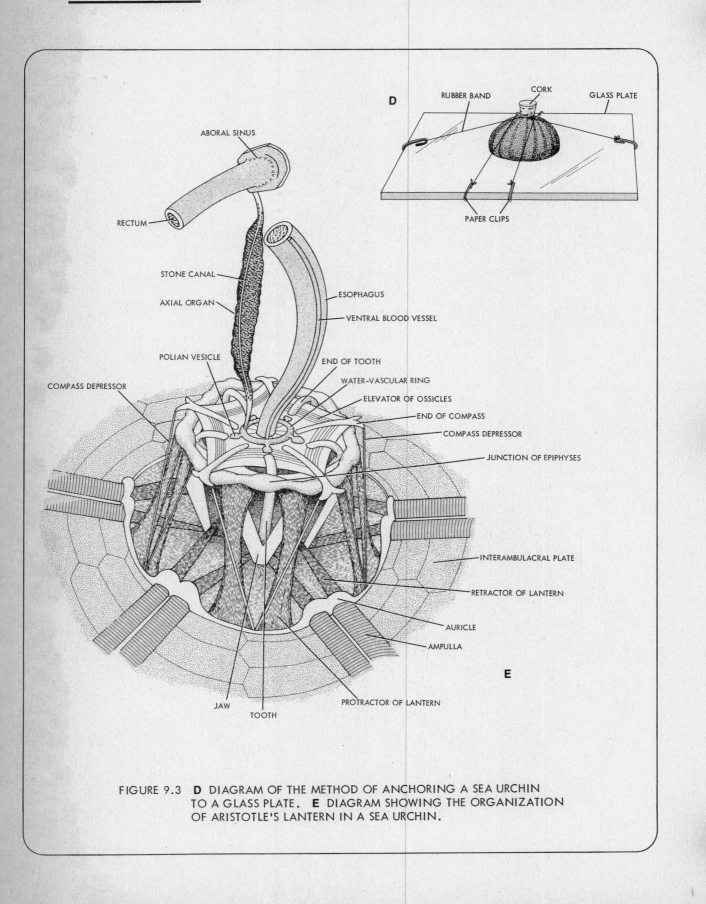

RUBBER BAND   CORK   GLASS PLATE

D

PAPER CLIPS

ABORAL SINUS

RECTUM

STONE CANAL

AXIAL ORGAN

ESOPHAGUS

VENTRAL BLOOD VESSEL

POLIAN VESICLE

END OF TOOTH

WATER-VASCULAR RING

COMPASS DEPRESSOR

ELEVATOR OF OSSICLES

END OF COMPASS

COMPASS DEPRESSOR

JUNCTION OF EPIPHYSES

INTERAMBULACRAL PLATE

RETRACTOR OF LANTERN

AURICLE

AMPULLA

E

JAW   TOOTH   PROTRACTOR OF LANTERN

FIGURE 9.3 **D** DIAGRAM OF THE METHOD OF ANCHORING A SEA URCHIN
TO A GLASS PLATE. **E** DIAGRAM SHOWING THE ORGANIZATION
OF ARISTOTLE'S LANTERN IN A SEA URCHIN.

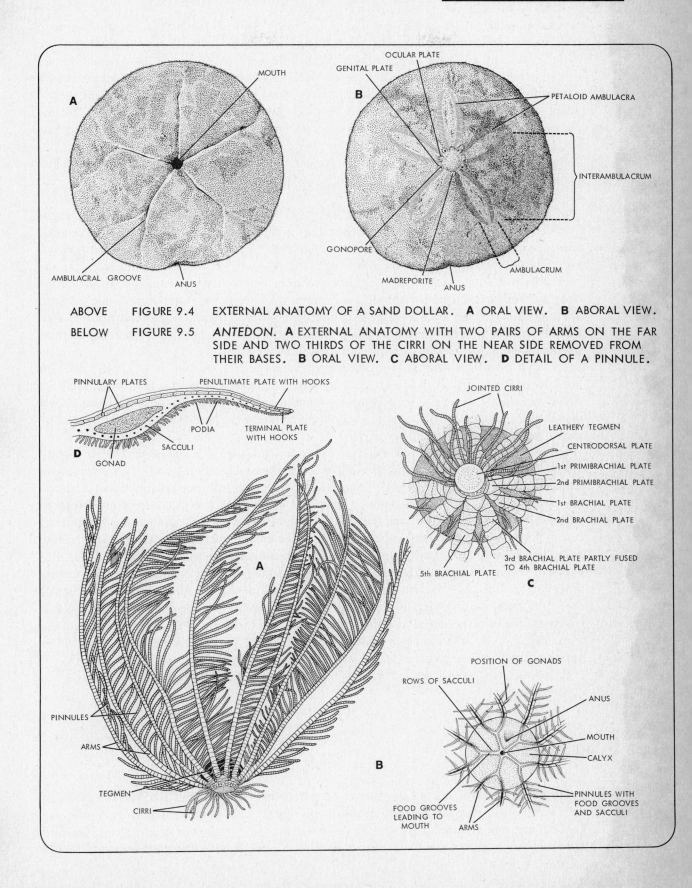

ABOVE   FIGURE 9.4   EXTERNAL ANATOMY OF A SAND DOLLAR. **A** ORAL VIEW. **B** ABORAL VIEW.
BELOW   FIGURE 9.5   *ANTEDON.* **A** EXTERNAL ANATOMY WITH TWO PAIRS OF ARMS ON THE FAR SIDE AND TWO THIRDS OF THE CIRRI ON THE NEAR SIDE REMOVED FROM THEIR BASES. **B** ORAL VIEW. **C** ABORAL VIEW. **D** DETAIL OF A PINNULE.

on alternate sides there arises a **pinnule** supported by a row of jointed ossicles. Ciliated food grooves run along the pinnules, with podia along both sides. The podia may be difficult to see as they probably will have been withdrawn into the food grooves. In life, food organisms are caught in the mucus produced by the glands on the podia and a strand of food-laden mucus is passed to the mouth by the ciliary currents in the food groove. The pinnules from the disc to about half way along the arms are swollen with **gonads.** There are no gonoducts, and the gametes are released by the rupture of the gonads. Notice the absence of spines and pedicellariae.

Since living specimens of crinoids are difficult to obtain and will not ordinarily be available, further details of crinoid anatomy will not be considered in this exercise.

v. HOLOTHUROIDEA. In adult holothuroids, or sea cucumbers, the secondary radial symmetry is obscured by a superimposed tertiary bilateral symmetry. You have observed that animal symmetry is related to the manner in which the organism orients itself in the environment; in general, radial symmetry is characteristic of sessile or sedentary animals, where the principal orientation is toward gravity. In holothurians, associated with burrowing habits, the body is cylindrical, cucumber-shaped, or vermiform and elongate in the aboral-oral direction. The ambulacra usually run along the length of the body, and their arrangement, together with that of the tube feet, has shifted so that a dorsoventral axis is established. Most holothurians habitually orient themselves with one surface toward the substrate; this **ventral** surface usually bears three ambulacra with all the locomotory tube feet (e.g., *Cucumaria*). However, tube feet need not be restricted to the ambulacral areas (e.g., *Thyone*). The tube feet of the opposite **dorsal** surface are in general reduced to warts and sensory papillae. Because the animal has a dorsal surface and a ventral surface, it is possible to establish anterior and posterior ends. The mouth is anterior and the anus posterior.

▶ Study the form of a preserved specimen of the sea cucumber *Cucumaria* or *Thyone* (Figure 9.6A). The endoskeleton is reduced to microscope **spicules** giving the body wall a tough and leathery quality. Run your fingers across the body surface of your specimen and note the absence of spines. The **mouth** may be identified at the center of a conspicuous crown of **tentacles,** which are modified tube feet. (These will be retracted if the animal was not relaxed before preservation. If so, pinch the body behind the mouth and tease out the tentacles.) How many tentacles are present? Locate the terminal **cloacal aperture** surrounded by its groups of papillate sensory tube feet. The arrangement of the ambulacral regions can easily be seen by looking at *Cucumaria* from one end. Is the interambulacral area the same size between each of the rows of tube feet? What does this imply about the symmetry of the organism? Can you find a madreporite or pedicellariae? Do the tube feet have suctorial discs? Does the structure of the tube feet vary in different regions of the body? If differences exist, how are they to be explained with respect to location and function?

▶ Based on your observations, correlate habit with external form in each of the classes of echinoderms.

9–2. THE SKELETON

The entire body surface of echinoderms is covered with an **epidermis** composed of **epithelial** cells that may or may not be ciliated, **neurosensory** cells, and **gland** cells. The epidermis also contains **pigment granules** that are responsible for the varied colors of echinoderms. Beneath the epidermis a thick **dermis** of fibrillar connective tissues secretes the **calcareous endoskeletal**

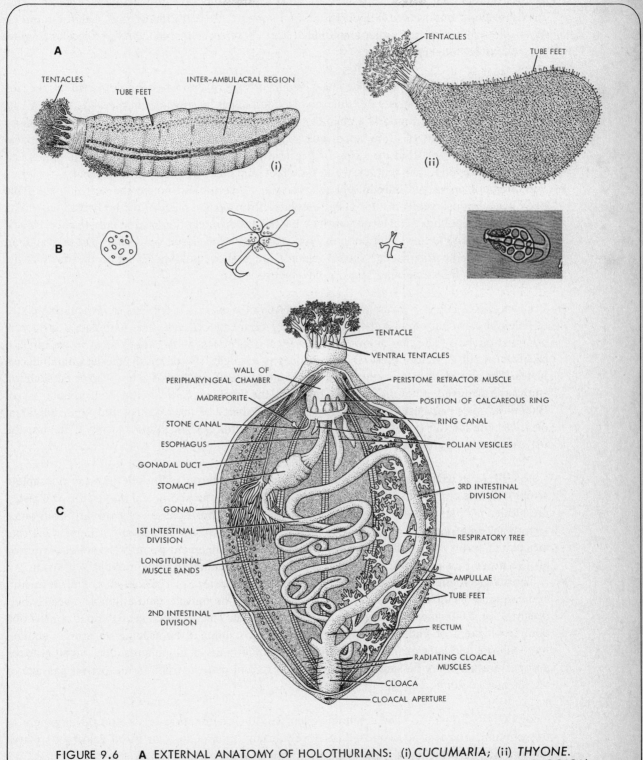

FIGURE 9.6   **A** EXTERNAL ANATOMY OF HOLOTHURIANS: (i) *CUCUMARIA*; (ii) *THYONE*.
B HOLOTHURIAN SPICULES. (PHOTOGRAPH COURTESY CAROLINA BIOLOGICAL
SUPPLY COMPANY.)   **C** INTERNAL ANATOMY OF A HOLOTHURIAN (*THYONE*).

plates or **ossicles** and their **spines,** and binds the plates more or less closely together. The dermis and its skeletal elements are mesodermal in origin and function principally in protection, but may also be of importance in locomotion and support. Beneath the dermal skeleton a **muscle layer** of outer **circular** and inner **longitudinal** fibers is separated from the internal **coelomic space** by a thin **flagellated epithelium.**

   i. HOLOTHUROIDEA.   The most reduced sort of endoskeleton is found in the sea cucumbers, where it consists of numerous isolated **plates** and **spicules** in the body wall. A ring of calcareous ossicles supports a chamber surrounding the pharynx; the ossicles and chamber may be the remnants of a jaw apparatus similar to that found in the echinoids. The cloacal aperture is also supported by a ring of five toothed calcareous plates. Calcareous plates and spicules also support the tentacles, the tube feet, the madreporite, and ring canal.

▶   Examine a preserved holothuroid *(Cucumaria, Thyone)* and notice the soft nature of the body wall. Endoskeletal spicules may be isolated from a small piece of the body wall by boiling it in sodium hypochlorite. The soft connective tissues disintegrate, leaving only the spicules. If the pieces of tissue are small enough, you may place the tissue on a slide and add sodium hypochlorite. The soft tissues should disintegrate without boiling. Examine the form of the spicules under the compound microscope (Figure 9.6B).

▶   ii. ASTEROIDEA.   Examine a dried skeleton of *Asterias*. It consists of articulating plates of calcium carbonate held together by leathery connective tissue. The ambulacral groove is bounded on both sides by a series of closely fitting plates, which meet to form an internal **ambulacral ridge.** Below these aboral ossicles is a single row of spine-bearing **adambulacral plates.** The aboral surface is supported by a framework of smaller ossicles. The dried skeleton shows the arrangement of the ossicles, but lacks the flexibility found in the living animal. Why? Examine a living starfish and note the limited flexibility of the ray. To study the individual ossicles, cut off the tip of one ray and boil it in some sodium hypochlorite. Compare the ossicles of the asteroid with those of the holothuroid.

   iii. OPHIUROIDEA.   The skeletal ossicles of the ophiuroid endoskeleton form a complete armor over the central disc. Note that in *Ophioderma,* all the aboral and part of the oral plates are hidden by pigmented granules. The skeleton of the arms consists of two parts—an outer superficial endoskeleton composed of aboral, lateral, and oral **shields** and a deeper internal articulated series of so-called **vertebral ossicles.** This arrangement permits the solidly armored arm to move by sinuous flexures and thus enables the animal to crawl rapidly and swim.

▶   The arrangement may be seen in the brittle star *Ophioderma* or *Ophioplocus*. Pull off an arm and examine it in cross section (Figure 9.2B). Locate the muscles and attempt to relate their organization to the movements made by a living animal. The vertebral articulations consist of a complex system of knobs and grooves; it may be examined by placing an arm in sodium hypochlorite and later examining the free plates under a dissecting microscope. Small grooves and perforations may be seen in the internal vertebral plates, which in life permit passage of nerves and coelomic vessels.

   iv. ECHINOIDEA.   The echinoids exemplify the culmination of skeletal development in this phylum. The test is composed of closely fitting, interlocking, or fused calcareous plates that are not movable as in the asteroids and ophiuroids.

▶   Examine a dried specimen* of a sea urchin (Figure 9.3B). Notice the points of attachment of

*A dry test may be prepared by placing a living or preserved animal in sodium hypochlorite solution.

the spines and the way in which each spine articulates with a **tubercle** on the test by a ball-and-socket arrangement. If the spines have not been removed from your specimen, clean it by scraping the surface free of spines with a scalpel or stiff brush. Observe that the plates of the test are arranged into 20 meridional rows organized into alternating double columns—five **ambulacral** and five **interambulacral.** The plates forming the ambulacra are perforated by **pore pairs** through which, in life, the tube feet pass. Examine the area of the periproct, and locate the madreporite and the genital and ocular plates.

▶ Compare the skeleton of the sea urchin to that of a sand dollar that has been similarly cleaned of spines (Figure 9.4). Can you detect perforations in the test of the sand dollar where the tube feet were located? In what zones are they found?

9–3. THE WATER VASCULAR SYSTEM, HYDRAULIC SKELETON, AND LOCOMOTION

The water vascular system of echinoderms is unique among the Metazoa and is perhaps the most distinctive feature of this phylum. The system is composed of a series of tubes, derived from the coelom, whose fluid contents are much like the seawater with which the system freely communicates via the madreporite. The system is evident externally as the blind ending branches that are the tube feet. The tube foot is an organ of multiple function and, depending on the group, may be used in locomotion, sensation, feeding, and/or respiration. In general the water vascular system is composed of **madreporite, madreporic (stone) canal, ring canal, radial canals, ampullae,** and **tube feet** (Figure 9.1**B**).

Before proceeding to study the water vascular system of echinoderms, review the general directions for the dissection of echinoderms at the end of this chapter.

▶ i. ASTEROIDEA   Obtain an anesthetized starfish and carefully remove the aboral surface of the central disc and of the three rays of the trivium. This is best done by inserting a scissor point into the body wall at the tip of a ray and cutting down both sides of the ray toward the central disc. Continue cutting so as to free the body wall in the region of the disc, being careful not to disturb the underlying organs. Separate the madreporite from the body wall, taking care not to damage it and making sure that it retains its connection with the rest of the water vascular system. Carefully loosen the underlying organs from the mesenteries by which they are suspended. Lift up the free aboral body wall and remove it completely from the rest of the body. Pin the opened specimen out in a wax-lined pan and cover with seawater.

▶ Locate the digestive tract (Figure 9.1**C**) and trace the **stone canal** that leads orally from the madreporite to the **ring canal** around the esophagus. The digestive tract may obscure this; attempt to identify the water vascular ring without removing the digestive tract, or identify the water ring when you come to dissect the digestive system (Section 9–6). Can you see the nine **Tiedemann's bodies** on the ring? The position of the tenth is occupied by the junction of the water vascular ring with the stone canal. The Tiedemann's bodies are organs of uncertain function; they may filter the water vascular system of debris and extraneous material with the aid of the amoebocytes found in them, performing a role much like the lymph nodes of vertebrates. Leading from the ring canal are five **radial vessels** which run one to each ray. Remove the hepatic (pyloric) caeca and gonads from the ray to expose the ambulacral ossicles of the skeleton that hide the **radial water vascular canal.** Lateral branches from the radial canal give rise to the tube feet on the oral surface and **ampullae,** which are bulbous projections situated at the aboral end of each tube foot (Figure 9.1**C**). A radial line of these bulbous ampullae can be seen along the internal surface of each ray.

▶ Using prepared slides of a cross section of the starfish ray (Figure 9.1**D**), identify the components of the water vascular system.

▶ You may demonstrate the anatomical connections of the water vascular system in the ray by severing a ray from its attachment to the central disc and removing a piece of the tip of the ray. Locate the radial vessel, and by means of a fine-tipped Pasteur pipet containing Evans blue dye, inject some dye into this vessel. What structures become colored by the blue dye?

▶ Allow a living starfish to walk along a glass plate. Observe the action of the tube feet through the glass. What is the sequence of activity in an individual tube foot as it moves? The tube foot is filled with fluid, and the walls of the ampulla and the foot itself are muscular, providing the basic components of a **hydraulic skeleton.** What muscles contract to make the tube foot turgid? Flaccid? What prevents back flow from the tube feet to the radial vessel? What is the function of the terminal sucker on the tube foot? If tube feet are damaged and the system leaks, how is loss of fluid compensated? Study the form of a tube foot in a burrowing asteroid such as *Astropecten*. What structure is absent? Correlate tube foot characteristics with behavior.

ii. ECHINOIDEA. The arrangement of the water vascular system in the sea urchin is essentially the same as that found in the asteroids, except that the ambulacra are closed and therefore the radial canals are internal to the skeleton.

▶ Recall the position of the madreporite of the sea urchin, or locate the madreporite on the aboral surface of a living sea urchin now. The anatomical details of this system will be seen when the gut and digestive system are dissected, or you may dissect a specimen now according to the directions given in Section 9–6. The water vascular system runs orally from the madreporite as the stone canal and forms a ring canal around the esophagus; five radial vessels run aborally from the ring canal and give rise to the tube feet (Figure 9.3**C**).

▶ Place a living sea urchin on a glass plate and observe the active spines and how the tube feet are used in locomotion. The processes are much like those seen in the asteroid, except that here the spines play a role in movement. This is easily observed by taking a sea urchin and turning the animal upside down (i.e., aboral surface in contact with the substrate). Are the spines used in the righting reaction? Are the tube feet? Do the tube feet have suckers?

iii. HOLOTHUROIDEA. Sea cucumbers show modifications not only of symmetry, but of the water vascular system. Holothurians, like echinoids, have a closed ambulacral system, so that the radial water vessels and other radial structures are internal to the body wall. The holothurian madreporite is also internal and is found at the end of the stone canal, lying free in the coelom or attached to the body wall by mesenteries.

▶ Obtain a relaxed specimen of *Cucumaria* or *Thyone,* and orient the specimen so that the three rows of podia (ambulacral areas) that are closest together face you. This is the functional **ventral** side of the animal (Figure 9.6**A**). Make a ventral incision from mouth to anus. Carefully open the incision and reflect the cut body wall laterally. Pin the opened specimen in place in a dissection pan and cover with water. The water vascular system has **corpuscles** containing **hemoglobin,** which give the system a pinkish appearance. Identify the dorsal madreporite and stone canal, which hang free in the coelom (Figure 9.6**C**). The stone canal connects to the water vascular ring canal, which surrounds the esophagus, and the ring has two elongate sacs called the **Polian vesicles,** which are presumed to be reservoirs of water vascular fluid. From the ring canal, radial canals run anteriorly to the oral tentacles, and then turn back as posterior branches that are embedded in the longitudinal muscle bands and give rise to the tube feet. Locate the five paired bands of **longitudinal muscles** running in the body wall. At the anterior end, these join powerful **retractors,** attached to the calcareous ring, which pull the peristomial area and tentacles into the body. **Circular muscles** line the body wall and form **oral** and **cloacal sphincters.**

▶ Sea cucumbers spend a considerable amount of time burrowing. Place a living *Cucumaria* or *Stichopus* (*Leptosynapta* or *Thyone*) on the surface of some sand, cover with seawater, and observe the behavior of the animal. Does a peristaltic wave pass along the organism during burrowing? What are the components of the hydraulic skeleton of the body? Pick up the animal and note the turgid state of the body. What similarities are there to other wormlike forms? Are there any groups of vermiform organisms that do not employ a hydraulic skeleton in burrowing? What would be the disadvantage of a rigid endoskeleton to burrowing in sea cucumbers?

iv. OPHIUROIDEA. The ophiuroids are often regarded as the most successful echinoderm group living today in terms of number of species. This is probably because of the smaller size and greater agility of its members. The tube feet of the water vascular system are reduced and are largely sensory in function; locomotion depends mainly on the well-developed skeleton of the arms and its associated musculature. The flexible arms are so constructed that movement is more or less restricted to the horizontal plane, but nevertheless the animals are swift-moving and energetic.

▶ If living specimens are available, observe their movements in an aquarium. Notice that one arm may lead or follow the others. Is it always the same arm? How are the other arms coordinated with each other in locomotion? Can the animal coil an arm around objects and thus pull itself along?

9–4. RESPIRATION AND CIRCULATION

Exchange of gases between the ambient environment and the body tissues depends on the amount of gas available, the rate of diffusion in the medium, and the method of transferring the gas to and from the internal tissues. Gaseous diffusion is much slower in water than in air, and to compensate for this in echinoderms, the ambient seawater is moved over the respiratory surface by ciliary action. Internally, the transport of gas, nutrients, and wastes to and from the tissues of animals with some bulk necessitates an active circulation. The derivatives of the coelom, that is, the water vascular system and the perivisceral coelom, play the dominant role in echinoderm respiration and circulation. Coelomocytes probably play a role in distribution of digested nutrients, but this has not been unequivocally demonstrated.

▶ i. ASTEROIDEA. Place a small drop of carmine suspension on the aboral surface of a living starfish (*Asterias, Pisaster*) in a bowl of seawater and observe how the particles move on the surface. In what direction are the ciliary currents? Blister-like bulges called **dermal papulae** or **branchiae** may be seen on the aboral surface. These are ciliated externally and are formed by protrusions of the coelom which emerge between the ossicles. Gaseous transfer occurs across the walls of these membranous branchiae as well as across the walls of the tube feet.

▶ Circulation of fluid internally in the spacious perivisceral coelom can be seen by injecting 0.5 to 1.0 ml of a heavy suspension of carmine in seawater into one of the rays of an active starfish by means of a hypodermic syringe fitted with a 20 gauge needle. Observe how the carmine color moves rapidly across the ray. Observe the dermal branchiae under the dissecting microscope. Can you see carmine particles within the branchiae? What causes the circulation? After a few minutes cut off the tip of a ray of this injected starfish, and allow the coelomic fluid to drain into a test tube. Examine a drop of the fluid under the compound microscope. Where are the carmine particles located? The carmine particles are moved by ciliary currents; these particles are ingested by amoebocytes, which are carried by the current in the coelom or move in amoeboid fashion through the coelom and into the dermal branchiae.

▶ Circulation time in asteroids may be determined by taking a live starfish, injecting 0.5 ml of carmine suspension into one ray, and recording the time taken for the particles to appear in a ray on the opposite side of the body from the site of injection. The distance divided by the time interval between injection and appearance of carmine gives an estimate of circulation time.

ii. ECHINOIDEA.   It is quite obvious that the hard test of echinoids is a barrier to gaseous diffusion, and in the absence of structures such as dermal branchiae, the tube feet serve as the main respiratory surface. The accessory gills at the edge of the peristome are thought to supply the respiratory needs of the jaw apparatus.

▶ Take a large living specimen of a sea urchin *(Strongylocentrotus, Arbacia)* and separate the oral from the aboral half by cutting through the equator with a saw so that the cut edges are smooth and even. Gently clean the inside of the aboral half (remove the gut and gonads). Remove the anal plates and plug the hole with a close-fitting cork or rubber bung. Put the test on a glass plate and seal the cut edge to the glass with a thick layer of petroleum jelly. The seal should be airtight. Further anchor the test to the plate with rubber bands and paper clips (Figure 9.3**D**). Fill the half-urchin through the anal opening with a thick suspension of luminescent bacteria, making sure that there are no air bubbles inside. Replace the cork and wash the external surface free of luminous bacteria. Immerse the entire apparatus in a bowl of well-oxygenated water and take it into a dark room. After your eyes have adapted to the dark, you should see five meridional rows of luminescence through the glass plate on the inside surface of the sea urchin. These correspond to the positions of the tube feet.

▶ The surfaces of the podia are ciliated, as is the epidermis that covers the test, and water is moved over the surfaces by the action of the cilia. Determine the direction of water movement on the test of a living sea urchin by placing a few drops of carmine suspension on various parts of the test of a specimen in a bowl of seawater.

▶ As in the asteroids, the perivisceral coelom of sea urchins is spacious, and the coelomic fluid circulates within. Inject a small amount (e.g., 0.5 ml) of carmine suspension into the soft tissue surrounding the mouth (peristome) by means of a hypodermic syringe fitted with a 25 gauge needle. A few minutes later, withdraw some coelomic fluid by means of a clean syringe and examine a drop under a compound microscope. Observe the large variety of **coelomocytes.** Do they contain carmine particles? The fluid will clot, but only if coelomocytes are present. Devise an experiment to verify this. Clotting can be accelerated by adding some clam juice. Why does the coelomic fluid not clot inside the animal? What are the functions of coelomocytes in echinoids?

iii. HOLOTHUROIDEA.   In sea cucumbers, as in other echinoderms, the coelomic and water vascular fluids fulfill a transport function. In some species, additional respiratory surfaces are provided by a pair of specialized organs lying in the coelom, the **respiratory trees,** and by the tube feet.

▶ Return to the specimen of *Cucumaria* or *Thyone* that you opened previously to identify the components of the water vascular system, or open an anesthetized specimen now, according to the directions given in Section 9–3. Locate the respiratory trees, which are the two large branching organs that lead out of the terminal end of the intestine, the **cloaca** (Figure 9.6C). If the specimen is still alive and has not been damaged too much during dissection, you may see rhythmic pulsations in the cloaca and in the terminal branches of the trees. Even if you cannot observe this in your specimen, attempt to work out the mechanics of circulation and the operation of the system in gaseous exchange. Contraction of circular muscles in the cloaca causes water to be expelled from the cloacal aperture if it is open, or forced into the respiratory tree if the aperture is closed. Contraction of radial cloacal muscles causes cloacal expansion, and water is sucked into the cloaca from the outside via the cloacal aperture or from the respiratory trees. Where must sphincters be located in this system? On occasion, the whole body contracts, and water spouts from the cloacal aperture.

▶ Observe the periodic opening and closing of the cloacal aperture in an intact animal. For a sea cucumber this is equivalent to breathing. You can estimate the respiratory rate of a sea cucumber by placing an animal in seawater (10 to 15°C) and counting the time interval between exhalent currents from the cloaca. Do this for 10 to 15 minutes. Now place this cucumber in a bowl containing seawater at 25°C. Allow the specimen to acclimate for 15 minutes, and then record the time intervals of the exhalent currents for another 10 to 15 minutes. What happens to the cloacal pumping rate with the increase in temperature?

Fluid also circulates in the perivisceral coelom, and in those forms without respiratory trees, e.g., *Leptosynapta,* the transfer of gases probably occurs across the body surface. The water vascular system contains **amoebocytes** which in some species may contain **hemoglobin (hemocytes)** e.g., *Thyone* and *Cucumaria.* In the intact animal the tube feet are red or pink because of the presence of these hemocytes. The circulation of hemocytes probably aids in respiration.

▶ Observe the tube feet of *Cucumaria* or *Thyone* under the dissecting microscope and note the movement of hemocytes within. To get a clearer view of these scrape a few of the tube feet from the surface, or scrape a few ampullae from the inner body wall surface of the specimen opened previously, and observe under the compound microscope. Are the hemocytes nucleated? Can you see ciliary activity within the tube feet?

▶ Extract the respiratory pigments from holothurians and compare them with others, e.g., human, as follows. Locate the water vascular ring in your specimen; snip off the bulbous red Polian vesicles, pick up with forceps, and place in a test tube. Add some distilled water so that the hemocytes lyse. Centrifuge at low speed to remove debris. The clear supernatant can be used for studying the absorption spectrum, (see Exercise 6, Section 6–5, for methodology). How does the absorption spectrum of holothurian hemoglobin compare with that of your own? Can you explain the presence of hemoglobin in some sea cucumbers, but its absence from other echinoderms?

iv. OPHIUROIDEA. The circulatory and respiratory devices of brittle stars are much the same as those of asteroids, but because they are difficult to dissect and study in the intact animal, they are omitted from consideration.

9–5. THE HEMAL SYSTEM

The hemal system of echinoderms is a unique and enigmatic series of ill-defined channels; because there are no definite walled vessels, the system is described as being **lacunar.** The hemal lacunae are enclosed by coelomic vessels referred to as **perihemal** or **hyponeural** vessels. The channels are best developed in association with the digestive system and may play a role in the uptake and distribution of digested materials, although this function is probably performed to a large extent by wandering amoebocytes. The hemal system reaches maximum development in echinoids and holothurians. It is thought by some authors to function in circulation; however, it is an open system with no true pumping device, so that any circulatory function is likely to be minimal.

The so-called **axial gland** and **sinus (axial complex),** which are absent in holothurians, are part of the hemal system and are situated near the aboral end of the stone canal. It has been suggested that the complex acts as a pumping device that circulates fluid in the hemal system;

for this reason the complex has been termed a heart, but together with the rest of the hemal system, it is more likely to be vestigial. Its function at the present time remains obscure.

▶   With the aid of Figure 9.3C, attempt to locate the axial complex and channels of the hemal system in an echinoid such as *Strongylocentrotus* or *Arbacia,* or wait until you dissect the digestive system (Section 9–6).

9–6. FEEDING AND THE STRUCTURE OF THE GUT

Echinoderms may be predacious carnivores, grazing herbivores, omnivores, or ciliary-mucoid detritus feeders. Correlated with their feeding habits are modifications of the gut.

i. ASTEROIDEA.   Starfish are principally voracious carnivores: they feed on a variety of material including gastropods, barnacles, and bivalves, which they open by exerting considerable and prolonged force on the valves with their tube feet. Others may extrude their stomach through the mouth and engulf smaller organisms.

▶   Return to the starfish that you opened previously to see the water vascular system, or open a specimen now according to the directions given in Section 9–3. Locate the digestive tract (Figure 9.1C). The short **esophagus** leads from the **mouth** and connects to the **stomach.** The stomach is divided into a globular protrusible **cardiac** portion and an aboral pentagonal **pyloric** portion. Running from the pyloric portion into each arm are a pair of **hepatic (pyloric) caeca** that function in digestion and food storage. The **rectum** runs to the aboral surface from the pyloric stomach and exits via a microscopic **anus.** Remove the pyloric caeca from one arm and identify the **cardiac retractor muscles** that retract the cardiac stomach.

▶   If you failed to identify the water vascular ring previously, (Section 9–3), remove the digestive tract by severing the connections at the mouth, anus, and pyloric caeca and attempt to locate the ring.

▶   In addition to pulling on the shell valves of bivalve molluscs, the starfish can protrude its stomach and enter it into a space of about 1 mm. If there is the slightest gap between the valves, the bivalve will be digested. If time permits, cut the adductor muscles of some *Mercenaria* or *Mytilus* and close the valves again by wrapping the shell with strong rubber bands. Vary the number of turns of rubber bands on the different shells so that it will require different forces to open the valves. Determine the number of turns of the rubber band needed to prevent a starfish from eating a clam. How much force is this, measured in pounds? (See Exercise 8, Section 8–6, for the methodology.)

ii. ECHINOIDEA.   Most species of sea urchin and sand dollar are principally herbivorous; some species, however, are omnivorous scavengers, carnivores, or even cannibals.

▶   Obtain an anesthetized specimen of the sea urchin *Strongylocentrotus* or *Arbacia,* and place the animal with the oral side down. With a fine saw or a pair of scissors, cut through the equator of the urchin, separating the oral and aboral regions. You may remove spines to facilitate cutting. Separate the two halves by cutting the mesenteries which hold the loops of the gut to the body wall and lay out as shown in Figure 9.3C.

▶   Orally, locate the powerful jaws of **Aristotle's lantern** that surround the mouth. The lantern, which is composed of muscles and calcareous plates, serves to grind and crush the food, to bore through the test of other urchins, into rocks, wooden pilings, or even steel pilings, or to scrape algae and other food organisms from rocks and other surfaces. A detailed description of this organ will not be given here, but the principal parts are identified in Figure 9..3E.

▶ Running around the oral surface of the lantern is the **water vascular ring.** Identify the spongy **Polian vesicles** attached to the ring opposite each interambulacrum. The **axial organ** and the **stone canal** also leave the ring and ascend aborally. Attempt to trace them. The **radial canals** that leave the ring will not easily be seen, but along the inner sides of the test you may be able to identify the radial canals and the **ampullae** of the tube feet.

▶ The **esophagus** ascends from the mouth and emerges from the water vascular ring from the oral surface of the lantern. It loops back on itself and proceeds in an anticlockwise direction as the **stomach,** followed by the **intestine.** After a complete circuit, it again loops back for another circuit and ascends aborally as the **rectum** to the **anus.** Running along the inner edge of the looping stomach is the **siphon,** which is a hollow ciliated tube that probably functions in removal of water from the gut contents and hence concentrates the feces. Digestive enzymes have been located in the digestive tract, and coelomocytes may function in food distribution.

iii. HOLOTHUROIDEA. Sea cucumbers are detritus feeders, employing the tentacles (buccal tube feet), or eating their way through the sand.

▶ Observe the feeding activity of a living *Cucumaria.* The arborescent tentacular crown expands (by what mechanism?) and sweeps over the sandy substrate, and small particles are trapped in the mucus of the expanded tentacles which are then withdrawn into the mouth and wiped clean.

▶ Return to the specimen of a sea cucumber that you opened to examine the water vascular system, or open a relaxed specimen now according to the directions given in Section 9–3. With the aid of Figure 9.6C, identify the parts of the alimentary tract. The **mouth** leads to the short **esophagus** via the **pharynx,** which is surrounded by a peripharyngeal chamber supported by a ring of **calcareous ossicles** that may be the remains of a jaw apparatus similar to the lantern of echinoids. The esophagus passes through the **water vascular ring canal,** and then gives rise to the **stomach** and looping **intestine.** The terminal portion of the intestine, the **rectum,** opens into the **cloaca** from which lead a pair of **respiratory trees.** Refer to Section 9–4 for a discussion of the function of the respiratory trees.

iv. OPHIUROIDEA. The ophiuroids are microphagous ciliary-mucoid feeders and may feed macrophagously as well. Mucous glands in the integument produce a network of mucus on the surface of the arms, and the mucus and food entrapped by it are moved by cilia, tube feet, and arms to the mouth. They also sweep their arms over the sandy bottom to capture larger particles, which may be brought to the mouth by flexion of the arms or by passage from podium to podium.

▶ If starved specimens of ophiuroids are available in the laboratory, attempt to observe them feeding in a dish of seawater. Offer pieces of chopped mussel of graded sizes as follows:

(a) 10 × 5 mm

(b) 1 to 2 mm square

(c) Finely ground mussel. In response to this, the animal should assume a feeding posture with raised arm tips and spread spines. The mucus produced forms a net between the spines, and one or more arms move to sweep the net across the bottom of the dish. Add a weak solution of toluidine blue to the dish and the mucus net should be rendered visible.

The digestive system of brittle stars is the simplest found in the phylum. The saclike stomach is restricted to the central disc, and there are no hepatic caeca, intestine, or anus.

▶ How does the length of the gut relate to the type of food eaten in echinoderms? Account for the presence of a reduced or microscopic anus in the carnivorous species.

9–7. THE NERVOUS SYSTEM

The nervous system of echinoderms is not easily demonstrated without the aid of sectioned material. Notable for its lack of centralization, it consists mainly of a more or less diffuse network of fibers beneath the epidermis all over the body, and ganglionated nerve cords arranged in the radiate pattern, i.e., a **circumoral ring** and **radial fibers.** Sense organs are not well developed in members of this phylum. Can you give any explanation for the absence of substantial cephalization in radiate animals?

i. ASTEROIDEA.   The asteroid nervous system consists of three interrelated parts, the **ectoneural** (sensory) system, the deeper **entoneural** (motor) system, and the deep accessory **hyponeural** system or **Lange's nerves,** which are also primarily motor.

▶   The ectoneural system is situated just below the epidermis; it consists of a pentagonal **circumoral ring,** which gives off a **radial nerve** to each ray, ending in a pigmented **eyespot, ocellus, or sensory cushion** at the tip of each ray. It is thought to function in photoreception by some authors, but this has not been unequivocally demonstrated. The response of starfish to light varies with the species, the body surface in general is photosensitive. An **ectoneural plexus** runs all over the body and innervates the appendages, spines, etc. Separate the tube feet of a ray of a preserved or anesthetized starfish, or hold the ray from which you removed the hepatic caeca and gonads (Section 9–3) up to the light, and locate the radial nerve. You may be able to trace the radial nerve to its junction with the circumoral nerve ring. Examine a prepared slide of a cross section of a ray and locate the radial nerve (Figure 9.1**D**). You will be unable to locate the epidermal plexus, as the fibers are too small to see.

The deeper entoneural system is composed of a **plexus** that lines the entire coelom, together with thickenings of the subepidermal plexus, which form marginal nerves that run along the outer edges of the ambulacra; branches innervate the muscles that run between the ossicles of the ambulacra.

The deeper hyponeural system comprises Lange's nerves, which innervate the muscles of the arms, where the nerves are arranged as a radial series of ganglionic concentrations. They continue from each arm to the peristomal region, where they form five aboral thickenings in the floor, or a **hyponeural sinus ring.** The nerves of both the hyponeural and entoneural system are difficult to see even in sectioned material.

ii. ECHINOIDEA.   The ectoneural system is dominant in echinoids; it consists of the **peripharyngeal** or **circumoral nerve ring** lying inside Aristotle's lantern at the oral end of the pharynx, the five **radial nerves,** and a **subepidermal plexus.** The radial nerves arise from the circumoral ring and run inside the body wall along the midline of the ambulacra. They send branches to the podia, pedicellariae, spines, and other structures of the test and body wall, making these effective sense organs and forming the subepidermal plexus. The circumoral ring also sends branches to the gut. The entoneural system is located in and innervates the lantern.

▶   In an opened specimen of a sea urchin (*Strongylocentrotus* or *Arbacia*), attempt to locate the radial nerves that run along the ambulacra and the branches that they give off to the ampullae of the tube feet.

iii. HOLOTHUROIDEA.   The arrangement of the nervous system in sea cucumbers is essentially similar to that in asteroids. Running from the radial nerves is a subepidermal plexus, which supplies the body wall with tactile receptors and chemoreceptors. Statocysts also occur in some species and are located in the radial nerve net near the calcareous ring. Their function as organs of equilibrium has not been completely proved. Entoneural and hyponeural components of the nervous system are also present.

▶ The components of the nervous system are difficult to locate in dissected material, but may be seen, if time permits, in histological preparations.

iv. OPHIUROIDEA.   The nervous system is highly developed in these active animals, with ganglionic enlargements in the arms along the radial nerve cords. However, it is difficult to see in dissected or stained material and will be omitted from this exercise.

9–8. BEHAVIOR

Behavioral studies similar to those outlined in this section for asteroids should be attempted using ophiuroids if time permits.

a. Movement, Righting, and Escape Responses

i. ASTEROIDEA.   Many investigators have shown that starfish walk, rather than pulling themselves along a surface (except on a vertical surface), that is, they push with the tube feet as well as pulling. Anchor a starfish on its aboral surface so the oral surface faces you (see p. 305 for directions), and place a small piece of aluminum foil or celluloid on the top of the tube feet; observe the stepping action of the tube feet. In which direction does the aluminum foil or celluloid move? Note that the tube feet of one of the arms swings parallel to the long axis of the arm and those of the other arms take up the swing in the same direction. Thus, one arm tends to lead: coordination with the other arms indicates the presence of central nervous control, probably effected by the nerve ring and the radial nerve. Coordination of podial movement makes possible directed movements.

▶ Release the anchored starfish and allow it to walk. Turn it over and with the oral side facing you, cut into the ambulacral ridge at the base of the ray. This severs the radial nerve. Allow the animal to walk. What is the effect of surgery on walking? Cut the nerve ring between another pair of arms to produce a two-armed portion and a three-armed portion. What is the effect on tube foot pointing? Note how two arms of the starfish now act as though they were leading arms—one in the two-arm portion and one in the three-arm portion. Sever the nerve ring connections between the remaining arms, one at a time, and note the effect of ambulatory movements after each cut is made. How many cuts must be made to arrest movement? Cut off one arm on the distal side of the nerve ring: note how it moves toward the disc as though it were a trailing arm. An isolated arm with a portion of the disc and nerve ring moves in the opposite direction, with the pointed end forward, as though it were a leading arm. This suggests that the center for imposing stepping direction is located at the junction of the radial nerve cord and the nerve ring. During the remainder of the laboratory period, check the movements of the isolated arm containing the nerve ring at regular intervals. Is the stepping direction maintained? Directional stepping in such a preparation ordinarily does not persist for long periods of time, suggesting that the center for imposing directional movement in the ray eventually loses dominance. In the intact animal, there is probably alternating dominance between the centers for the different rays. Of what adaptive value is this to a radially symmetrical animal, with respect to its ability to meet the environment?

▶   Asteroids have a strong tendency to keep the oral surface in contact with the substratum. Turn a living asteroid on its back and note the righting reaction. How many rays are attached to the substratum before the rest of the animal turns? Repeat, and see if the same ray leads. If you make an incision between rays so that the nerve ring is cut, what happens to coordination in righting? Does an isolated ray right itself? If an isolated ray can right itself, what does this indicate about the nervous center for initiation of righting?

▶   Using a glass plate which can be inclined, test the geotactic responses of a starfish. Allow the animal to crawl on the plate, and then lift one edge. Does the animal change direction? Now reverse the lifted edge. Does the animal now change its direction? Does this animal show geotaxis?

▶   ii. ECHINOIDEA.   Test the righting response of a sea urchin when placed on its aboral surface in contact with a horizontal glass plate. Are spines as well as podia involved in the righting response? Using forceps or a scalpel, remove a number of spines to make a fixed point of orientation on the animal's body. Repeat the preceding experiment and determine whether the same ambulacrum is always turned first. Repeat again, but as soon as you observe the animal adhering to the glass plate, move the plate to the vertical position. Does the animal continue to right itself?

Note that sea urchins will climb vertical surfaces by the use of podia, as do the asteroids. The organs of equilibrium in some urchins are modified bean-shaped spines known as **sphaeridia.** These are usually found on the ambulacral areas around the mouth (Figure 9.3B), but some urchins have them along the entire ambulacra. Carefully remove the tube feet at the edge of the peristome and attempt to remove the sphaeridia, located between each double row of tube feet in the ambulacra; again test the righting response. If the response is not abolished in the operated urchins, is it delayed? What is the cause of delay in righting in urchins lacking sphaeridia?

▶   Determine the coordinating center for the righting reaction as follows. Anchor an urchin, with its aboral surface down, to a wire screen by means of rubber bands (see p. 305 for directions). Remove all the large spines that can reach the region of the peristome. At the edge of the peristome locate the two buccal tube feet that mark the ambulacrum and beneath which runs the radial nerve. Make a shallow incision between test and peristome in this area to sever the radial nerve. Do this under water and make sure you have adequate lighting. Take ten animals and divide them into five groups of two each. Sever one nerve, two nerves, three nerves, four nerves, or all five nerves in each group, respectively, and place the pairs of operated urchins in separate bowls. Observe the righting reaction in each pair. How does this compare with asteroid behavior?

▶   iii. HOLOTHUROIDEA.   Study the reactions of a living *Cucumaria* or *Thyone* to mechanical stimulation. Using a pencil, stroke the animal on the tentacles, on the body surface, and in the region of the anus, and determine the areas with the greatest sensitivity. Under what circumstances does the entire animal react? Are podia sensitive to touch?

▶   Holothurians generally right themselves when turned over. Attempt to study the righting reaction in available sea cucumbers. First, identify the functionally ventral podia, and then turn the animal so these podia are no longer in contact with the substratum. Does the animal right itself? Are podia involved? Are muscles involved?

▶   Test the geotactic responses of the organism by placing the animal on a horizontal glass plate. Raise one edge of the plate. Does the animal reorient itself? Repeat, with the organism facing in the opposite direction initially. Is the organism positively or negatively geotactic?

(*Cucumaria* is most suitable for this experiment, because *Thyone* does not show a geotactic response.)

▶ Bury one end of an animal in moist sand or mud and note the effect on burrowing. Then remove the animal and bury the other end. Is there polarity in burrowing?

b. Spines, Pedicellariae, and Podia

i. ASTEROIDEA. Pincer-like pedicellariae (Figure 9.7), which are modified spines, are variously distributed over the external surface of some starfish and function primarily in cleaning the surfaces and ambulacral grooves, protection of the papulae, and capture of small prey. The pedicellariae are composed of three ossicles; the basal ossicle forms a stalk and the two distal ossicles work against each other with a shearing action, like the blades of a pair of scissors.

▶ Study the aboral and oral surfaces of a living starfish (*Asterias, Pisaster*) in a bowl of seawater under the dissecting microscope, and observe the form and distribution of pedicellariae around the spines and edges of the ambulacra. The operation of pedicellariae may be easily observed by drawing a fine thread across them or touching with a fine paintbrush. How do the jaws react? Remove a few, mount on a slide, and observe under the compound microscope. Greater detail may be obtained by adding a drop of sodium hypochlorite solution. Are there different types of pedicellariae? Notice the minute muscles attached to the ossicles that open and close the jaws.

▶ ii. ECHINOIDEA.   Spines and pedicellariae are probably the principal weapons of defense available to sea urchins. Each spine articulates with a raised tubercle on the test by a **ball and socket joint** (Figure 9.7). Refer to a cleaned test to see this more clearly. The spine is held or locked in position by an inner ring of muscles called **cog muscles.** Under normal conditions the spines are kept erect by the contraction of these muscles; if they relax, the spine falls limply toward the surface of the test. Carbon dioxide is capable of anesthetizing the cog muscles. Blow a jet of CO_2 gas on an area of the test. What is the reaction of the spines? Repeat, using a jet of air or nitrogen to prove the effect is really anesthesia by CO_2 and not the mechanical effect of the stream of gas. To feel the cog muscle in operation, hold on to the tip of a spine and try to move it. Can you feel the cog muscles lock the spine into position? What do you think is the significance of positioning in this way? Now touch the surface of the test gently with the tip of a pencil or a probe. In what direction do the spines point? Stimulation of the spines with greater intensity causes them either to lock rigidly into place or to point away from the source of stimulation, presumably exposing the pedicellariae for action. Continued or very strong stimulation should be avoided or the animal may take flight.

The muscles involved in directional pointing of the spines are the outer ring of muscles at the base of each spine. In order for spines to point when an area of the test is stimulated, the cog musculature must be inhibited. Which muscles contract in spine pointing?

▶ To examine the role of the nervous system in mediating the spine movements, circumscribe an area on the test with a scalpel so that you cut down to the calcareous plates, isolating the area together with the subepidermal plexus from the rest of the body. Apply a stimulus within the circle. What is the reaction? Apply a stimulus outside the circumscribed area. What happens to the spine pointing? Remove a piece of the test with the spines intact and test the response of the isolated portion to tactile stimulation. What does this indicate about spine coordination and the nervous system? Now take a sea urchin and circumscribe a number of

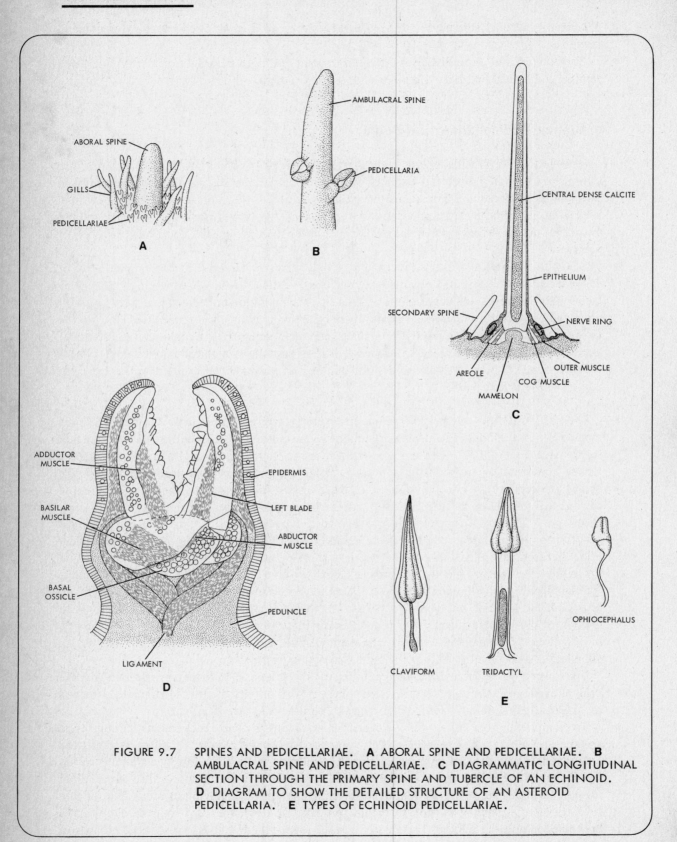

FIGURE 9.7   SPINES AND PEDICELLARIAE.   **A** ABORAL SPINE AND PEDICELLARIAE.   **B**
AMBULACRAL SPINE AND PEDICELLARIAE.   **C** DIAGRAMMATIC LONGITUDINAL
SECTION THROUGH THE PRIMARY SPINE AND TUBERCLE OF AN ECHINOID.
D DIAGRAM TO SHOW THE DETAILED STRUCTURE OF AN ASTEROID
PEDICELLARIA.   **E** TYPES OF ECHINOID PEDICELLARIAE.

circles on the test in the manner just described, and see if locomotion is impaired. What then is the role of the subepidermal plexus in locomotory behavior?

▶ Echinoid pedicellariae come in a variety of forms (Figure 9.7) and are variously distributed over the body surface. They will react to mechanical and chemical stimuli even on isolated portions of the test. Study the form of pedicellariae on the test of a living sea urchin in a bowl of seawater under the dissecting microscope. How do they differ from asteroid pedicellariae? Using a fine paintbrush or a fine glass needle, test the reaction of pedicellariae to tactile stimuli on various parts, e.g., the jaws and stalk. Do all intensities of touch produce the same reaction? The calcareous jaws may be studied in greater detail by removing pedicellariae, placing them on a slide with a drop of sodium hypochlorite, and examining under the compound microscope. What are the functions of echinoid pedicellariae?

▶ Podia react to light, chemicals, and touch. Test the reactions of podia to light mechanical disturbance, and then to stronger stimulation. What is the reaction of podia to jarring of the urchin? What is the adaptive significance of this reaction?

c. Evisceration in Holothuroidea

▶ One of the behavioral defense mechanisms of some sea cucumbers is the ability to eviscerate almost all the internal organs, which can then be regenerated. *Thyone* and *Stichopus* are most suitable for these experiments and can be induced to eviscerate by placing in 0.1% NH_4OH in seawater for 1 to 2 minutes and then removing the animal. If the organism is held up by the posterior end, most of the internal organs will be extruded. Complete regeneration may take months, but some regeneration can be seen in 2 weeks. What internal organs are shed? Of what advantage is evisceration to a sea cucumber?

9–9. REPRODUCTION

Echinoderms are dioecious, and the number of gonads present varies from class to class. In the crinoids there are many gonads situated in the arms or the pinnules of the arms, from where the gametes are shed by rupture of the walls. In ophiuroids the variable number of gonads are usually restricted to the disc. In asteroids there are two or more gonads in each arm; in regular echinoids there are five gonads and four in irregular ones; in holothuroids there is a single gonad. The gonads, located in coelomic sacs, are composed of masses of sex cells which fill the sac; these cells are produced from a central rachis or from the epithelial lining. In all classes except the crinoids the gametes are shed through the gonopores, to which the gonads are attached by a short gonoduct. Fertilization and development usually occur in the sea, and breeding is generally seasonal. There is no good way of distinguishing between the sexes by external characters.

i. ASTEROIDEA.   The testes or ovaries of *Asterias* and *Pisaster* occur as paired organs in each of the rays. Each gonad is surrounded by a coelomic sinus and attached by a gonoduct that opens to the exterior via a gonopore located at the junction of the rays. Using the starfish that you opened previously (Section 9–3) or another specimen, remove the aboral surface of a ray and locate the pair of gonads (Figure 9.1**C**). The ovary is usually reddish in color, whereas the testis is white or yellow.

ii. HOLOTHUROIDEA.   Return to the specimen of a sea cucumber that you opened previously, or open another specimen according to the directions given in Section 9–3. Locate the single gonad, which is composed of a cluster of finger-like parts and suspended in the body cavity by the dorsal mesentery. The gonoduct opening is located on a papilla between the two tentacles in the dorsal interambulacrum (Figure 9.6C). The sex of the organism can be determined by microscopic examination of crushed gonad filaments.

▶   iii. ECHINOIDEA.   The five gonads seen in regular echinoids are located in the interambulacral regions of the animal. In the specimens of *Strongylocentrotus* or *Arbacia* that you hemisected previously (Section 9–6) or in a freshly opened specimen, locate the gonads (Figure 9.3C). The ovary is orange-yellow (red in *Arbacia*), whereas the testis is gray. Is your specimen male or female?

▶   If the organisms are in breeding condition, spawning may be induced by a variety of methods. The simplest and most successful method is by injecting 0.5 ml of 0.5 M KCl into the perivisceral cavity via the peristome, using a hypodermic syringe fitted with a 25 gauge needle. Invert the animal over a 250 ml beaker of seawater and allow the animal to shed. Sperm are best collected "dry," that is, without being diluted with seawater. Collect with a Pasteur pipet from the surface of the test after allowing the animal to sit on a damp towel during shedding. Eggs are orange or yellow and sperm are white; observe their release from the genital pores located in the aboral genital plates. Directions for studying echinoid development using the gametes procured here will be treated in the following section.

9–10. DEVELOPMENT

Echinoderm eggs are easily obtained in the laboratory in large quantities; although breeding is seasonal, the season itself varies from species to species so that it can be said that there is never a time of year when echinoderm eggs of some sort are unobtainable. Cleavage of the echinoderm egg is **total, indeterminate,** and **radial,** the blastopore forms the anus (i.e., echinoderms are **deuterostomes**), and development of the coelom is **enterocoelic.** Cleavage and development thus resemble the chordate pattern rather than the spiral, determinate, schizocoelic pattern of the protostomatous invertebrates. For these reasons echinoderms have been extensively used in the labortatory for experimental studies of embryology and development, and you should acquaint yourself with the main features of this aspect of echinoderm biology. See the reference by Costello et al. (1957) for further details.

a. Fertilization and Cleavage*

Instructions for obtaining eggs and sperm from a representative echinoid are given in the previous section.

▶   Take a small (250 ml) beaker of unfertilized eggs in seawater, and, using a Pasteur pipet, transfer a sample to a slide and examine it under the compound microscope. The egg is spherical and surrounded by a **jelly coat** (Figure 9.8A); if mature, eggs will show a small, clear **nucleus** and uniformly distributed pale **yolk granules.** Slightly larger red granules may be present, containing **echinochrome pigment** (*Arbacia*).

*This section should be started at the beginning of the laboratory period to allow time for completion.

▶ Take a clean toothpick and dip the tip into concentrated sperm. Swirl the toothpick in the egg suspension and gently stir once or twice with a glass rod. DO NOT ADD TOO MUCH SPERM. A microscopic amount is plenty, as multiple fertilization (**polyspermy**) and abnormal development may result when too many sperm are present. Immediately take a sample on a slide and examine under the microscope. Under the high power the sperm will be seen clustering around the egg, and may even cause it to revolve by the lashing of their tails (Figure 9.8A). One sperm head will penetrate, leaving the tail behind, and within a few seconds the egg responds. The **vitelline membrane** surrounding the egg rises rapidly from the surface within 5 seconds and helps to prevent polyspermy; observe the raising of the membrane. A **perivitelline space** is left around the egg. The membrane hardens and thickens in about 5 minutes, after which it is called the **fertilization membrane.** Ten minutes after penetration, a clear **hyaline layer** is formed at the surface of the egg, the **hyaline plasma membrane.** It is composed of a sticky substance, which holds the blastomeres together after cleavage.

The first division of the egg occurs at about 50 to 60 minutes after fertilization at 23°C, second cleavage at 1½ hours, third cleavage at 1¾ hours, and the blastula is formed at approximately 6 hours.

▶ Leave your fertilized eggs on a laboratory bench at room temperature and examine at intervals. Examine samples from cultures set up earlier by the instructor; they should show the different cleavage stages and young larvae (Figure 9.8A).

b. Larval Development

Although most echinoderms are radially symmetrical sessile creatures, the majority of the larvae are bilaterally symmetrical free-swimming organisms. The free-swimming larva distributes the species and undergoes a complex metamorphosis at the end of its larval life. Larvae of the various classes have evolved independently of the adult and of each other, and the larval form varies accordingly, although in the early stages all are similar. The basic larval type has been dubbed the **dipleurula** ("little two sides").

▶ Using prepared slides and with the aid of Figure 9.8**B**, examine the developmental stages of the starfish; if time permits and material is available, compare the **bipinnaria** and **brachiolaria** larva of the asteroid with the **pluteus** of the echinoid.

References

Barnes, R. D. *Invertebrate Zoology,* 3rd ed. Philadelphia: W. B. Saunders Co., 1974.

Binyon, John. *Physiology of Echinoderms*. Oxford: Pergamon Press, 1972.

Brown, F. A., Jr. (ed.). *Selected Invertebrate Types*. New York: John Wiley & Sons, 1950.

Boolootian, R. A. (ed.). *Physiology of the Echinodermata*. New York: John Wiley & Sons, 1966.

Bullough, W. S. *Practical Invertebrate Anatomy*. New York: St. Martin's Press, 1962.

Costello, D. P., M. E. Davidson, A. Eggers, M. H. Fox, and C. Henley. *Methods for Obtaining and Handling Marine Eggs and Embryos*. Woods Hole, Mass.: The Marine Biological Laboratory, 1957.

Ferguson, J. C. Cell production in the Tiedemann bodies and haemal organs of the starfish *Asterias forbesi, Trans. Amer. Microsc. Soc.,* **85**(2):200–209, 1966.

Fontaine, A. R. The feeding mechanism of the ophiuroid *Ophiocomina nigra, J. Mar. Biol. Ass. U.K.* **45**:373, 1965.

Giudice, G. *Developmental Biology of the Sea Urchin*. New York: Harcourt Brace Jovanovich, 1973.

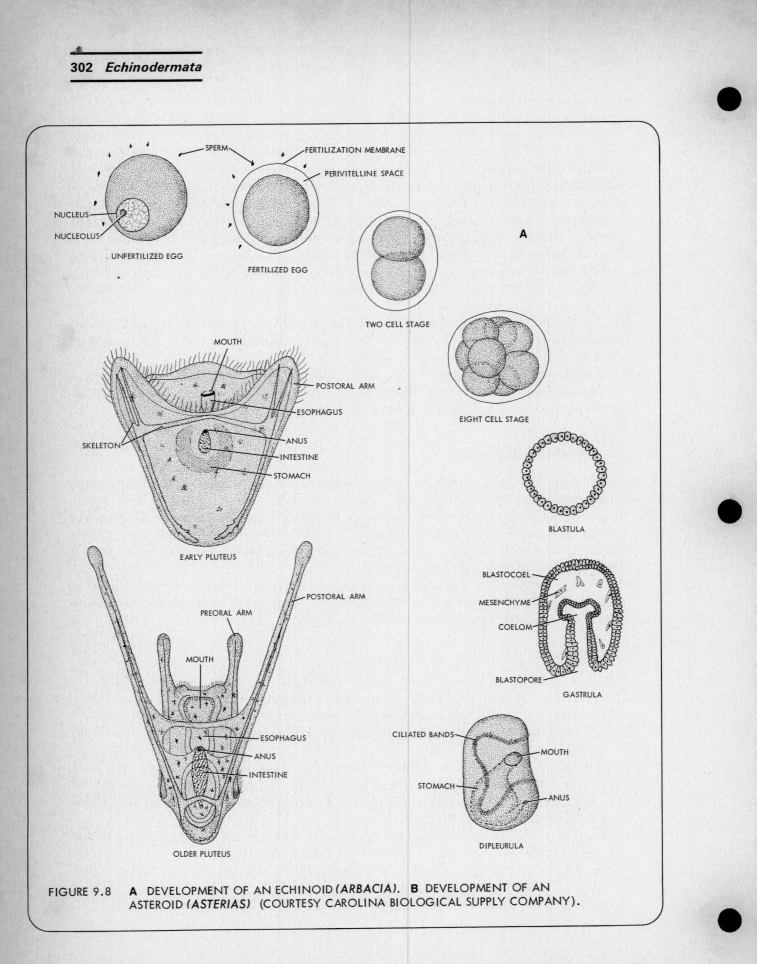

FIGURE 9.8   **A** DEVELOPMENT OF AN ECHINOID *(ARBACIA)*.   **B** DEVELOPMENT OF AN
ASTEROID *(ASTERIAS)* (COURTESY CAROLINA BIOLOGICAL SUPPLY COMPANY).

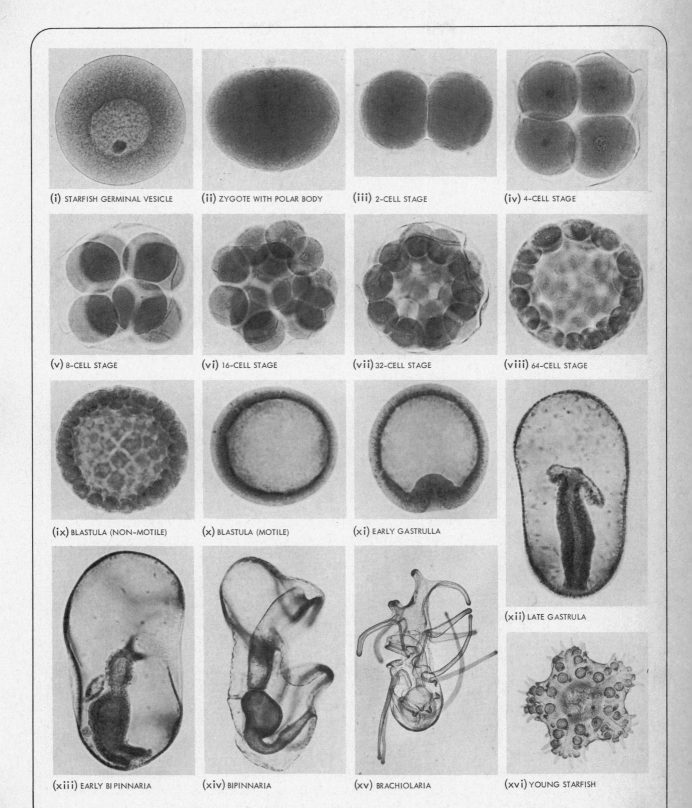

(i) STARFISH GERMINAL VESICLE   (ii) ZYGOTE WITH POLAR BODY   (iii) 2-CELL STAGE   (iv) 4-CELL STAGE

(v) 8-CELL STAGE   (vi) 16-CELL STAGE   (vii) 32-CELL STAGE   (viii) 64-CELL STAGE

(ix) BLASTULA (NON-MOTILE)   (x) BLASTULA (MOTILE)   (xi) EARLY GASTRULLA   (xii) LATE GASTRULA

(xiii) EARLY BIPINNARIA   (xiv) BIPINNARIA   (xv) BRACHIOLARIA   (xvi) YOUNG STARFISH

B

Hyman, L. H. *The Invertebrates,* Vol. IV. *Echinodermata.* New York: McGraw-Hill Book Co., 1955.

Johnson, Phyllis T. The coelomic elements of sea urchins *(Strongylocentrotus),* I. The normal coelomocytes; their morphology and dynamics in hanging drops. *J. Invert. Pathol.* **13**(1):25, 1969.

Meglitsch, P. *Invertebrate Zoology.* New York: Oxford University Press, 1967.

Millott, N. *Echinoderm Biology. Symp. Zool. Soc. London,* No. 20, 1967.

Nichols, D. *Echinoderms.* London: Hutchinson and Co., 1962.

Terwilliger, Robert C., and Kenneth R. H. Read. The hemoglobins of the holothurian echinoderms *Cucumaria miniata* Brandt, *Cucumaria miniata* Stimson and *Molpadia intermedia* Ludwig, *Comp. Biochem. Physiol.* **36**(2):339, 1970.

An Abbreviated Classification of the Phylum Echinodermata (After Hyman and Nichols)

Phylum **Echinodermata**

> ### *Class I* **Crinoidea**
>
> Stalked or stalkless, oral surface directed upward, cuplike calyx (theca) houses lower part of body, and domed flexible roof (tegmen) houses rest of body and bears mouth at its center. Pentamerous, mobile arms, e.g., *Antedon, Heliometra.*
>
> ### *Class II* **Holothuroidea**
>
> Elongate in aboral-oral axis, secondary bilateral symmetry, mouth surrounded by tentacles, endoskeleton reduced to microscopic spicules, locomotory podia in five ambulacral areas or spread over body surface, e.g., *Cucumaria, Thyone, Leptosynapta, Stichopus.*
>
> ### *Class III* **Echinoidea**
>
> Spheroidal or disclike, movable spines, endoskeleton forms close-fitting plates (test), ambulacral plates pierced by pores for podia, e.g., *Strongylocentrotus, Arbacia, Echinus, Echinarachnius, Dendraster, Lytechinus.*
>
> ### *Class IV* **Asteroidea**
>
> Flattened, starlike with arms radiating from a central disc, oral surface held downward, locomotory podia on oral surface, endoskeleton flexible or separate ossicles, e.g., *Pisaster, Asterias, Astropecten, Patiria.*
>
> ### *Class V* **Ophiuroidea**
>
> Flattened; flexible arms distinctly set off from central disc, podia not used for locomotion, anus and intestine absent, e.g., *Ophioderma, Gorgonocephalus, Ophioplocus.*

General Directions for Laboratory Work on Echinoderms

1. Animals may be **relaxed (anesthetized)** by immersion of the entire organism for 40 minutes in 7% $MgCl_2$ made up in tap water. This relaxation is reversible, and anesthetized animals may be easily returned to full activity by bathing in running seawater. Holothurians may be relaxed by injection of 5 to 10 ml of $MgCl_2$ into the coelomic cavity, 40 to 60 minutes prior to dissection.

2. Small animals are the most favorable material for dissection. They should be relaxed first or immobilized by anchoring to a wire mesh screen by means of a few rubber bands. The mesh can be anchored to a wax pan with pins. This will permit you to dissect with both hands, and the animal's movements will not hinder your efforts.

3. For **examination under seawater,** place the relaxed or immobilized animals in a wax-lined pan or finger bowl and completely cover the organism with seawater to at least 1 cm above the surface of the animal.

4. **Lighting** is as important in dissection as in microscopy. Direct the microscope lamp so that enough light shines on the animal and at the same time does not hurt your eyes or reflect from the surface of the water. During dissection, coelomic fluid, eggs, sperm, etc., may obscure your view. These materials can be removed by application of small jets of clean seawater from a pipet and frequent changes of the covering water.

Equipment and Materials

Equipment

Dissecting microscope
Compound microscope
Cover slips and slides
Dissecting instruments and trays
Pipets
Wire mesh
Rubber bands
Paper clips
Glass plate
Hypodermic syringe
20 and 25 gauge needles
Colorimeter
Celluloid or aluminum foil
250-ml beaker
Syracuse watch glass
Finger bowls
Bunsen burner

Solutions and Chemicals

Carmine powder
Evans blue (1%)
Sodium hypochlorite
Carbon dioxide gas
0.1% NH_4OH
0.5 M KCl

Petroleum jelly
Toluidine blue

Prepared Slides

Starfish ray, cross section
Asterias larvae

Preserved Material

Gorgonocephalus
Antedon
Asterias
Echinarachnius

Living Material*

Mercenaria or *Mytilus* (9,12)
Luminescent bacteria
  (*Photobacterium fischeri*) (3)
Pisaster or *Asterias* (9,12)
Strongylocentrotus or *Arbacia*
  (7,9,12)
Ophioderma or *Ophioplocus* (7,9,12)
Cucumaria (7,9,12)
Thyone (7,9,12)
Parastichopus, Stichopus (7,9,12)
Leptosynapta (7,9,12)
Dendraster, Mellita, Echinarachnius
  (7,9,12)

*Numbers in parentheses identify sources listed in Appendix B.

PROTOCHORDATAES

Contents

Considerable understanding of the chordate phylum and its relationship to the invertebrates may be gained by studies of the more primitive chordates and related groups, collectively known as **protochordates** (*proto,* first; *chorda,* cord; L.). The protochordates are a diverse group of invertebrates that share some characteristics with chordates and vertebrates. Chordates are **enterocoelous deuterostomes,** which at some stage in their life history possess **pharyngeal gill slits,** a **dorsal hollow tubular nerve cord,** and a **postanal metamerically segmented tail.** The advanced chordates are vertebrates, but the chordate phylum is invertebrate in origin and many of its representatives are without a vertebral column.

Three groups of organisms, all marine, are commonly called protochordates: the Hemichordata or acorn worms, the Urochordata, which includes the sea squirts and salps, and the Cephalochordata or lancelets (see Classification of Protochordates, p. 325). There is considerable variability in the adult form, and many of the affinities of this group are based on larval and developmental considerations.

10–1. BODY ORGANIZATION

i. HEMICHORDATA.   The relationships of the Hemichordata have been uncertain for as long as they have been known. They are sometimes included as a subphylum of the Chordata, along with the Urochordata, Cephalochordata, and Vertebrata. However, because of their doubtful affinities they may be placed in a phylum of their own.

Hemichordates are common marine animals of wide distribution. The most common representatives are the soft, wormlike **acorn** or **tongue worms,** which generally live in burrows in the shallow waters of the intertidal zone. These are the Enteropneusta; the most common genus of the Atlantic and Pacific coasts is *Saccoglossus.* The deep-water Pterobranchia (e.g., *Rhabdopleura* and *Cephalodiscus*) are also hemichordates, but because of their rarity they will not be studied in the laboratory.

▶ Obtain a specimen of *Saccoglossus* (Figure 10.1A). Living specimens smell mildly of iodoform. Examine the softness of the body and note its wormlike form. The epidermis is ciliated and plentifully supplied with mucus-secreting cells, which are important in burrowing and feeding. The body of the animal is divided into three regions: an anterior **proboscis,** a short **collar,** and, posteriorly, an elongate **trunk.** In living specimens the proboscis is cream-colored or pink with a blunt tip and narrows posteriorly to form the **proboscis stalk.** A tiny **proboscis pore** on the stalk allows the passage of water in and out of the proboscis coelom during burrowing. The permanently open **mouth** is ventral, located at the base of the proboscis, and obscured from view by the overlapping anterior edge of the cylindrical collar (orange in living animals). The remainder of the body, the trunk, is divisible into three regions: an anterior **branchial region** containing **branchial pores,** a **genital region** that may be extended into **genital ridges** because of the distension produced by the enlarged gonads, and, posteriorly, the **abdominal region** (greenish yellow in living specimens), containing the **intestine** and lateral pouches of the **hepatic caeca.** The **anus** is terminal. Keep this specimen in a finger bowl of seawater for future study.

ii. UROCHORDATA.   Ascidians, or tunicates, are the most common representatives of the subphylum Urochordata. They are exclusively marine and sessile and are covered by a rather firm **test** or **tunic** from which they get their name. The "simple" ascidian (Figure 10.2) consists of a saclike body (*ascus,* sac; Gk.) attached to the substratum by its base; at the free end it bears two openings: a **mouth** or **branchial opening** and an **atrial opening.**

▶ In a finger bowl of seawater, observe a living specimen of a solitary tunicate (*Ciona,* Figure 10.3A, or *Molgula,* Figure 10.3B). Make sure your specimen is completely covered by water, and allow it to relax and extend its two siphons. The siphons are of unequal length: the **exhalent** or **atrial siphon** that terminates in the atrial opening marks the dorsal side of the animal, and the **inhalent, branchial,** or **oral siphon** that terminates in the mouth is anterior. In *Ciona* there are eight lobes to the oral siphon, and in *Molgula* there are six; the atrial opening of *Ciona* has six lobes, that of *Molgula* has four.

▶ Adjust the lighting and place your specimen in such a way that, by using the dissecting microscope, you can see through the body of the animal. The body is divided into two regions: the **thorax** and **abdomen.** The thorax is occupied chiefly by the **branchial** or **pharyngeal** portion of the gut. The space between the pharynx and the inner body wall is the **atrium** or **peribranchial chamber.** Into this space, near the base of the exhalent siphon, open the **reproductive ducts,** the **intestine,** and the **gill slits** or their derivatives. Material is discharged from the atrium via the exhalent siphon. The abdomen contains the **viscera** (**gonads, gut,** and **heart**).

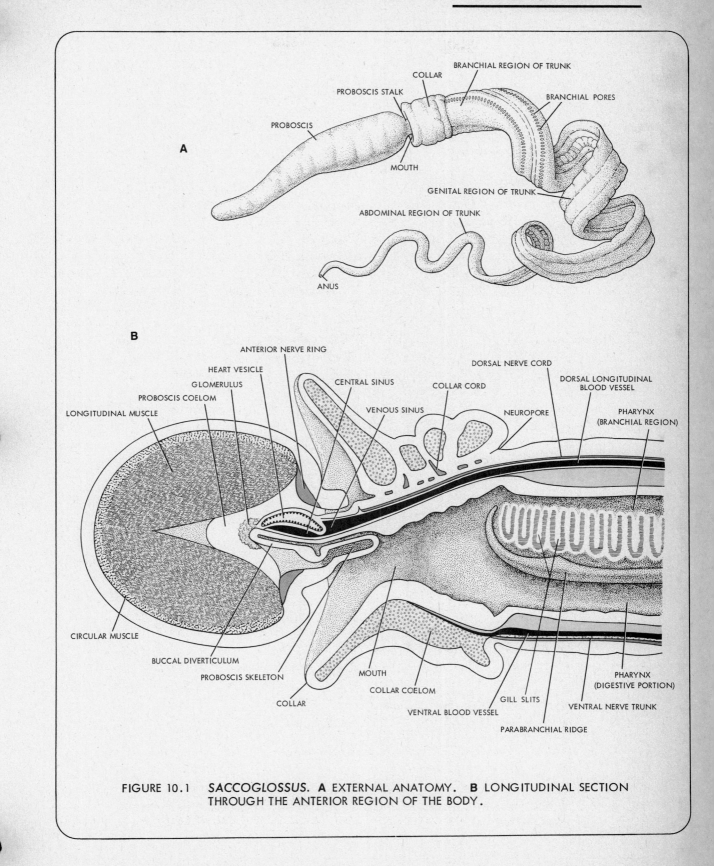

FIGURE 10.1 *SACCOGLOSSUS.* **A** EXTERNAL ANATOMY. **B** LONGITUDINAL SECTION THROUGH THE ANTERIOR REGION OF THE BODY.

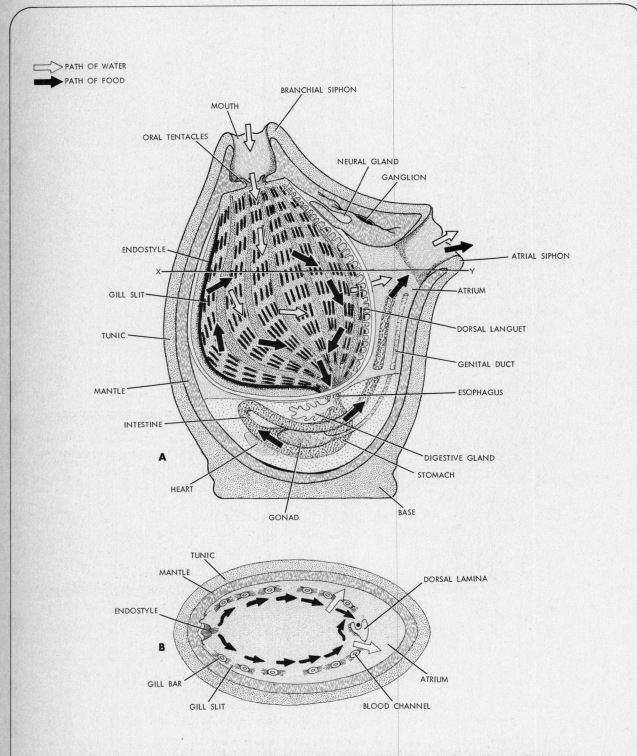

PATH OF WATER
PATH OF FOOD

BRANCHIAL SIPHON
MOUTH
ORAL TENTACLES
NEURAL GLAND
GANGLION
ENDOSTYLE
X
GILL SLIT
TUNIC
MANTLE
INTESTINE
A
HEART
GONAD
ATRIAL SIPHON
Y
ATRIUM
DORSAL LANGUET
GENITAL DUCT
ESOPHAGUS
DIGESTIVE GLAND
STOMACH
BASE

TUNIC
MANTLE
ENDOSTYLE
B
GILL BAR
GILL SLIT
DORSAL LAMINA
ATRIUM
BLOOD CHANNEL

FIGURE 10.2 **A** DIAGRAMMATIC REPRESENTATION OF A TUNICATE SHOWING THE PHARYNX AND MOVEMENT OF WATER AND FOOD. **B** CROSS SECTION AT LEVEL X–Y.

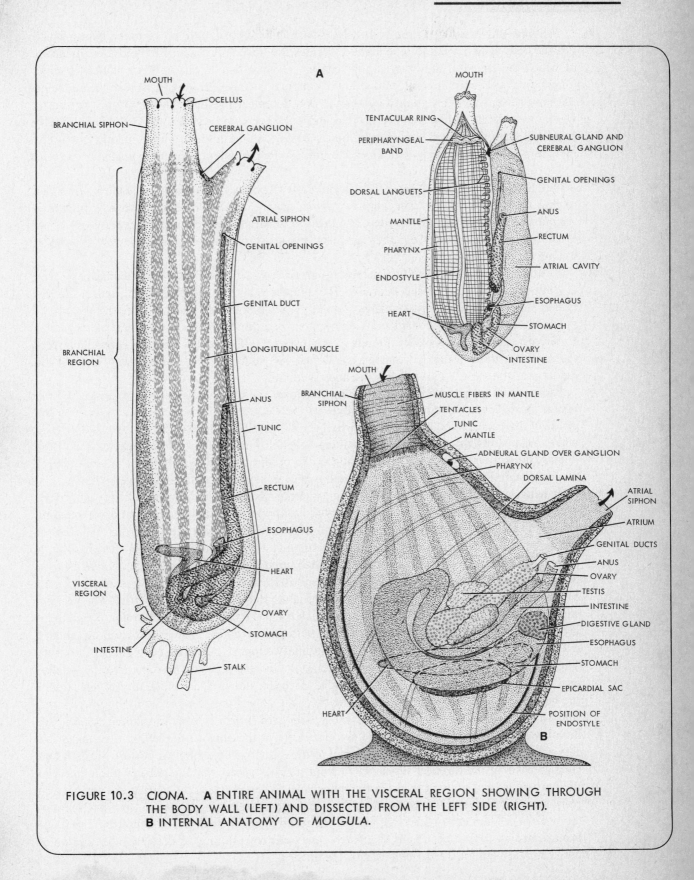

FIGURE 10.3 *CIONA.* **A** ENTIRE ANIMAL WITH THE VISCERAL REGION SHOWING THROUGH THE BODY WALL (LEFT) AND DISSECTED FROM THE LEFT SIDE (RIGHT). **B** INTERNAL ANATOMY OF *MOLGULA.*

▶   Examine the structure of the body wall, which consists of a firm **test** or **tunic,** below which lies the body wall or **mantle.** The mantle secretes the tunic and contains the muscles by means of which the shape of the body can be altered. Using a pair of fine scissors, carefully make longitudinal cuts into the tunic, and strip the cut portions from the animal. Return the animal to the bowl of seawater. Examine a wet mount of a thin piece of tunic. About 90% of the tunic is water, associated with which are protein and polysaccharide (cellulose). Cellulose normally occurs in plant tissues; its presence in the tunic of urochordates is probably unique among the members of the animal kingdom. Does the tunic contain blood vessels, blood cells, or nerves?

iii. CEPHALOCHORDATA.   The origin of the Cephalochordata may have been very close to that of the vertebrates. The only two living genera, *Branchiostoma* (formerly known as *Amphioxus*) and *Asymmetron,* are often used to illustrate the fundamental features of verte-brate organization. However, they possess many invertebrate features as well as several unique to themselves.

▶   Examine a whole mount of the lancelet *Branchiostoma* (Figure 10.4A). The animal is pointed at both ends and compressed laterally. The adult is a bottom-dwelling form, and in life lies buried in fine sand with only a small portion of the anterior end exposed. Notice the absence of a distinct head, and the conspicuous chevron-shaped **muscle bands (myotomes)** of the body. Externally, a **median dorsal fin** extends over most of the body and a **median ventral fin** covers the posterior third; a **tail fin** connects the dorsal and ventral fins at the posterior end. At the junction of the ventral and tail fins the **anus** is found.

▶   The more blunt anterior end of the animal encloses the **vestibule,** a cavity surrounded by the **buccal (oral) hood.** Hanging from the edges of the buccal hood are tentacles or **cirri.** On the inside of the buccal hood is the ciliated **wheel organ,** which in life creates a vortex of water. The wheel organ consists of several dark-staining finger-like projections; at the anterior end of the longest projection is **Hatschek's pit,** which in life produces mucus. Behind the wheel organ is the diaphragm-like **velum.** The **mouth** is an opening through the velum and fringed with sensory **velar tentacles,** leading into the large and conspicuous **pharynx.** The pharynx is easily identified; it consists of many lateral sloping **gill bars** that support **gill slits** opening into a surrounding space, the **atrium.** The atrium opens to the exterior via the **atriopore.** The ventrolateral edges of the atrium hang down as **metapleural folds.** Identify the ventral **endostyle** in the pharynx, a midventral ciliated groove with thick, mucus-secreting walls. Anteriorly, right and left ciliated **peripharyngeal bands** pass dorsally from the endostyle and meet in the midline at the ciliated **hyperpharyngeal groove,** which passes back along the dorsal wall of the pharynx. The skeletal supports of the ciliated gill bars are fused to each other dorsally but free ventrally. Strengthen-ing cross bars or **synapticula** join alternate gill bars (known as the **primary gill bars). Secondary** gill supports have grown down between the primary supports, dividing each gill slit into two. They pass external to the synapticula. Posterodorsally, the pharynx narrows into the **esopha-gus,** which leads in turn to the wider **stomach.** From the stomach, a blind-ending diverticulum, the **hepatic caecum,** projects forward along the right wall of the pharynx into the atrial cavity. The stomach leads to the **intestine** at the junction of which is a deep-staining region, the **iliocolonic ring.** The intestine terminates posteriorly at the **anus** to the left of the caudal fin. The **nerve cord** is a tubular structure dorsal to the intestine. The anterior portion of the nerve cord consists of the **cerebral vesicle,** prominant **pigment spots,** and an **olfactory pit.** Ventral to the nerve cord but above the gut is the **notochord.** When present, the **gonads** are arranged in pairs in two lateroventral rows in the pharyngeal and postpharyngeal region. In cross section, the **ovaries** contain cells with large nuclei, the **testes** appear streaky. The excretory organs or **protonephridia** consist of tubular flagellated cells **(solenocytes),** located dorsally in the region of the secondary gill bars and opening into the atrium.

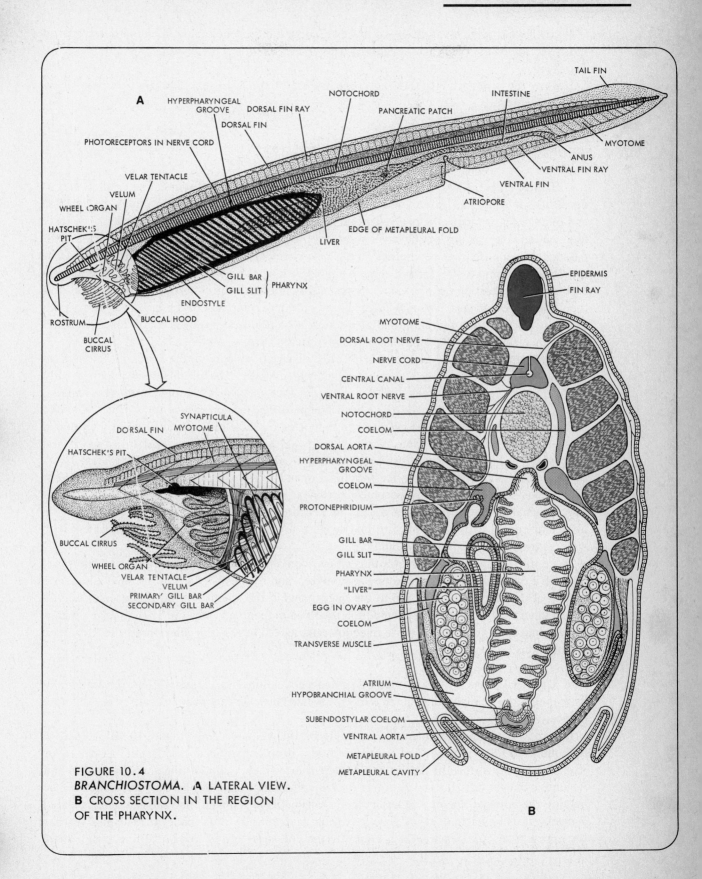

FIGURE 10.4
BRANCHIOSTOMA. **A** LATERAL VIEW.
B CROSS SECTION IN THE REGION
OF THE PHARYNX.

► If you have trouble locating any of these structures in the whole mount, examine a prepared cross section of the pharyngeal region (Figure 10.4B) in which the anatomy should be clearly displayed.

► Review the chordate characters present in each of the representative protochordates you have examined.

10–2. FEEDING AND THE STRUCTURE OF THE GUT

i. HEMICHORDATA.   The hemichordates are **ciliary-mucus** feeders. The adult enteropneust lives in sand flats; it occupies a mucus-lined burrow from which it extrudes the anterior end of its body. It moves the proboscis and collar about on the surface of the sand and collects food particles by means of the mucus secretion of the epidermis. The mucus and food are passed to the mouth by the epidermal cilia. The alimentary canal is a straight tube running from mouth to anus, and food is passed along it largely by ciliary action. From the mouth the mucus-entrapped food particles travel via the **buccal cavity** into the ventral portion of the **pharynx** and then into the **esophagus.** The esophagus is the only muscular part of the gut; it compacts the food. In the **intestine,** secretion and absorption take place. Undigested materials, sand, and debris are ejected from the anus and extruded from the burrow, usually as a spiral sand cast.

Pharyngotremy, that is, perforation of the pharynx, is one of the cardinal features of chordates, and the hemichordates possess this characteristic. The gill slits do not open directly to the exterior in most species of *Balanoglossus;* each one leads to a pouch or **branchial sac,** which in turn opens to the outside by a **branchial pore.** The branchial sacs are in effect a divided atrium. The septa between the gill slits are supported by skeletal bars and fused at their dorsal ends; each gill slit is divided into two by the downgrowth of secondary skeletal supports and processes. The pharynx thus resembles that of Cephalochordates and Urochordates. However, the hemichordate pharynx does not function in the trapping of food particles to the extent that it does in urochordates and cephalochordates. It has been suggested that pharyngotremy may originally have arisen as a means of permitting the escape of water that enters the alimentary canal with the food. Can you suggest any other function in which a pharynx with such a structure might be important?

► Place a living *Saccoglossus* in a finger bowl containing seawater and a layer of fine sand. Prepare a thick suspension of carmine in seawater, and by means of a Pasteur pipet add a drop or two of the carmine suspension to the finger bowl in the vicinity of the proboscis. Can you detect ciliary currents? Is muscular action involved in feeding? Does the animal have a means of rejecting large particles? How is rejection effected? Where is the **ciliary organ** and what are its functions?

ii. UROCHORDATA.   The method of food collection in ascidians differs fundamentally from that of the hemichordates in that the specialized urochordate pharynx is used as a filtering device.

► Return to the living specimen of *Ciona* (Figure 10.3A) or *Molgula* (Figure 10.3B), and allow the animal to relax in a bowl of seawater. Add a drop or two of a suspension of carmine or Aquadag to the surface of the water and determine which siphon is inhalent (branchial) and which is exhalent (atrial). Orientation of the animal is important and the following guides will be helpful: the **inhalent siphon** is **anterior,** the **exhalent siphon** is **dorsal,** and the glandular, ciliated **endostyle** is a conspicuous white groove running midventrally in the **pharynx.** The pharynx has the appearance of a sievelike bag. Opposite the endostyle in the middorsal line is the **dorsal**

lamina, a ciliated fold (*Ciona* has **dorsal languets,** which are ciliated projections from the lamina). Anteriorly, ciliated **peripharyngeal bands** lead from the dorsal lamina to the endostyle on each side. The wall of the pharynx is perforated by slits **(stigmata)** and supported by longitudinal gill bars. Strengthening cross bars join the longitudinal bars to each other, and the surfaces are heavily ciliated and may be covered with ciliated **papillae.** In animals that are fairly transparent, it may be possible to trace the movement of particles through part of the alimentary tract. Water enters the inhalent siphon at the **mouth,** which is tubular and consists of an inturned portion of the tunic. From here, it flows through a grid of branched **velar tentacles** into the pharynx, where the particles become trapped in mucus secreted by the endostylar cells. This mucus sheet with entrapped particles is moved dorsally by means of cilia on the pharyngeal wall and peripheral bands to the dorsal languets in *Ciona* (dorsal lamina in *Molgula*) and then backward to the **esophagus** leading out of the pharynx posterodorsally. Observe the location of accumulated graphite or carmine particles in the pharynx. Water and rejected particles pass through the pharyngeal slits **(stigmata)** into the **atrium,** and exit via the atrial siphon (Figure 10.2).

▶ If you cannot see these details in your specimen, dissect the animal* as follows: insert a pair of scissors into the oral siphon and cut downward slightly to the right of the median line. Continue this cut around the base of the body, cutting through the tunic, mantle, and pharyngeal wall. Spread the halves of the body as you would the pages of a book. Pin the animal out, cover with water, and identify the parts of the body already referred to (Figure 10.3). If a piece of the pharyngeal wall is removed from a fresh specimen, placed in a syracuse watch glass or mounted on a slide in a drop of seawater and then viewed with the microscope, you will see the stigmata, the activity of the lateral cilia that create the water current, and the blood vessels. What is the function of the blood vessels?

▶ The remainder of the gut is ciliated and consists of a short **esophagus** leading to a saclike **stomach;** from the stomach the posterior arm of the **intestine** leads around to make a semicircular loop, which runs anteroventrally and terminates in the **anus** at the base of the atrial siphon. The anus opens into the atrium and feces are carried out with the exhalent stream of water. The external surface of the stomach may be covered by the brownish **digestive gland.**

▶ *Determination of filtration rate in a tunicate.* Place one or more living ascidians (e.g., *Ciona* or *Molgula*) in a graduated vessel containing 1 to 2 liters of filtered seawater at 15°C, and aerate gently and continually. Prepare a fresh suspension of graphite (Aquadag) in distilled water as follows. Shake about 10 to 20 mg of the graphite paste vigorously in a test tube with distilled water for 1 to 2 minutes, and then pour into a beaker. After 5 minutes, decant the supernatant into a bottle, make up to 50 ml with distilled water, and allow to stand for 5 minutes. Use 1 to 2 ml of stock per liter of seawater. Avoid precipitated carbon and do not use the suspension more than 5 hours after its preparation. Add the Aquadag to the vessel containing the tunicate(s), and, as a control, add an equal amount of Aquadag suspension to another vessel containing the same amount of seawater but without animals. The initial concentration of Aquadag should be 0.2 to 0.5 mg per liter. Record the total volume in each container, and periodically (e.g., at intervals of 5, 15, 30, and 60 minutes) sample the water. Using a colorimeter or a spectrophotometer, determine the absorption of the samples at 570 or 650 nm. Graphically represent the results of this experiment by making a plot of concentration (ordinate) versus time (abcissa). Calculate the volume of water filtered using the following equation:

---

*Animals that are contracted and difficult to open may be narcotized first; see page 326 for methodology.

$$m = -\frac{M}{(t)}\left(2.303 \log \frac{P_t}{P_0}\right)$$

where m = volume of water cleared
M = total volume in container
P_t = concentration of Aquadag at time t
P_0 = concentration of Aquadag at time 0

Filtration rate may be calculated from m/t where m is the volume of water cleared and t is the time taken $(t_1 - t_2)$. For further details see the references by Jørgenson (1953, 1966).

iii. CEPHALOCHORDATA. Feeding in *Branchiostoma* depends on the current of water that enters the mouth, passes through the pharynx to the atrial cavity, and exits via the atriopore (Figure 10.4A). The buccal cirri are folded over one another and act as a screen. Mucus is discharged from Hatschek's pit onto the wheel organ and swept toward the mouth. As in ascidians, the main current is produced by the beating of lateral cilia on the gill bars. Mucus secreted by the endostyle is driven forward and upward over the pharyngeal wall and peripharyngeal bands, collected dorsally with entrapped food particles and passed backward along the hyperpharyngeal groove into the alimentary tract.

▶ If you have not worked out the structure of the pharynx from a prepared slide (Section 10–1), do so now. Is the pharynx substantially different in structure or function from that of *Ciona*, *Molgula*, or *Balanoglossus*? Of what adaptive advantage is an atrium?

10–3. NEUROMUSCULAR SYSTEM

▶ i. HEMICHORDATA. *Saccoglossus* is a burrower, and the organ chiefly employed during burrowing is the proboscis. Observe a living specimen in a bowl of seawater and note the extension and contraction of the proboscis. In what direction does the peristaltic wave pass? How is contraction and extension of the proboscis effected? Place the animal in a finger bowl containing seawater and a layer of sand and observe the burrowing process. Disturb the animal during burrowing and observe the characteristic reverse peristalsis.

▶ To gain some idea of the muscular arrangement in the proboscis, study prepared slides of a longitudinal section through the proboscis region (Figure 10.1B). Is there a coelomic space? How are the circular and longitudinal muscles arranged? The proboscis is supported in part by a **skeletal rod** located at the base of the proboscis where it joins the collar. An anterior diverticulum of the buccal cavity, the **stomochord** or **buccal diverticulum** has been regarded as a notochord, but this view is not confirmed by considerations of its structure, function, and anatomical position. Verify this latter opinion by comparing the longitudinal section with cross sections showing the notochord of *Branchiostoma* (Figure 10.4B).

▶ The muscular movements require coordination, which is obtained through the nervous system. The organization of the enteropneustan nervous system corresponds very closely to the ectoneural system of echinoderms. Throughout the body there is an **epidermal plexus,** which is locally concentrated into **longitudinal cords;** the most prominent are the middorsal cord of the proboscis and collar and the middorsal and midventral cords of the trunk. Only the collar cord is hollow and can be regarded as chordate in character. Return to the longitudinal section of *Saccoglossus* and identify as many of the components of the nervous system as possible.

It is not easy to study the coordination between nerve and muscle in the acorn worm, and much of the analysis has been achieved by making simple incisions or by studying the behavior of isolated body parts. The following studies (from Knight-Jones, 1952) should be performed only if time and animals are readily available.

▶ Using a living animal, make a dorsal incision that severs only the dorsal proboscis cord and note the behavior of the peristaltic wave. In a second animal cut only the ventral half of the proboscis. Make similar cuts through the trunk cords and observe the effects on burrowing and retreat peristalsis. What conclusions can you draw about the role of the cords in neuromuscular coordination? To locate the center responsible for initiation of locomotory behavior, divide an animal into two pieces by making a cut between proboscis and collar. Sever the body of another animal between collar and trunk. What portions of the body of these two animals continue to show burrowing behavior? What conclusions can you draw about the initiation center for locomotory activity in enteropneusts?

ii. UROCHORDATA. Ascidians possess a reduced nervous system that has not been well studied and is incompletely understood.

There is a single **cerebral ganglion,** located between the siphons, which is closely associated with a **subneural gland.** (In *Molgula* the gland is dorsal to the ganglion.) A duct leads from the gland and opens into the dorsal anterior end of the pharynx. The function of the subneural gland is unknown. It has been variously reported to be an endocrine organ and a possible forerunner of the vertebrate pituitary, to secrete mucus, or to have excretory or lymphatic functions. A pair of anterior nerves innervates the oral siphon, and other branches supply the body wall.

▶ Return to your specimen of *Ciona* (Figure 10.3**A**) or *Molgula* (Figure 10.3**B**), and attempt to locate the cerebral ganglion–subneural gland complex. In opened specimens of *Ciona* and *Molgula* you may be able to trace the nerves that run from their origin at the cerebral ganglion.

▶ The body wall musculature of *Ciona* (Figure 10.3**A**) consists of two groups of longitudinal muscles arising from the base and passing forward to the siphons. Using a needle or a glass rod drawn out to a fine tip, touch various parts of the test of a living animal in a bowl of seawater. Use a variety of pressures and record the responses. Which portions of the body are most sensitive? Stimulate first the exterior and then the interior of the atrial siphon and note the responses. Repeat on the oral siphon. What is the reaction? What do you think is the adaptive value of this response? Now take a small grain of sand and drop it into the oral opening; observe the reaction.

▶ The role of the cerebral ganglion in regulating the previously demonstrated **direct** and **crossed responses** can be determined by extirpation. Anesthetize the animal by placing it in ice-cold seawater. With a pair of fine scissors, carefully cut through the tunic and mantle in the intersiphonal area and remove the ganglion. (The subneural gland is dorsal to the ganglion in *Molgula.*) Allow the animal to recover and then test the response to touch. Which response is dependent on the presence of the cerebral ganglion?

iii. CEPHALOCHORDATA. Because living amphioxids are not ordinarily available, you will be able to study only the anatomy of the nervous and muscular systems.

▶ Recall the arrangement of myotomes in the whole mount of *Branchiostoma* or re-examine it now (Figure 10.4**A**). The myotomes are separated by **myosepta,** and the muscle fibers within each myotome run horizontally between myosepta. The myotomes of the left side alternate with those of the right side so that when the muscle fibers of one side contract, they bend the animal in that direction; during locomotion the animal bends its body in first one direction and then the other.

▶   Recall the position of the dorsal nerve cord in *Branchiostoma*. Examine cross sections through the pharyngeal region (Figure 10.4**B**), and note the relationship of the dorsal nerve cord to the notochord, myotomes, and epibranchial groove.

10–4. VASCULAR SYSTEM AND EXCRETION

▶   i. HEMICHORDATA.   The vascular system of hemichordates consists of **vessels** and **lacunae.** Examine a longitudinal section through the collar and proboscis region of *Saccoglossus* (Figure 10.1**B**) and identify the **heart vesicle, venous sinus,** and **central sinus.** What course does the blood take in *Saccoglossus?* At this time, locate the **glomerulus** and note its relationship to the previously named structures and to the buccal diverticulum. The glomerulus has been thought to perform some function in excretion.

    ii. UROCHORDATA.   The tunicate vascular system is of the open type, consisting of **hemocoelic lacunae** which ramify throughout the body, and few real vessels. The **heart** is an evagination of the wall of the **pericardial cavity,** forming an elongate, tubular U-shaped structure. Located on the right side of the body within a loop of the intestine (in *Perophora* it lies beside the pharynx rather than posterior to it), it is easily recognized in the living animal by its peristaltic beat.

▶   Place a living *Ciona* (or *Molgula*) in a wax dissecting pan, anchor with pins, and cover the animal with seawater. If your view is obscured, carefully remove portions of the tunic below the pharyngeal region so that this portion of the body can be seen clearly. Blood vessels leave the heart and run to the test, pharynx, and viscera. One of the pecularities of the tunicate heart is its periodic reversal of beat. Attempt to observe how the peristaltic wave begins at one end, and, after a number of pulsations in one direction, pauses and reverses. What is the cause of this reversal of beat?

    Tunicate blood is interesting because of the variety of blood cell types and the ability of these cells to leave the bloodstream and migrate in the body, to play a role in regeneration, and to concentrate rare metals such as vanadium. Students interested in these aspects of urochordate biology should consult the references by Endean (1960), Freeman (1964), or Carlisle (1968).

▶   Paired outpushings from the pharynx, so-called **epicardial sacs,** end blindly on either side of the heart. They allow seawater to circulate around the heart and may function in excretion and possibly in respiration. In some species, including *Molgula,* the connection with the pharynx is lost, and the sacs store uric acid. In some Urochordates, the sac wall gives rise to cells involved in budding and regeneration. Attempt to identify the epicardial sacs in your specimen (Figure 10.3).

    iii. CEPHALOCHORDATA.   The vascular system of *Branchiostoma* will not be dealt with here. Students are advised to read Young (1962), pp. 33–35, for a brief discussion.

    You have already identified the nephridia in prepared slides of *Branchiostoma* (Section 10–1). If not, do so now. Note that the only other animal group to possess excretory organs of this kind (with solenocytes) are the annelid worms.

10–5. REPRODUCTION AND DEVELOPMENT

a. Sexual Reproduction

    i. HEMICHORDATA.   *Saccoglossus* is dioecious. In living animals, the reproductive organs (gray in the female and yellow in the male) are located in the genital region of the trunk.

Some specimens may show gonadal tissue in the branchial region. The gonads open to the exterior by **genital pores,** which cannot be distinguished externally. Fertilization is external and the egg develops into a pelagic **tornaria** larvae, which closely resembles the **bipinnaria** larva of starfish.

▶ Examine prepared slides of tornaria and bipinnaria larvae, and compare their structure. (See also Exercise 9, p. 303.)

ii. CEPHALOCHORDATA

▶ *Branchiostoma* is dioecious. The 26 pairs of gonads lack gonoducts, the gametes are shed into the atrium by rupture of the atrial wall; they escape via the atriopore with the exhalent current and fertilization is external. If you have not already done so, attempt to locate the gonads in prepared slides of *Branchiostoma* (Section 10–1 and Figure 10.5). Gonads will not be present in immature animals, frequently used for preparing whole mounts.

▶ iii. UROCHORDATA. Most ascidians are hermaphroditic. *Ciona* is oviparous and the **ovaries** (yellowish-green) and **testes** (white) are found in the intestinal loop (Figure 10.2A). In living *Molgula manhattensis* the translucent green ovary is located in the loop of the intestine; the testis is opaque white and overlaps the ovary (Figure 10.3). The gonads open by ducts (oviduct and sperm duct) into the atrial cavity near the base of the excurrent, atrial siphon. Return to your opened specimens or examine new ones and locate the gonads.

▶ Early development of *Ciona* or *Molgula manhattensis* may be obtained in the laboratory. It is advisable to use gametes procured from different individuals. Slit open the test of living animals and cut the superficial layer of muscles so that the animals can be extended. By means of Pasteur pipets, remove eggs and sperm from the oviduct and sperm duct, respectively. Make a suspension of sperm in seawater. The mature *Ciona* egg is peach-colored and remains viable for 18 hours after removal from the animal. The eggs should be transferred through a few changes of seawater prior to insemination. Add a drop or two of sperm suspension to eggs in a syracuse watch glass of seawater so that a milkiness is imparted to the seawater and allow to stand for 15 minutes, after which the eggs should be washed free of excess sperm. Development into tadpole larvae occurs within 24 hours. Details for other species can be found in the references by Costello et al. (1957) and Milkman (1967).

b. Tadpole Larva

The larval stage of ascidians, commonly called a **tadpole larva** because of its superficial resemblance to the amphibian tadpole, is important in studies of protochordate and vertebrate evolution; at this stage of development the organism possesses body structures that clearly identify its chordate nature (Figure 10.5). In general, two types of tadpoles are formed. The "simple" tadpoles of *Ciona* and *Molgula manhattensis* are 1 mm long, with three adhesive papillae, and have a simple trunk containing a cerebral vesicle and the beginnings of a branchial basket. They usually swim freely for 6 to 24 hours. The more complex larvae of *Botryllus* and *Amaroucium* have an elaborate internal structure, and their free-swimming existence is reduced to minutes or a few hours at most.

▶ To study the organization of the tadpole larva, use the living material developed previously from *Ciona* or *Molgula* or obtain larvae from *Amaroucium* or *Botryllus* as follows.

▶ Larvae are normally released from sexually mature colonies of the brooding tunicate *Amaroucium constellatum* at dawn. However, they can be obtained at a more suitable hour by holding the colonies overnight in shrouded containers or a dark room or drawer. Swarms of active tadpoles should appear 15 to 20 minutes after exposure of the colonies to the light.

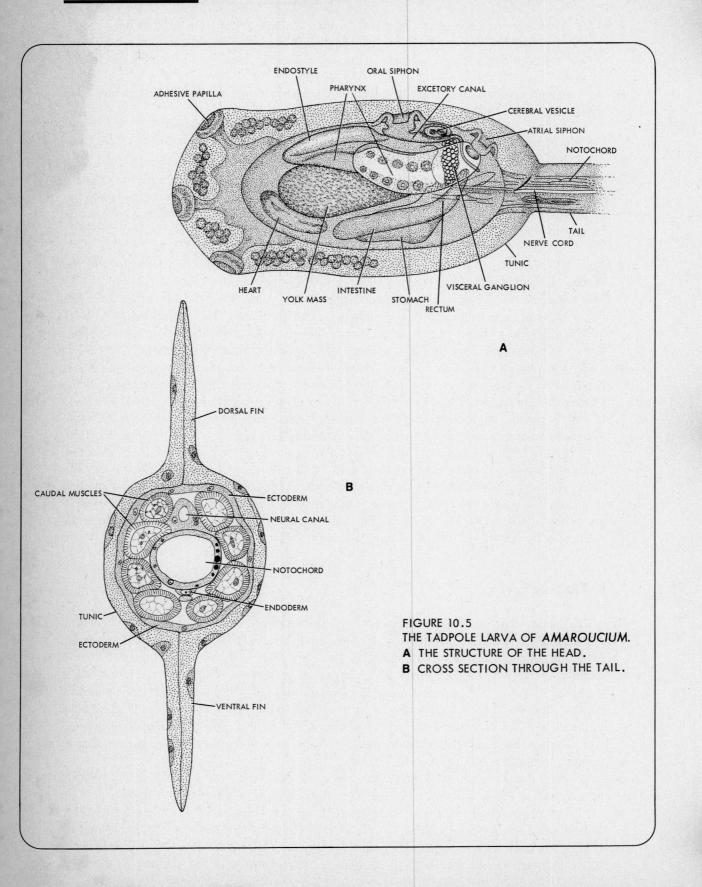

ADHESIVE PAPILLA

ENDOSTYLE

PHARYNX

ORAL SIPHON

EXCETORY CANAL

CEREBRAL VESICLE

ATRIAL SIPHON

NOTOCHORD

TAIL

NERVE CORD

TUNIC

VISCERAL GANGLION

RECTUM

STOMACH

INTESTINE

YOLK MASS

HEART

A

B

DORSAL FIN

CAUDAL MUSCLES

ECTODERM

NEURAL CANAL

NOTOCHORD

TUNIC

ENDODERM

ECTODERM

VENTRAL FIN

FIGURE 10.5
THE TADPOLE LARVA OF *AMAROUCIUM.*
A THE STRUCTURE OF THE HEAD.
B CROSS SECTION THROUGH THE TAIL.

Alternatively, squeeze a piece of the colony over a dish of seawater to express the tadpole larvae together with eggs, developmental stages, and zooids (members of the colony). Using a dissecting microscope, you should be able to identify the tailed larvae by their swimming motion. Transfer some of these to clean seawater by means of a Pasteur pipet.

▶ Developmental stages can be obtained from the atrial cavity of *Botryllus schlosseri* by dissection of colonies in which fertilization has occurred. Free-swimming larvae may also be obtained by placing adult colonies in large finger bowls of fresh seawater near a window between 10 A.M. and 1 P.M., out of direct sunlight. If the colonies are sexually mature, they may begin to release larvae within a few minutes.

▶ Make wet mounts of the developmental stages or examine prepared slides of whole mounts (Figure 10.5A).

▶ The **tadpole larva** consists of a **body** and **tail** enclosed in a **tunic.** At the anterior end of the body are **adhesive papillae,** and the tail is posterior. Locate the dorsally situated **cerebral (sensory) vesicle** with its pigmented sense organs, the **eye spot** and **statolith.** The conspicuous **pharynx** occupies the anterodorsal portion of the body, and the **endostyle** is located in its dorsal wall. Ventral to the pharynx are located the **heart, stomach,** and **intestine.** The intestine loops posterodorsally and terminates in an **anus** near the base of the atrial siphon. The **notochord** extends along the anterior two thirds of the tail and passes into the body below the sensory vesicle, terminating anteriorly in contact with the pharynx. The **nerve cord** runs dorsally above the notochord and connects with the sensory vesicle. Two bands of **muscle,** one dorsal and one ventral, surround the nerve cord and notochord.

▶ It difficulty is encountered in locating any of these structures, examine the diagram of a transverse section across the tail of a tadpole larva (Figure 10.5B).

After a brief (15 minute to 24 hour) free-swimming existence, the larva attaches to the substrate by its adhesive papillae. The tail is resorbed, there is rapid growth between the point of attachment and the oral region so that the oral and the atrial siphons are shifted until they point upward from the region of fixation, and the larva undergoes **metamorphosis** into the adult form. For further details concerning metamorphosis see the references by Cloney (1961, 1963, 1966).

▶ Review the chordate features of the larval and adult tunicate. Can you suggest how the vertebrates might have arisen from this group? (See the reference by Berrill, 1955.)

c. Colony Formation in Urochordata

Many urochordates have a high capacity for asexual reproduction. Budding produces **daughter zooids,** which remain in close association with the parent through a common tunic, and this results in the formation of colonies. Many ascidians show an alternation of phases of asexual reproduction (winter) and sexual reproduction (summer). Some species have little or no capacity for asexual reproduction and remain solitary (e.g., *Ciona, Molgula*).

Colonial ascidians can be divided into two types: (1) social ascidians where the individuals tend to be separated from one another but connected by a **stolon,** and (2) compound ascidians composed of individuals embedded in a common tunic.

▶ Study the organization of the colony of the social ascidian *Perophora* (Figure 10.6B). Examine a small portion of a colony in a syracuse watch glass containing seawater. Observe the arrangement of individuals along the stolon. Do the stolons branch? Does there appear to be any regularity in the way buds arise from the stolons? Is the stolon hollow? Is the stolon compartmentalized? Is there a free circulation of blood from individual to individual? Attempt to detect the beating of the heart and time the duration of beating in each direction. Are the

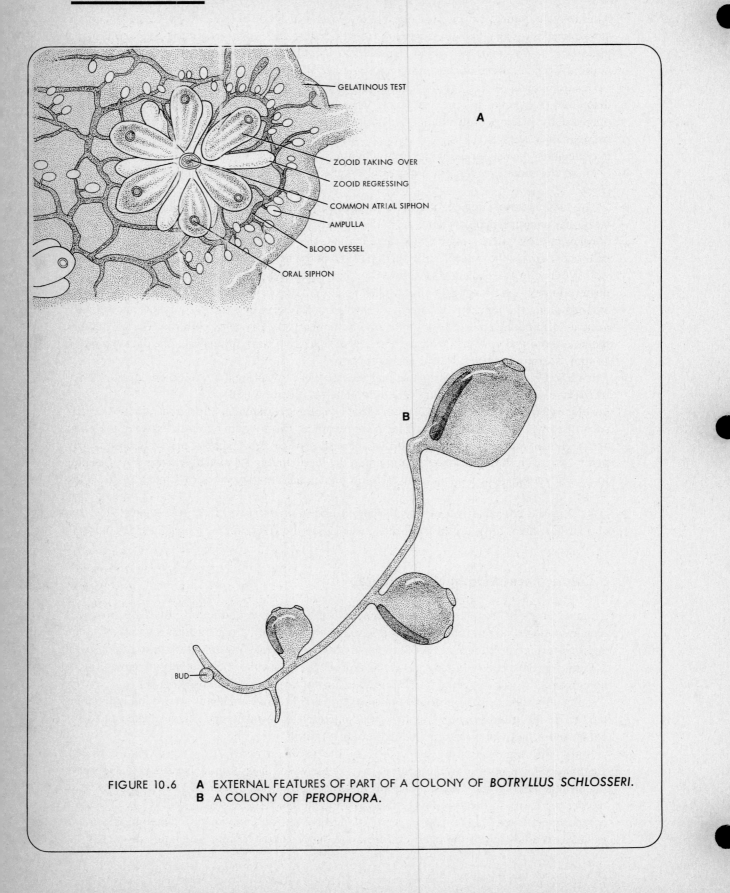

FIGURE 10.6 **A** EXTERNAL FEATURES OF PART OF A COLONY OF *BOTRYLLUS SCHLOSSERI*. **B** A COLONY OF *PEROPHORA*.

heartbeats of individuals on the same stolon synchronized? Examine a stolon with buds of different ages and compare the degree of development of different buds. Compare the orientation of organs in *Perophora* with that in *Molgula* or *Ciona*.

▶   Observe living specimens of the compound tunicate *Botryllus* (Figure 10.6A). Colonies may be removed from the substrate on which they are growing and attached to microscope slides with petroleum jelly or small rubber bands. After a few days the colonies become firmly attached to the glass. Place slides on which young colonies of *Botryllus* are growing into a shallow dish containing enough seawater to cover the colonies, and examine under the dissecting microscope. Each *Botryllus* colony is composed of a rosette containing 5 to 18 **zooids** embedded in a common translucent **tunic.** The **oral siphons** are peripheral and open directly to the exterior, whereas the **atrial siphons** open into a common **atrial chamber.** In and around the zooids is a **vascular system** consisting of **blood vessels** and **ampullae.** Identify these structures and the **endostyle, pharynx, gut,** and other parts identified previously in the solitary ascidian.

▶   Directions for obtaining tadpole larvae from an *Amaroucium* colony were given previously (Section 10–5b). Using the same technique, attempt to remove unbroken zooids from the common tunic, and examine them with the dissecting microscope (Figure 10.7). The body is divisible into three regions: **thorax, abdomen,** and **postabdomen.** The thorax contains the **pharynx,** the abdomen the **digestive tract** and **gonoducts,** and the postabdomen the **gonads** and **heart.** The postabdomen may be regarded as a stolon in which the gonads and heart are located and from which buds are formed. Locate the above-named parts of the body and those previously identified in the solitary ascidian.

d. Regeneration

▶   In general, high regenerative powers are characteristic of organisms that reproduce asexually by budding. Obtain specimens of *Botryllus* growing on glass slides (Figure 10.6A) and remove everything but the peripheral vascular system. Replace the slides in a seawater aquarium, and during the next few weeks look for regeneration of new zooids. What does this tell you about the source of the regenerating materials?

▶   *Perophora viridis* forms colonies on a vascular stolon (Figure 10.6B). To determine whether new zooids arise by septal proliferation or from blood cells and dedifferentiated somatic cells, perform the following experiment. Isolate a stolon with two zooids in a petri dish of seawater. Amputate or injure one zooid with a needle. Observe the regenerative pattern over the next two weeks and do not feed the organisms during this time. What is the source of material for the regenerate?

References

Anderson, M. Electrophysiological studies on initiation and reversal of the heart beat in *Ciona intestinalis, J. Exp. Biol.* **49**(2):363, 1968.

Barrington, E. J. W. *The Biology of the Hemichordata and Protochordata.* San Francisco: W. H. Freeman and Co., 1965.

Berrill, N. J. *The Tunicata.* London: Ray Society, 1950.

Berrill, N. J. *The Origin of the Vertebrates.* New York: Oxford University Press, 1955.

Berrill, N. J. *Growth Pattern and Development.* San Francisco: W. H. Freeman and Co., 1961.

Brown, F. A., Jr. (ed.). *Selected Invertebrate Types.* New York: John Wiley & Sons, 1950.

Bullough, W. S. *Practical Invertebrate Anatomy.* New York: St. Martin's Press, 1962.

Carlisle, D. B. Vanadium and other metals in ascidians, *Proc. Roy. Soc.* **171**(1022):31, 1968.

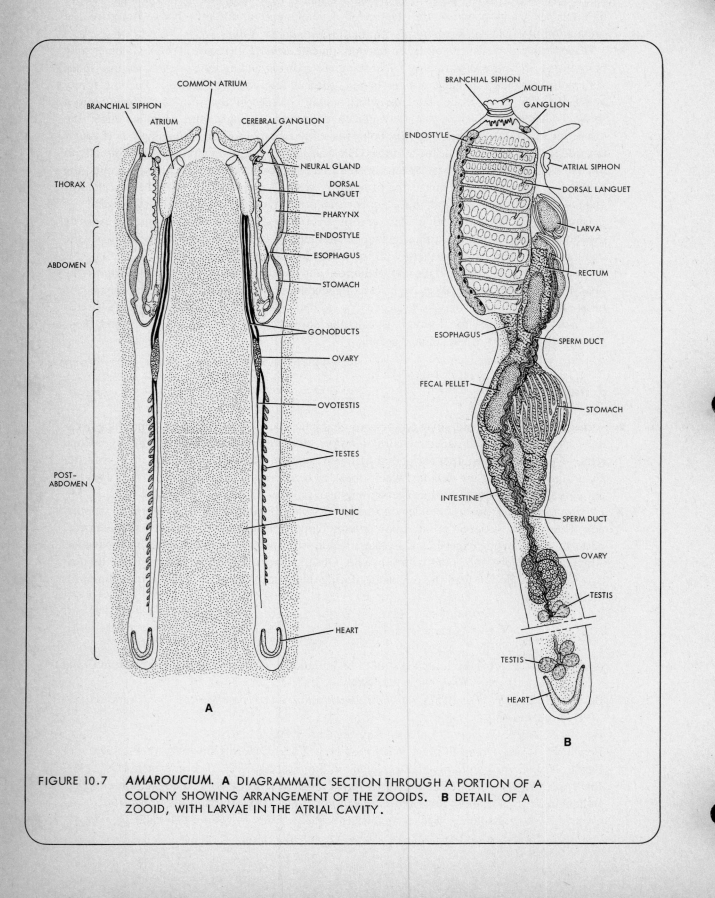

FIGURE 10.7 **AMAROUCIUM. A** DIAGRAMMATIC SECTION THROUGH A PORTION OF A COLONY SHOWING ARRANGEMENT OF THE ZOOIDS. **B** DETAIL OF A ZOOID, WITH LARVAE IN THE ATRIAL CAVITY.

Cloney, R. A. Observations on the mechanism of tail resorption in ascidians, *Amer. Zool.* **1:**67–87, 1961.

Cloney, R. A. The significance of the caudal epidermis in ascidian metamorphosis, *Biol. Bull.* **124:**241–53, 1963.

Cloney, R. A. Cytoplasmic filaments and cell movements: Epidermal cells during ascidian metamorphosis, *J. Ultrastruct. Res.* **14:**300–28, 1966.

Colwin, A. L., and L. H. Colwin. Behavior of the spermatozoon during sperm blastomere fusion and its significance for fertilization (*Saccoglossus kowalevskii:* Hemichordata), *Zeitschr. f. Zellforsch.* **78:**208–20, 1967.

Costello, D. P., M. E. Davidson, A. Eggers, M. H. Fox, and C. Henlry. *Methods for Obtaining and Handling Marine Eggs and Embryos.* Woods Hole, Mass.: The Marine Biological Laboratory, 1957.

Endean, R. The blood cells of the ascidian *Phallusia mamillata, Quart. J. Microscop. Sci.* **101:**177–198, 1960.

Freeman, G. The role of blood cells in the process of asexual reproduction in the tunicate *Perophora viridis, J. Exp Zool.* **156:**157–183, 1964.

Hyman, L. H. *The Invertebrates,* Vol. V. *Smaller Coelomate Groups.* New York: McGraw-Hill Book Co., 1959.

Jørgenson, C. B., and E. D. Goldberg. Particle filtration in some ascidians and lamellibranchs, *Biol. Bull.* **105:**447–89, 1953.

Jørgenson, C. B. *Biology of Suspension Feeding.* Oxford: Pergamon Press, 1966.

Knight-Jones, E. W. On the nervous system of *Saccoglossus cambrensis, Phil. Trans. Roy. Soc., Ser. B.* **236:**315–54, 1952.

Kriebel, M. E. Studies on the cardiovascular physiology of tunicates, *Biol. Bull.* **134:**434–55, 1968.

Milkman, R. Genetic and developmental studies on *Botryllus schlosseri, Biol. Bull.* **132:**229–43, 1967.

Millar, R. H. *Ciona. L.M.B.C. Memoirs XXXV.* Liverpool: University of Liverpool Press, 1953.

Packard, A. Asexual reproduction in *Balanoglossus* (Stomochordata), *Proc. Roy. Soc.* **171** (1023):261–72, 1968.

An Abbreviated Classification of the Protochordates (After Hyman and Others)

Phylum **Hemichordata**

Enterocoelous coelomate animals with tripartite body and coelom, epidermal nervous system, without nephridia, preoral gut diverticulum; gill slits present or absent and without tentaculated arms.

Class 1 **Enteropneusta**

Vermiform, solitary, with numerous gill slits, straight intestine, e.g., *Saccoglossus, Balanoglossus.*

Class 2 **Pterobranchia**

Aggregated or colonial, digestive tract U-shaped, two tentaculated arms; gill slits absent, e.g., *Cephalodiscus, Rhabdopleura.*

Phylum **Chordata**

Enterocoelous, bilateral, coelomate animals with at some stage in the life history a dorsal hollow tubular nerve cord, pharyngeal gill slits, and a postanal, metamerically segmented tail.

Subphylum 1 **Urochordata**

Unsegmented, notochord restricted to tail (present in larval stage); cellulose-like test, central nervous system reduced in adult.

Class 1 **Ascidiacea**

Adult is sedentary and lacks a tail, degenerate nervous system and atrium opens dorsally; gill clefts divided into stigmata by longitudinal bars, e.g., *Ciona, Molgula, Perophora, Botryllus, Amaroucium, Ascidia.*

Class 2 **Thaliacea**

Pelagic tunicates, adult has no tail, degenerate nervous system, atrium opens posteriorly, gill clefts divided into stigmata by longitudinal bars.

Class 3 **Larvacea**

Pelagic tunicates without metamorphosis; adult retains the larval tail.

Subphylum 2 **Cephalochordata**

Segmented, fishlike body; notochord and nerve cord, many gill slits, well-developed coelom, without specialized head, e.g., *Branchiostoma, Asymmetron.*

Subphylum 3 **Vertebrata**

Backbone, or vertebral column; distinct head including brain, olfactory organs, eyes, internal ears and a skull (cranium); ventral heart divided into several chambers; paired appendages, when present, originate from several segments; hepatic portal system.

Directions for Narcotizing Urochordates

Animals may be narcotized and thus relaxed for dissection by sprinkling crystals of menthol on the surface of the vessel of seawater containing the organism; allow the animal to sit for 1 to 4 hours. Alternatively, place the animal in 1% chloroform in seawater and allow to sit until relaxed.

Equipment and Materials

Equipment

Microscope and lamp
Microscope slides and cover slips
Finger bowls
Syracuse watch glasses
Petri dishes
Glass rod
Sand
Dissecting instruments
Colorimeter
Pipets
Depression slide
Beakers
Rubber bands
Petroleum jelly

Solutions and Chemicals

Carmine powder
Aquadag A

Prepared Slides

Ascidia
Saccoglossus, longitudinal section, anterior region
Saccoglossus, cross section, anterior region
Tadpole larvae
Tornaria larva
Bipinnaria larve (starfish)
Branchiostoma, whole mount
Branchiostoma, cross section, pharyngeal region

*Living Material**

Amaroucium constellatum (7,9,12)
Botryllus schlosseri (7,9,12)
Molgula manhattensis (7,12)
Branchiostoma (7)
Saccoglossus (12)
Perophora (7,12)
Ciona (9,12)

Preserved Material

Saccoglossus
Branchiostoma
Ciona
Molgula
Amaroucium

*Numbers in parentheses identify sources listed in Appendix B.

APPENDIX

A. PREPARATION OF SOLUTIONS AND MEDIA

NOTE: Use distilled water in all solutions except where indicated.

ACETIC ACID, 5%: 5 ml 3 N (99 to 100%) acetic acid; 95 ml water.

ACIDIFIED METHYL GREEN: 1 g methyl green; 1 ml acetic acid; 100 ml water.

AMMONIUM HYDROXIDE, 0.1%: 1 ml ammonium hydroxide; 280 ml seawater.

AMPHIBIAN RINGER'S SOLUTION: 6.5 g sodium chloride; 0.2 g sodium bicarbonate; 0.14 g potassium chloride; 0.16 g calcium chloride ($CaCl_2 \cdot 2H_2O$); 1000 ml water.

CHLORETONE (1,1,1-Trichloro-2-methyl-2-propanol; chlorobutanol): 2 g chlorobutanol; 1000 ml water.

CONGO-RED-STAINED YEAST SUSPENSION: 3 g compressed yeast; 30 mg Congo red; 10 ml water. Boil gently 10 minutes.

CYANOL–CONGO RED: 500 mg cyanol; 500 mg Congo red; 100 ml water.

EVANS BLUE: 1 g Evans blue; 100 ml water.

GIEMSA STAIN: Air dry film. Fix for 1 minute in absolute methanol. Air dry. Stain in Giemsa stain (1 ml stain, 10 ml water, buffered to pH 7.2 with phosphate buffer) for 30 minutes. Rinse in distilled water and air dry.

GLUTATHIONE (Reduced), 10^{-5} M: 3 mg reduced glutathione; 10 ml water. Dilute 1 to 10 to give 10^{-5} M.

HYDROCHLORIC ACID, 0.1 N: 8.6 ml of 11.6 M (36%) hydrochloric acid; 1000 ml water.

HYDROGEN PEROXIDE SOLUTION, 1%: 33 ml hydrogen peroxide (3%); 66 ml water.

INDIGO CARMINE, 1%: 1 g indigo carmine; 100 ml water.

INSECT RINGER'S SOLUTION: 9 g sodium chloride; 0.2 g potassium chloride; 0.27 g calcium chloride ($CaCl_2 \cdot 2H_2O$); 4 g dextrose. Dissolve in distilled water and add sufficient sodium bicarbonate to bring the pH to 7.2. Bring total volume to 1000 ml with water.

IODINE SOLUTION, 1%: 1 g iodine; 100 ml water.

JANUS GREEN–NEUTRAL RED: 100 ml 95% ethanol; 1 g neutral red; 0.5 g Janus green. Dip slide in stain and air dry.

LITHIUM CHLORIDE, 2%:   2 g lithium chloride, 100 ml water.

LUGOL'S IODINE: 6 g potassium iodide; 4 g iodine; 100 ml water.

MAGNESIUM CHLORIDE, 2%:   2 g magnesium chloride ($MgCl_2 \cdot 6H_2O$); 100 ml water.

MAGNESIUM CHLORIDE, 7.2%:   7.2 g magnesium chloride ($MgCl_2 \cdot 6H_2O$); 100 ml water.

METHOCEL (Methyl cellulose), 10%: 10 g methyl cellulose; 45 ml boiling water. Allow to soak 30 minutes at 10°C; add 45 ml cold water and stir until smooth. Bubbles disappear on standing.

METHYLENE BLUE, 0.5%:   0.5 g methylene blue; 100 ml water.

NEUTRAL RED, 0.1%:   10 mg neutral red; 10 ml water.

NEUTRAL RED, 0.25%:   25 mg neutral red; 100 ml water.

NEUTRAL RED, 1%:   10 mg neutral red; 1 ml water.

NIGROSIN, 10%:   10 g nigrosin; 100 ml water.

PEPTONE–ACETATE MEDIUM: 1 g trypticase, 100 mg yeast extract, 100 mg sodium acetate, 100 ml water. Tube in 5 ml volumes and after autoclaving add 1 drop of 95% ethanol to each tube.

PHOSPHATE BUFFER, 0.01 M, pH 6.8:   4.583 g potassium phosphate; 4.646 g sodium phosphate ($NaH_2PO_4 \cdot H_2O$); 1000 ml water. Dilute 1 to 5 to give 0.01 M.

POTASSIUM CHLORIDE, 0.5 M: 37 g potassium chloride; 1000 ml water.

PROTEOSE PEPTONE, 1%:   1 g proteose peptone; 100 ml water. Adjust pH to 7.0 to 7.2.

SEAWATER, 30%:   30 ml seawater, 70 ml distilled water.

SILVER NITRATE, 0.01 M: 1.7 g silver nitrate; 1000 ml water.

SODIUM CHLORIDE, 0.85%:   8.5 g sodium chloride; 1000 ml water.

SODIUM CHLORIDE, 10%:   10 g sodium chloride; 100 ml water.

SODIUM CHLORIDE IN PHOSPHATE BUFFER (0.125 M sodium chloride in 0.01 M phosphate buffer), pH 6.5 to 7.0:   0.7 g sodium chloride; 100 ml 0.01 M phosphate buffer.

SODIUM HYPOCHLORITE, 5%:   Commercial Clorox is about 6% sodium hypochlorite.

SUCROSE, 5%:   5 g sucrose; 100 ml water.

B. SOURCES OF MATERIALS

1. Algal Culture Collection
   Indiana University
   Bloomington, IN 47401

2. Ariolimax
   College Biological Supply
   P. O. Box 1326
   Escondido, CA 92025

3. Carolina Biological Supply
   Company
   Powell Lab Division
   Gladstone, OR 97027
         or
   Carolina Biological Supply Company
   Burlington, NC 27215

4. Macmillan Science Co. Inc.
   (Turtox-Cambosco)
   8200 S. Hoyne St.
   Chicago, IL 60620

5. Connecticut Valley Biological
   Supply House
   Valley Road
   Southampton, MA 01073

6. Dahl Company
   P. O. Box 566
   Berkeley, CA 94701

7. Gulf Specimen Company Inc.
   P. O. Box 237
   Panacea, FL 32346

8. Ricarimpex
   33 Audenge (GDE)
   France
   (For living *Hirudo medicinalis*)
      or
   Mogul-Ed
   Oshkosh, WI 54901
   (For living *Macrobdella*)

9. Pacific Bio-Marine Supply
   Company
   P. O. Box 285
   Venice, CA 90293

10. Nasco-Steinhilber
    Fort Atkinson, WI 53538
       or
    P. O. Box 3837
    Modesto, CA 95352

11. Wards Natural Science
    Establishment, Inc.
    P. O. Box 1712
    Rochester, NY 14603
       or
    P. O. Box 1749
    Monterey, CA 93940

12. The Supply Department
    The Marine Biological
    Laboratory
    Woods Hole, MA 02543

C. SOURCES OF ILLUSTRATIONS

The illustrative material in this manual has been selected in the hope that the student will obtain an accurate picture of the anatomy of the organism under study without having to resort to long and detailed written descriptions of morphology. The authors and illustrator are deeply indebted to the many biologists and artists whose work has served as a basis for the illustrations used in this laboratory guide. The following figures were adapted from the sources indicated.

Alexopoulos, C. J., and H. C. Bold. *Algae and Fungi*. New York: Macmillan Publishing Co., 1967. 1.3**A**, 1.24**A**.

Barnes, R. D. *Invertebrate Zoology*. Philadelphia: W. B. Saunders Co., 1974. 7.1**A** (iii) after Cuenot, 7.2**A** after Kaestner, 7.13 (i) after Rolleston, 10.2.

Barrington, E. J. W. *Invertebrate Structure and Function*. Boston: Houghton Mifflin Co., 1967. 7.20**A** after Parry.

Batham, E. J., and C. F. A. Pantin. *J. Exp. Biol., 27*:264, 1950. 3.15**B**.

Bayer, F. M., and H. B. Owre. *The Free-living Lower Invertebrates*. New York: Macmillan Publishing Co., 1968. 2.2**C** after Fjerdingstad and Rasmont, 3.3**A**, 3.4**D** after Hyman, 3.6**C** after Naumos, 3.11**A**, 4.1**C** after Wilhemi.

Beck, D. E., and L. F. Braithwaite. *Invertebrate Zoology Laboratory Workbook*. Minneapolis: Burgess Publishing Co., 1968. 1.5**B**.

Borradaile, L. A., and F. A. Potts. *The Invertebrata*. London: Cambridge University Press,

1961. 4.1**A** after Shipley and Macbride, 7.3**B** (i) after Koch, 7.20**B** after Parker and Haswell, 7.24**A, C,** and **D** after Dietrich, Faxon, and Bate, 9.3**B,** 10.3A after Shipley and Macbride.

Brill, B. *Z. Zellforsch.*, **144:**231, 1973. 2.2**B.**

Brooks, W. K. *Handbook of Invertebrate Zoology*. Boston: Bradlee, Whidden, 1897. 7.4**C,** 8.10**A,** 9.8**A.**

Brown, F. A., Jr. (ed.). *Selected Invertebrate Types*. New York: John Wiley & Sons, 1950. 3.12 after Parker, 3.13**B** and **C** after Schulze, 5.1**B,** 6.2**C** after Turnbull, 6.2**D,** 6.5**A** and **B,** 7.14, 8.4**B,** 8.7**F,** 8.8**A,** 9.2**A,** 9.2**B** after Delage and Herouard, 9.3**A,** 9.3**C** (i).

Buchsbaum, R. *Animals Without Backbones*. Chicago: University of Chicago Press, 1956. 7.23 after Hesse.

Bullough, W. S. *Practical Invertebrate Anatomy*. New York: St. Martin's Press, and London: Macmillan and Company Ltd., 1958. 8.2, 8.4**D,** 9.5**B, C,** and **D.**

Burtt, E. T. *Proc. Roy. Soc. Lond.*, **124B:**13, 1937. 7.27.

Chandler, A. C., and C. P. Read. *Introduction to Parasitology*. New York: John Wiley & Sons, 1961. 4.5.

Clark, R. B. *A Practical Course in Experimental Zoology*. New York: John Wiley & Sons, 1966. 7.11, 8.5**B.**

Coe, W. R. *Conn. Geol. and Nat. Hist. Survey,* 1912. 9.1**B,** 9.6**A** (ii).

Cohen, M. *Proc. Roy. Soc. Lond.*, **152B:**30, 1960. 7.22.

Crofton, H. D. *Nematodes*. London: Hutchinson and Co., 1966. 5.6 after Fairbairn, 5.7.

Dales, R. P. (ed.) *Practical Invertebrate Zoology*. London: Sidgwick and Jackson, 1969. 9.5**A.**

Dawson, J. A., and W. Etkin. *Basic Exercises in College Biology*. New York: Thomas Y. Crowell Co., 1951. 6.4**A,** 7.5**E,** 8.6**A.**

De Laubenfels, M. W. *A Guide to the Sponges of Eastern North America*. Coral Gables, Fla.: University of Miami Press, 1953, 2.6.

Finley, H., et al. *J. Protozool.*, **11:**266, 1964. 1.9**B.**

Fjerdingstad, E. J. *Z. Zellforsch.*, **53:**645, 1961. 2.2**C.**

Gibbons, I. R., and J. V. Grimstone. *J. Biophys. Biochem. Cytol.,* **7:**697, 1960. 1.2.

Gilbert, S. G. *Atlas of General Zoology*. Minneapolis: Burgess Publishing Co., 1966. 7.5**B, C,** and **D,** 7.17**A,** 8.7**E.**

Grassé, P. *Traité de Zoologie,* vols. IV, V, and XI. Paris: Masson et Cie, 1948. 5.2**B** and **C,** 5.3, 6.2**F** and **G,** 8.3, 10.1**B** after Spengel, 10.5 after Lahille, 10.5 after Grave, 10.6**A,** after Delage and Herouard, 10.6**B** after Franz, 10.7**A** after Herdman.

Grove, A. J., and G. E. Newell. *Animal Biology*. London: Univ. Tutorial Press, 1942, 6.2**A** and **B.**

Guthrie, M. J., and J. M. Anderson. *General Zoology*. New York: John Wiley & Sons, 1957. 4.3**E,** 4.7**A** after Stempel.

Hall, R. P. *Protozoology*. Englewood Cliffs, N.J.: Prentice-Hall, 1953. 1.19**A** after Loefer, Kudo, 1.21**A** (i) and (ii) after Hollande and Dunihue.

Hayashi, T. *Sci. Amer.,* **205**(3):184, 1960. 1.15**A.**

Hegner, R. W. *Invertebrate Zoology*. New York: Macmillan Publishing Co., 1933. 1.14**A** after Mast, 9.3**E** after Macbride.

Hegner, R. W., and J. G. Engemann, *Invertebrate Zoology,* 2nd ed. New York: Macmillan Publishing Co., 1968. 1.14**A** after Mast, 2.5**A,** 3.4**A** and **B** after Hyman, 3.4**C** after Kepner and Miller, 3.5**B,** 3.6**A,** 3.8**A,** 4.2**A** after Faust, 4.4 after Cort, 7.3**B** (ii) after Borradaile and Potts, 7.10**A** and **C** after Woodruff and Carter, 7.19 after Kennedy, 7.21**A** after Snodgrass.

Hegner, R. W., and K. A. Stiles, *College Zoology*, 7th ed. New York: Macmillan Publishing Co., 1959. 4.3**A, B, C,** and **D,** 4.7**B** after Newman, 10.4**B.**

Hertwig, R. *A Manual of Zoology*. New York: Holt, 1912. 7.12, 8.8**C** after Haller.

Hickman, C. P. *Biology of the Invertebrates*. St. Louis: C. V. Mosby Co., 1967. 7.1**B** after Lang.

Holter, H. *Sci. Amer.*, **205**(3):170, 1960. 1.18.

Hyman, L. G. *The Invertebrates*. New York: McGraw-Hill Book Co. Vol. I. *Protozoa Through Ctenophora*. 1940. 2.1**A**, 2.7 after Metschnikoff and Hammer, 2.8**A** and **B** after Brien and Evans, 3.1**A** and **B**, 3.5**C**, 3.6**B** and **D**, 3.7**A** and **B**, 3.11 **A**, 3.13**A**, 3.16**A** and **B**, 3.17**A**. Vol. II. *Platyhelminthes and Rhynchocoela*. 1951. 4.7**C**. Vol. III. *Acanthocephala, Aschelminthes and Entoprocta*. 1951. 5.2**A** after Hickernell, 5.8**A**. Vol. VI. *Mollusca I*. 1967. 8.8**C**.

Jakus, M. A., and C. E. Hall. *Biol. Bull.*, **91**:141, 1946. 1.22**C**.

Jones, A. R. *The Ciliates*. New York: St. Martin's Press, 1974. 1.23.

Jurand, A., and G. G. Selman. *The Anatomy of **Paramecium aurelia***. London: Macmillan, 1969. 1.22**A**.

Kaestner, A. *Invertebrate Zoology*, Vol. I. New York: Interscience Publishers, 1967. 3.11**B** after Fowler, 3.15**C** after Hyman, 6.2**E** after Nicoll, 6.3**B** after Kukenthal, 6.5**A** after Ashworth, 6.5**E** after Nicoll.

Lackey, J. B. *Trans. Amer. Microsc. Soc.*, **78**:205, 1959. 2.3**A**.

Lane, C. E. *Sci. Amer.*, **202**(3):158, 1960. 3.10.

Lee, D. *The Physiology of Nematodes*. Edinburgh: Oliver and Boyd Ltd., 1965. 5.8**B**, 5.7 (ii) after Wallace.

Mackinnon, D. L., and R. S. J. Hawes. *An Introduction to the Study of Protozoa*. New York: Oxford University Press, 1961. 1.6 after Kirby and Wenyon, 1.7**A**, 1.9**A** after Stein, 1.13**A**, 1.16 after Pierson and other sources, 1.20**B** after Chen, 1.24**B, E,** and **F** after various sources.

Mann, H. K. *Leeches (Hirudinea). Their Structure, Physiology, Ecology and Embryology*. Oxford: Pergamon Press, 1962. 6.3**A** and **D**.

Mavor, J. W. *General Biology*. New York: Macmillan Publishing Co., 1950. 7.21**B** after Huxley.

Meglitsch, P. *Invertebrate Zoology*. New York: Oxford University Press, 1967. 7.2**B**, 7.3**A** after Koch, 7.5**F** after Snodgrass and others, 7.8**B** after Demoli and Versluys, 8.1.

Metcalf, C. L., and W. P. Flint. *Destructive and Useful Insects*. New York: McGraw-Hill Book Co., 1962. 7.10**E**.

Milkman, R. *Biol. Bull.*, **132**:229, 1967. 10.6**A**.

Millar, R. H. *Ciona. L.M.B.C. Memoirs XXXV*. Liverpool: University of Liverpool Press, 1953. 10.3**A** and **C**.

Morton, J. E. *Molluscs. An Introduction to Their Form and Function*. London: Hutchinson and Co., 1958. 8.1

Nichols, D. *Echinoderms*. London: Hutchinson and Co., 1962. 9.7 (iii).

Parker, T. J., and W. A. Haswell. *Textbook of Zoology*, Vol. I. New York: Macmillan Publishing Co., 1951. 4.3**F** after Shipley, 6.5**D** after Claparéde, 7.24**B** after Macbride, 8.10**B** after Küpfer, 8.11**A** after Patten.

Patterson, J. T., and W. S. Stone. *Evolution in the Genus **Drosophila***. New York: Macmillan Publishing Co., 1952. 7.25.

Prosser, C. L., and F. A. Brown, Jr. *Comparative Animal Physiology*. Philadelphia: W. B. Saunders Co., 1961. 7.26 after Matsumoto.

Ramsay, J. A. *Physiological Approach to the Lower Animals*. London: Cambridge University Press, 1952. 3.14, 3.15**A**.

Rasmont, R. L'Ultrastructure des Choanocytes d'Eponges. *Ann. Des. Sci. Nat. Zool.*, 12th Series, **1**:253, 1959. 2.2**A**.

Raven, P. *Morphogenesis. The Analysis of Molluscan Development*. Oxford: Pergamon Press, 1966. 8.11**B**.

Reisch, Donald J. *Laboratory Manual for Invertebrate Zoology*. Long Beach: California State College, 1969. 3.17**B**.

Rowett, H. G. Q. *Dissection Guides*. Vol. V. *Invertebrates*. New York: Holt, Rinehart and Winston, and London: John Murray Publishers, 1957. 6.4**B** and **D**, 7.5**B, C,** and **D,** 7.13 (ii), 7.16, 7.17**B** and **C**, 8.5**B, D,** and **E**.

Russell-Hunter, W. D. *A Biology of Lower Invertebrates*. New York: Macmillan Publishing Co., 1968. 2.1**B**, 2.5**A** and **C**.

Sasser, J. N., and W. R. Jenkins. *Nematology*. Chapel Hill: University of North Carolina Press, 1960. 5.1**B**, 5.5 after Hirschmann.

Sayles, L. P. *Manual for Comparative Anatomy*. New York: Macmillan Publishing Co., 1950. 10.4**A**.

Schechter, V. *Invertebrate Zoology*. Englewood Cliffs, N.J.: Prentice-Hall, 1959. 3.3**B, C,** and **D**, 7.9**B**, 7.10**B**, 8.5**A**, 8.6**D**, 9.3**C**, 9.4, 9.6**A** (i).

Simpson, G. G., and W. S. Beck. *Life*. New York: Harcourt Brace Jovanovich, 1965. 3.1**C**.

Smith, J. E. *Phil. Trans. Roy. Soc.,* **227B**:111, 1937. 9.1**D**.

Smith, R. I., et al. *Keys to Marine Invertebrates of the Woods Hole Region*. Contribution 11. Systematics-Ecology Program. Woods Hole, Mass.: Marine Biological Laboratory, 1964. 3.7**C**, 3.8**B** and **C**, 3.9**A** and **B**.

Smyth, J. D. *The Physiology of Trematodes*. San Francisco: W. H. Freeman and Co., 1966. 4.6.

Snodgrass, R. E. *Textbook of Arthropod Anatomy*. Ithaca, N.Y.: Cornell University Press, 1952. 7.2**B**, 7.3**A** (ii), 7.5**A** and **F**, 7.6, 7.7, 7.8**A**.

Stephenson, W. *Parasitol.,* **38**(3):128, 1947. 4.2**B**.

Storer, T. I., and R. L. Usinger. *General Zoology,* 3rd ed. New York: McGraw-Hill Book Co., 1957. 3.3**A**, 3.5**A**, 4.1**B**, 6.1, 6.6, 7.1**A** (i) and (ii) after Snodgrass, 7.4**A** and **B**, 8.8**D**, 9.1**A** and **C**, 9.7 (i), 9.8**A**.

Suzuki, S. *J. Sci. Hiroshima Univ.,* **15**:206, 1954, 1.10.

Tartar, V. *The Biology of Stentor*. Oxford: Pergamon Press, 1961. 1.12**B** and **C**.

Threadgold, L. T. *Quart. J. Microbiol. Sci.* **104**:505, 1963. 4.2**C**.

Tonapi, G. T., and V. A. Ozarkar. *Turtox News* **44**:242, 1966. 7.18.

Van Name, W. The North and South American Ascidians. *Bull. Amer. Mus. Nat. Hist.* **84**:39, 1945. 10.7**B**.

Vickerman, K., and F. E. G. Cox. *The Protozoa*. Boston: Houghton Mifflin Co., 1967. 1.4**A**, 1.8**C** after Ehret and Powers and Grell, 1.12**A** after Tartar, 1.20**A**.

Villee, C. A., W. F. Walker, and F. E. Smith. *General Zoology,* 2nd ed. Philadelphia: W. B. Saunders Co., 1963. 3.16**B**, 10.2**A** and **B**.

Weichert, C. K. *Anatomy of the Chordates,* 2nd ed. New York: McGraw-Hill Book Co., 1958. 10.3**B**.

Wells, G. P. *J. Mar. Biol. Ass. U.K.,* **28**:465, 1949, 6.8.

Wells, G. P., and R. P. Dales. *J. Mar. Biol. Ass. U.K.,* **29**:661, 1951. 6.8.

Welsh, J. H., R. I. Smith, and A. E. Kammer. *Laboratory Exercises in Invertebrate Physiology*. Minneapolis: Burgess Publishing Co., 1968. 7.28**B** after Edwards.

Whitehouse, R. H., and A. J. Grove. *The Dissection of the Cockroach*. London: University Tutorial Press, 1957. 7.13 (iii).

Wichterman, R. *The Biology of Paramecium*. New York: Blakiston Division, McGraw-Hill Book Co., 1951. 1.8**A**, 1.17 after Lund.

Wilmoth, J. H. *Biology of Invertebrates*. Englewood Cliffs, N.J.: Prentice-Hall, 1967. 6.7 after Richards.